D0074482

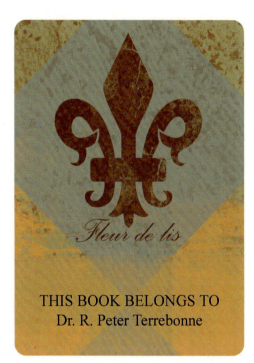

Fleur de lis

THIS BOOK BELONGS TO
Dr. R. Peter Terrebonne

STATICS

Lawrence E. Goodman
Professor of Mechanics
Department of Aeronautics and Engineering Mechanics
University of Minnesota

and

William H. Warner
Associate Professor
Department of Aeronautics and Engineering Mechanics
University of Minnesota

DOVER PUBLICATIONS, INC.
Mineola, New York

Copyright

Copyright © 1961, 1963, 1964 by Lawrence E. Goodman and William H. Warner

All rights reserved under Pan American and International Copyright Conventions.

Published in Canada by General Publishing Company, Ltd., 895 Don Mills Road, 400-2 Park Centre, Toronto, Ontario M3C 1W3.

Published in the United Kingdom by David & Charles, Brunel House, Forde Close, Newton Abbot, Devon TQ12 4PU.

Bibliographical Note

This Dover edition, first published in 2001, is a corrected republication of *Statics* from the combined printing of *Statics and Dynamics,* originally published by Wadsworth Publishing Company, Inc., Belmont, California, in 1964. An Author's Addendum for the Dover Reprint Edition has been added on p. vii.

Library of Congress Cataloging-in-Publication Data

Goodman, Lawrence E. (Lawrence Eugene), 1920–
    Statics / by Lawrence E. Goodman and William H. Warner.
        p. cm.
    Originally published: Belmont, Calif., : Wadsworth Pub. Co., [c1963]
    Includes bibliographical references and index.
    ISBN 0-486-42005-1 (pbk.)
     1. Statics. I. Warner, William H. (William Hamer), 1929–   II. Title.

QA821 .G58 2001
531'.12—dc21

                         2001047004

Manufactured in the United States of America
Dover Publications, Inc., 31 East 2nd Street, Mineola, N.Y. 11501

PREFACE

For some years we have presented a first course in mechanics for engineering students who have previously studied calculus. *Statics* and its companion volume, *Dynamics*, are intended to supplement this course by providing material to which students may refer for details of proofs of theorems, for ancillary illustrative material, and for exercises. By this means the lecturer will be freed, it is hoped, from the need to present every proof and pertinent illustration in the course of his lectures. He will be better able to emphasize important, difficult, or especially interesting aspects of the subject.

We believe that a first course in mechanics should provide engineering students with the professional background needed to deal with mechanical analyses that are not limited to two spatial dimensions. The approach we have adopted, therefore, is "vectorial," so that the three-dimensional theory may be presented with reasonable completeness and rigor. At the same time, care has been taken to smooth the transition from vector symbolism to the representation of vectors in specific coordinate systems. The engineer, when coming to grips with specific problems, usually finds that there is some symmetry present which makes a special coordinate system advantageous.

A glance at the Table of Contents will tell the reader that the order and coverage of topics is that of a standard syllabus in professional engineering training, except, perhaps, for the complete discussion of the principles of statics prior to the consideration of applications. (Chapter I, dealing with vector algebra, may be omitted by those who have studied that subject in earlier training.) We usually devote about one-third of an academic year to the material of *Statics*, two-thirds to that of *Dynamics*. On the other hand, virtual work is treated in dynamics rather than in statics. We believe that a knowledge of kinematics is essential for an understanding of the virtual work principle. In presenting the applications of the principles of statics we have tried to bear in mind that for the student of technology the study of statics is normally a forerunner to a course in "strength of materials" or deformable-body mechanics, as well as to dynamics. The rather full discussion of statical determinacy and indeterminacy in Chapter III is intended to provide background for such a course. So is the presentation of distributed force applications—hydrostatics, aerostatics, loaded cables, and the computation of resultant bending moments and shear forces—in Chapter IV. In the fifth chapter we have ventured to present, in necessarily abbreviated form, some of the leading ideas of modern research into the friction of rubbing solids.

The text is reasonably self-contained. There is, we feel, a sufficiently wide choice of applications to permit the individual lecturer to select those he considers most interesting and important. This choice is bound to vary from lecturer to lecturer and even from year to year. Instructors who devote more than thirty lecture hours to statics should have no difficulty finding material for a longer course. Sections and paragraphs of the text set in smaller type are considered to be not essential to the main development. The solved Examples, also in smaller type, are an integral part of the text material and, we trust, illuminate the theory. A conventional numbering system has been used for easy cross-reference; for example, Eq. 2.4-5 is the fifth equation in Section four of Chapter II. The exercises, for which no special originality can be claimed, have in many cases been set as examination questions at the University of Minnesota. They are graduated in difficulty somewhat, the first in each set being the simplest; most, we feel, are of moderate difficulty.

Again we are happy to acknowledge the assistance of many who have helped us, both in the preparation of manuscript and in the development of our ideas: the staff members of Wadsworth Publishing Company, who have dealt patiently with what must at times have seemed a rather exacting pair of authors; the instructors who have taught recitation sections using our notes; Mrs. Helen Woodward, who has typed the various versions of the manuscript; Professor B. J. Lazan, head of the Department of Aeronautics and Engineering Mechanics at the University of Minnesota, and our colleagues who regularly teach the courses in mechanics; Professor Harold De Groff of Purdue University, for his comments on the final manuscript; the delegates of the Clarendon Press, Oxford, for permission to quote the material of Table I, Chapter V; the Cambridge University Press for permission to include among the exercises a number of problems set in examinations at the University of Cambridge; and all of the students of engineering at the University of Minnesota whom we have taught.

L.E.G.

W.H.W.

AUTHOR'S ADDENDUM
FOR THE DOVER REPRINT EDITION

Soon after this book was published originally, Professor Goodman left the Department of Aerospace Engineering and Mechanics for the Department of Civil Engineering, serving there first as head, and then as professor until his retirement. Unfortunately, his death last year (in 2000) came before we knew of Dover's decision to reprint our books, *Statics*, and *Dynamics* (0-486-42006-X).

I am pleased that Dover is making them available again, though I am sure they could have been made better if Professor Goodman or I had ever had the time to complete our planned revision. I thank all of those who helped persuade Dover that these were worth reissuing, especially Dr. Carl Nelson (one of Professor Goodman's students), whose idea this was and who did all the work in seeing it through.

June 2001

WILLIAM H. WARNER
Professor Emeritus
Aerospace Engineering and Mechanics

CONTENTS

Contents

Contents

HISTORICAL INTRODUCTION

La Dynamique est la science des forces accélératrices ou retardatrices et des mouvements variés qu'elles doivent produire. Cette science est due entièrement aux modernes, et Galilée est celui qui en a jeté les premiers fondements.... Les découvertes des satellites de Jupiter, des phases de Venus, des taches du Soleil, etc., ne demandaient que des télescopes et de l'assiduité; mais il fallait un génie extraordinaire pour démêler les lois de la nature dans des phénomènes que l'on avait toujours eus sous les yeux, mais dont l'explication avait néanmoins toujours échappé aux recherches des philosophes.... La Mécanique devint une science nouvelle entre les mains de Newton, et ses *Principes mathématiques*, qui parurent, pour la première fois, en 1687, furent l'époque de cette révolution.   *J. L. Lagrange*

The aim of theoretical mechanics is to provide quantitative predictions of the motions of material objects.   The practical applications of such predictive power are obvious, and a mastery of the subject has been a longstanding objective of men.   In succeeding chapters matters are treated from a logical approach which ignores the false starts and obscurities so common in the history of science.   The history of mechanics is, however, of considerable cultural and intellectual interest of itself; its influence may be detected even in modern treatments of the subject.   For this reason it seems worthwhile to sketch that history briefly.

Although ancient civilizations must have had a grasp of practical mechanics for the erection of their impressive structural monuments, it is to the Greeks that we owe the first recorded systematic efforts to provide a theoretical basis for the subject. Among them, the great name is that of Archimedes (287?–212 B.C.). He appears to have been trained in Alexandria, that center of hellenistic culture where Euclid (c. 330–c. 260 B.C.) had been master. In his works *On the Equilibrium of Planes* and *On Floating Bodies*, Archimedes begins, in the manner of Euclid, by setting out a number of postulates considered as self-evident. With these as foundation he derives a variety of propositions on the equilibrium of levers and on centers of mass, as well as the celebrated theorem in hydrostatics that bears his name. The inspiration for these propositions cannot be known to us with certainty, but, judging from the evidence available, it would seem that the scaffolding for the finished logical structure was provided by experiment. For example, Archimedes is reported to have cut out and weighed a segment of a parabola while investigating the formula for its area. The proofs themselves have been criticized—after two thousand years—on the grounds that they embody rather subtle unstated assumptions. Their cogency, however, is undeniable. Later, the idea of the moment of a force was generalized by Hero of Alexandria (c. 150), and other simple machines were analyzed.

No understanding of dynamics comparable to Archimedes' grasp of statics was achieved for almost sixteen hundred years. We can only conjecture why this long hiatus should have occurred. It has been attributed to the fact that the Roman economy was a slave economy, which provided little incentive for the development of technology. Many of the Roman wars were simply large-scale slave raids. The feudal economy that followed was based on the labor of the serf, whose condition was scarcely better. Partly, too, progress was retarded by the fact that, among Greek thinkers, it was the great logician Aristotle (384–322 B.C.), rather than Archimedes, whose work was accepted by the schoolmen of the early middle ages. In natural science Aristotle's primary interest lay in biology; he was led to identify motion with growth and change generally. Every body had a "natural" place in the universe. Motion toward this position was "natural" and would persist; other motion was "unnatural" or "violent" and would decay. These ideas did not promote the development of mechanics. Also, the completely geometrical approach of the Greeks was a handicap in the treatment of kinematical questions.

It must not be thought, however, that the period prior to the Renaissance was void of activity in science. Modern historical scholarship has brought to light the names of many who preceded Galileo. Robert Grosseteste, Chancellor of the University of Oxford and Bishop of Lincoln in 1235, questioned the mechanics of Aristotle. He inspired John Duns Scotus (c. 1266–1308), Roger Bacon (1214?–1294), and William of Occam (c. 1270–1349). In the period 1325–1350, a group associated with Merton College, Oxford, led by Thomas Bradwardine, William Heytesbury, Richard Swineshead, and John Dumbleton, worked out the kinematics of rectilinear motion. In France, Jean Buridan, who was in 1325 rector of the University of Paris, advanced the idea that a body in motion possessed a certain *impetus*, a quantity very like present-day momentum. His pupil, Nicole Oresme (?–1382), devised a semi-graphical analysis of the kinematics of rectilinear motion in which, essentially, velocity was plotted against time and the area of the diagram used to determine distance traversed. In Spain, Dominic Soto (1494–1560) worked out the laws of freely falling bodies. Finally Simon Stevin (1548–1620), a Netherlands engineer, brilliantly matched the statical achievements of Archimedes. Starting from an analysis of the inclined plane which was based on the impossibility of perpetual motion, he proved the vectorial addition of forces. In hydrostatics he was first to conceive the notion of the principle of solidification, which permitted him to find the pressure exerted by a fluid on its container by considering that any portion of the fluid may be regarded as a frozen solid, without disturbing equilibrium or the pressures in the remainder of the fluid.

At first slowly, and then more rapidly, progress developed. The year 1543 saw the publication of Copernicus' *On the Revolution of the Heavenly Spheres* (and also of Vesalius' book *On the Fabric of the Human Body*, which freed medicine from dependence upon Galen). Galileo Galilei (1564–1642), whose writings and whose discoveries with the telescope did much to secure acceptance for the Copernican theory, may be regarded as the first of the "moderns" in mechanics. It was he who introduced that combination of experiment and analysis that is the hallmark of the best scientific work. He established the elementary theory of projectile motion and can be credited with a rudimentary grasp of the idea of inertia. His writings exerted wide influence. They read well, even today, and we may recall the visit that the young Milton paid him in his old age and to which Milton refers in several places in his poetry. After the time of Galileo it was no longer completely respectable to draw conclusions about natural phenomena solely from authority. Huyghens (1629–1695) perfected and completed Galileo's mechanical discoveries and developed the analysis of the motion of the pendulum.

The year of the death of Galileo saw the birth of Isaac Newton (1642–1727), whose great achievements in mechanics may be judged by the quotation from Lagrange that begins this Introduction: "Mechanics became a new science in the hands of Newton, and his *Principia*, which appeared in 1687, broke the way for this revolution." Indeed, it is no exaggeration to say that the *Principia* and Locke's *Essay on the Human Understanding* (1690) marked the opening of that movement which its followers termed the Age of Reason. Newton introduced the concepts of force and mass and stated the laws of motion in the form in which we use them today. Furthermore, he developed the mathematical calculus by means of which the planetary laws so painfully extracted by the dedicated labor of Kepler from the observations of Tycho Brahe could be seen to follow directly from the laws of motion. The principle of universal gravitation is also his, though here credit must be shared with Robert Hooke (1635–1703) and others. The material of the present text is that upon which he set the mark of his genius, and it may properly be denominated *newtonian mechanics*.

The treatment of statical problems by the principle of virtual work was at this time systematized by John Bernoulli (1667–1748). In 1717 he enunciated the first comprehensive statement of this principle (sometimes termed the principle of virtual velocities or virtual displacements). The basic idea can be traced to hellenistic mechanics; it was known to Stevin in a primitive form. Galileo recognized that only the component of the velocity in the direction of the force is effective. Bernoulli also took as basic the product of the force and the virtual velocity component in the direction of the force, but he appreciated that the sum of all such products must vanish for any possible small displacement, if the force system is in statical balance.

The eighteenth and early nineteenth centuries saw the rapid extension of newtonian mechanics. The era was not one of uniform progress, however. There was, for example, an extended controversy between those who felt that the proper measure of a force was the change in kinetic energy produced by it and those who preferred the change in momentum. Leibniz, who had independently developed the calculus, played a role in this dispute, which we now recognize as arising from alternative ways of integrating the second law of motion. The leaders in the main stream of the consolidation of newtonian mechanics during this period were Leonhard Euler (1707–1783) and Jean le Rond d'Alembert (1717?–1783).

Euler's is certainly one of the greatest names in mathematical physics. A prolific writer, he undertook a series of fundamental texts covering not only the mechanics of particles but hydrodynamics and the mechanics of deformable solids as well. He was one of the first to make full use of calculus in solving Newton's differential equations of motion—Newton, in the *Principia*, employed geometrical methods, which he felt would be more understandable to his readers. Euler's treatises served to train most European workers in applied mechanics well into the nineteenth century. In the mechanics of rigid bodies, we owe to Euler, among other things, the angular coordinates used to describe spatial position; the fundamental kinematical theorem, which asserts that any motion of a rigid body can be decomposed into a translation followed by a rotation; the extension of Newton's equations of motion from the single-particle system to the rigid body (an extension which required an appreciation of the inertia tensor); and the solution of those extended equations to describe free precessional motion. The calculus of variations was in large part created by him as a result of the stimulation presented by the early development of variational principles in mechanics. Many of the fundamental ideas in the development of mechanics that are now associated with other names find their earliest lucid expression and rigorous proof in his work.

D'Alembert was also the author of a text celebrated in the history of mechanics. His *Treatise on Dynamics*, first published in 1743, amended in 1758, contains many illustrations of the method he devised for analyzing the motion of systems of rigid bodies subject to constraints. He also is to be credited with the first introduction of the idea of a vector *field*, and with showing that conservation of mechanical energy was a consequence of Newton's laws of motion. Like Newton, d'Alembert took an active part in the political and intellectual life of his day. For a long time he was joint editor of Diderot's *Encyclopedia*, that remarkable expression of the humanistic philosophy which did so much to discredit well-established institutions such as human slavery, colonialism, religious intolerance, and war, and which emphasized the importance of industry, the value of technical knowledge, and the dignity of labor.

The contribution of d'Alembert was subsumed into an approach to the equations of motion due to J. L. Lagrange (1736–1813). The *Analytical Mechanics* of Lagrange was published in 1788, one hundred years after Newton's *Principia*. Lagrange's inspiration was to use the scalar quantities work and energy rather than the vector quantities force and acceleration to determine the motion of mechanical systems. Combining d'Alembert's principle with the principle of virtual work, he derived a form of the equations of motion that is today the starting point for most advanced treatments of mechanics. The lagrangian approach does not, indeed, give us the solution of any particular problem that could not equally well be treated by direct reference to Newton's laws of motion (from which it is derived), but it does make it possible to draw conclusions about the behavior of large classes of systems. An analysis for one system will hold for any other whose energy depends upon the position coordinates in the same way. When the concept of energy was enlarged to include electric and magnetic effects, the motion of electromechanical systems became a natural subject for treatment by Lagrange's method.

During the nineteenth century, in the hands of Maxwell, Joule, Clausius, Carnot, Gibbs, and W. Thomson, thermodynamics developed along lines suggested by the progress in mechanics. Continuum mechanics—stress analysis, fluid dynamics, and the theory of sound—became well-developed disciplines in their own right, underlying much of today's engineering. Among the leaders in these fields were Navier, Cauchy, Kirchhoff, and de St. Venant; Helmholtz, Stokes, Joukowsky, and Rayleigh. Meanwhile, two whose names are familiar to students of elementary mechanics were Gaspard Coriolis (1792–1843), who completed the kinematic analysis of particle acceleration referred to moving axes, and J. B. L. Foucault (1819–1868), who devised the gyroscope and pendulum that bear his name. Foucault succeeded, where Galileo had failed, in devising a laboratory demonstration of the rotation of the earth. It should perhaps be noted that Coriolis and Foucault were men of diverse attainments: the former was director of studies at the prestigious École Polytechnique; the latter was a science reporter for the Paris press. The mathematical theory in which the Coriolis component of acceleration explains the behavior of Foucault's pendulum followed, rather than preceded, the experiments. In 1846 newtonian mechanics achieved what was, in some respects, its crowning success with the discovery of the planet Neptune in precisely the position that U. J. J. LeVerrier and J. C. Adams, after analysis of small anellipticities of the orbit of Uranus, predicted it should occupy.

All of the foregoing work was based upon the three laws of motion due to Newton and upon the principle of virtual work. Is there a single principle from which Newton's laws, and, consequently, all other mechanical effects, may be derived? As we have seen, Lagrange took a long step toward achieving this goal, which occupies an important place in the history of mechanics. As early as 1740, Pierre Louis Moreau de Maupertuis (1698–1759) asserted that the line integral of the quantity $mv \cdot d\mathbf{s}$, which he termed the *action* of the particle, is always a minimum.   In the integrand, m denotes the mass of the particle, $\mathbf{v}$ its velocity, and $d\mathbf{s}$ an element of arc-length along the path of the particle.   Maupertuis had the misfortune to fall out with Voltaire, then a fellow member of the Prussian Academy of Arts and Science, and to be hilariously caricatured in *The Diatribe of Dr. Akakia*.   Euler, who retained a high opinion of his colleague, while remaining in the good graces of Voltaire, showed, in 1744, that the orbit of a particle moving under the influence of a central force would indeed be one in which the action would have a stationary value. Lagrange, in the *Analytical Mechanics*, presented a more general proof, in which he showed that the action would be a minimum for a particle in a conservative force field, subject to time-independent constraints, and would have a stationary value in other cases.   The proof of the *principle of least action* used today is due to K. G. J. Jacobi (1804–1851) and was published posthumously in his *Lectures on Dynamics* (1866).   Newton's equations of motion can be derived from the principle of least action; the converse is also true.   By starting with an extremal principle of this type, it is possible to dispense with the concept of force as a fundamental one.

The principle of least action differs from the newtonian approach to particle motion by dealing with an integrated effect taken over a length of path rather than with the relation between force and acceleration at any instant.   Methods of this sort are known as variational methods.   They occupy a distinguished position in mechanical analysis and have an extensive history of their own.   The most important such principle is *Hamilton's Principle*, due to W. R. Hamilton (1805–1865), which asserts that, for a particle moving in a conservative force field, among the various possible neighboring paths the particle may take between any two points the time average of the difference between the kinetic and potential energies has a stationary value over the actual path traversed.   (The principle requires modification if the force field is not conservative or if velocity-dependent constraints are present.)   Both Newton's and Lagrange's forms of the equations of motion may be derived from Hamilton's principle.   It is this formulation of the laws of classical mechanics that has proved most fruitful for the purposes of quantum mechanics.

With technological and scientific progress, increasingly accurate measurements become possible and increasingly broad scientific theories are demanded.   In recent decades, minute departures from the predictions of newtonian mechanics have made men realize that, on a macroscopic scale, newtonian mechanics is inappropriate for bodies moving with speeds approaching the speed of light just as, on a microscopic scale, it is inappropriate for describing motions in the nucleus. On the other hand, the variational formulation of the principles of mechanics, being invariant with respect to coordinate transformations, could be modified by Einstein, Poincaré, and Lorentz into modern relativistic mechanics. This development in no way diminishes the lustre of conventional or, as it is called, *classical* mechanics—which, in fact, has played a leading role in the development of the modern einsteinian and quantum mechanics and which forms the basis of present-day technology.

Suggestions for Further Reading

Clagett, M., *Science of Mechanics in the Middle Ages* (University of Wisconsin Press, 1959).

Dugas, R., *A History of Mechanics* (New York: Central Book Co., 1955).

McKenzie, A. E. E., *The Major Achievements of Science*, Vol. I (Cambridge University Press, 1960).

STATICS

CHAPTER I

Vector Algebra and Allied

Geometrical Concepts

1.1 Introduction

Mechanics is concerned with the motion of material systems of all kinds. The subject is conventionally divided into statics and dynamics, of which the former deals with those structures and machine elements that are at rest—or, more precisely, whose parts all have zero acceleration—and the latter with accelerating objects. From a broad point of view, statics is only a special case of dynamics, a case in which all accelerations have the particular value zero. In engineering and technology, however, the importance of this special case is so great that it demands a separate treatment. Most of the structures designed by the engineer are intended to remain in equilibrium; even moving elements of machinery, if accelerations are not large, will be found to be designed on a statical basis. Continuum mechanics of deformable solids, stress analysis of structures and machinery, hydrostatics, and soil mechanics all take an understanding of statics as the starting point of engineering design and analysis. Statics is, in fact, the basis of much day-to-day engineering design, and the well-trained engineer must be able to use the principles of statics, in all their various forms, with the same effortless skill displayed by any good craftsman in the handling of his tools.

1.2 Vector and Scalar Quantities

It is desirable to begin by establishing an efficient notation for the description of the quantities encountered in mechanics (and in other branches of science as well).   Many of these quantities involve the idea of direction as well as magnitude.   For example, the position of a point P relative to a point O may be specified by means of the directed line segment drawn from O to P.   If O' and P' are any other two points such that the direction and the distance from O' to P' is the same as the direction and the distance from O to P, we say that

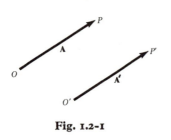

Fig. 1.2-1

the position of P' relative to O' is the same as the position of P relative to O. Another way of putting these statements is to say that a particle moving from O to P has experienced a *displacement* or change of position given by the directed line segment $\overline{OP}$.   Such a directed line segment is conveniently represented by an arrow drawn from O to P—that is, with its initial point at O and its terminus at P.   From our present point of view the displacement $\overline{O'P'}$ would be represented by an arrow drawn from O' to P'.   This arrow has the same direction and the same length as the arrow $\overline{OP}$, which corresponds with the fact that the displacement from O to P is the same as the displacement from O' to P' or that the position of point P relative to O is the same as the position of point P' relative to O'.   Quantities like displacement, which can be represented by a directed line segment drawn to uniform scale and which combine physically according to the rules or elementary operations of the next section, are known as *vector quantities*.   The arrow that represents the quantity is called its *vector*.   The length of the arrow is called the *magnitude* of the vector.   Note that from our present point of view a vector has magnitude and direction but is not localized in space; e.g., the vector which represents $\overline{OP}$ is regarded as the same as the one that represents $\overline{O'P'}$.   When we wish to emphasize this aspect of the vector quantity we speak of it as a *free* vector quantity; later on, in the consideration of moment and force vectors, the concepts of bound and sliding vectors will be introduced.

Vectors are of such common occurrence that it is convenient to have a special symbol for them. It is common practice, in print, to use boldface type for vectors.* For example, we denote the directed line segment $\overline{OP}$ by the special symbol **A** and the directed line segment $\overline{O'P'}$ by the symbol **A**'. It must be remembered that such a symbol represents more than an ordinary number; it stands for a quantity whose complete specification involves direction as well as magnitude. We may, however, still borrow the equality sign from algebra and write

$$\mathbf{A} = \mathbf{A}' \qquad\qquad \text{I.2-I}$$

provided we interpret this expression as meaning "the vector **A** has the same magnitude and the same direction as the vector **A**'." An advantage accruing from the use of boldface type (or the overbar) is that ordinary type can then serve to denote the magnitude of the vector. The magnitude of the vector **A** is written simply as A. In Fig. I.2-I, A represents the distance from O to P. It follows from Eq. I.2-I that

$$A = A'. \qquad\qquad \text{I.2-2}$$

Notice, however, that this is only one of the conclusions that may be drawn from Eq. I.2-I. Furthermore, the symbols A and A' denote ordinary numbers whereas Eq. I.2-I is a *vector equation*. A variation in notation sometimes encountered is the use of the symbol $|\mathbf{A}|$ in place of A to emphasize the fact that it is the magnitude of a vector that is being represented, and not some other numerical quantity.

While many of the important quantities encountered in mechanics have directional properties, there are others that require only a single number for their complete specification. These are called *scalar quantities*. We may say, for example, that a box has a volume of twelve cubic feet. The word "volume" indicates the kind of quantity we have in mind, the words "cubic feet" denote the *unit*, and the word "twelve" denotes the *measure*, that is, the number of units contained in this example of the quantity under discussion. When we deal with a scalar quantity, attention tends to center on the

---

* In manuscript or typescript, the use of an overbar on the symbol for the vector is one way of indicating a vector quantity.

measure, since this is the number that appears in mathematical equations. The ordinary real number provides the typical example of a scalar quantity, just as displacement typifies the vector quantity. Ordinary weight type is used for the measure numbers of scalar quantities.

Not all quantities of physical interest are represented by scalars or vectors; the student may well inquire about quantities dependent on two or more directions. These general *tensor* quantities are also important in mechanics; scalars and vectors are special cases of them. In the elementary courses in dynamics and deformable-body mechanics, the student will become acquainted with a few such tensor quantities: the inertia tensor, the stress tensor, and the strain or strain-rate tensor, for example. Other physical quantities, such as finite rotations of bodies, do not fit into the scalar-vector-tensor pattern at all and are described by mathematical entities behaving differently from those we treat here.

1.3 Elementary Operations

If the introduction of special vector symbols is to be justified, we must be able to manipulate these symbols in meaningful ways, just as the symbols representing ordinary numbers are handled in

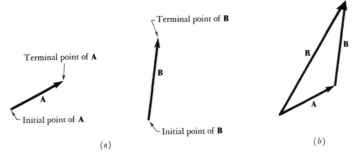

Fig. 1.3-1

conventional algebra. To accomplish this, certain operations are defined. The first is called *vector addition*. To form what is known as the sum of two vectors, such as **A** and **B** shown in Fig. 1.3-1a, we must first shift the vectors, without changing their magnitudes or directions, until the initial point of **B** coincides with the terminal point of **A**, as shown in Fig. 1.3-1b. Then the sum of **A** and **B** is defined as the vector represented by the directed line segment from the initial point of **A** to the terminal point of **B**. This vector is

denoted **R** in Fig. 1.3-1b.   One could equally well form **R** by making
the initial points of **A** and **B** coincide and completing the parallelogram, as shown in Fig. 1.3-2, to locate the terminal point of **R**.   The
method of construction shown in Fig. 1.3-1b is known as the triangle
law of addition, while that shown in Fig. 1.3-2 is known as the parallelogram law.   The two procedures are, of course,
completely equivalent, since the directed line segment $\overline{PQ}$ in Fig. 1.3-2 is (vectorially) equal to the
vector **B**.   It is customary to denote the operation
of vector addition by borrowing the + sign from
algebra and to write

$$\mathbf{A} + \mathbf{B} = \mathbf{R}. \qquad \text{1.3-1}$$

The vector **R** is known as the *sum* or the *resultant*
of **A** and **B**.   The vectors **A** and **B** are called
components of **R**.

Fig. 1.3-2

We can form the sum of more than two vectors by adding the
first two, then adding their resultant to the third, and so on.   It is
simpler to omit the intermediate steps and to form the vector polygon,
as shown in Fig. 1.3-3.   (Note that point S is not necessarily in the
plane of points OPQ.)   The sum of the vectors **A, B,** and **C** is the
vector **R** and we write

$$\mathbf{A} + \mathbf{B} + \mathbf{C} = \mathbf{R}. \qquad \text{1.3-2}$$

It may happen, of course, that the point S in Fig. 1.3-3 coincides
with the point O.   In that case we say that the
resultant is a *null vector*.   This null vector is represented by the symbol **o** or, where no confusion
between vectors and scalars is likely to result,
simply by the ordinary symbol for zero, o.   The
null vector may be regarded as a vector of zero
magnitude and arbitrary direction.

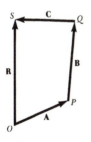

Next we define the negative of a vector as
another vector having the same magnitude as the
original one, but reversed in direction.   We borrow
the minus sign from algebra to indicate this and
denote the negative of **A** by $-\mathbf{A}$.   Then the sub-

Fig. 1.3-3

traction of **A** from **B** consists in the addition of $-\mathbf{A}$ to **B**.   This is
pictured in Fig. 1.3-4.

The borrowing of symbols such as +, −, = from the algebra
of ordinary real numbers (scalars) requires justification.   The algebra

of ordinary real numbers is based on certain fundamental laws of which two apply to the operations of addition and subtraction. The first of these, the commutative addition law, asserts that, for any two real numbers a and b, $a+b=b+a$. It follows from the definition of

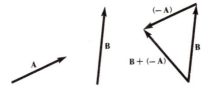

Fig. 1.3-4

vector addition that we may make a like statement for any two vectors, **A** and **B**. In Fig. 1.3-2 the vector $\overline{SQ}$ is the same as **A** and $\overline{PQ}$ is the same as **B**, so that we reach the same point Q whether we add **B** to **A**, as in triangle OPQ, or **A** to **B**, as in triangle OSQ. The second addition law of algebra, the associative addition law, asserts that, for any three real numbers a, b, c, the equation $a+(b+c)=(a+b)+c$ holds. This rule also holds for vector addition. To prove it we note that $(\mathbf{B}+\mathbf{C})$ in Fig. 1.3-5 is represented by $\overline{PS}$, so that

$$A+(B+C) = R. \qquad \textbf{1.3-3}$$

Similarly $(\mathbf{A}+\mathbf{B})$ is represented by $\overline{OQ}$ so that

$$(A+B)+C = R. \qquad \textbf{1.3-4}$$

It follows at once that

$$A+(B+C) = (A+B)+C. \qquad \textbf{1.3-5}$$

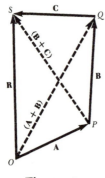

Fig. 1.3-5

The argument is readily extended to any number of terms. We conclude that the sum of any number of vectors is independent of the grouping of the terms.

If we add to the vector **A** another vector **A** we have, according to the rule for vector addition, a resultant vector which has the same direction as **A** and whose magnitude is twice as great as that of **A**. It is only natural to want to describe this fact symbolically by writing $\mathbf{A}+\mathbf{A}=2\mathbf{A}$. But the symbol 2 denotes a typical scalar quantity, and so we are led to define the product of a vector, **A**, and a positive scalar quantity, m, as a vector having the same direction as **A** but with magnitude m times as great. If m is negative, m**A** is again a vector whose magnitude is that of **A** multiplied by the absolute value of m but whose direction, in accordance with the definition of the negative of a vector, is reversed. That is, for negative values of m we would write m**A** $= -(-m$**A**$)$. This operation is known as multiplication by a scalar.

We must verify the fact that the operation of multiplication by a scalar conforms to the fundamental laws of the algebra of real numbers. The so-called "commutative" multiplication law m**A** $=$ **A**m is satisfied if we define post-multiplication (**A**m) of a vector by a scalar just as we have defined pre-multiplication. The "associative" law $n(m$**A**$) = m(n$**A**$)$ follows at once from the definition of multiplication by a scalar and the ordinary laws of arithmetic, as does the "distributive" law in the form $(m+n)$**A** $= m$**A** $+ n$**A**. It remains to prove that

$$m(\mathbf{A}+\mathbf{B}) = m\mathbf{A}+m\mathbf{B}. \qquad \textbf{1.3-6}$$

To show this, let $\overline{OP}$ in Fig. 1.3-6 represent **A** and $\overline{PQ}$ represent **B**.

The directed line segment $\overline{OR}$ may be taken to represent m**A** while $(\mathbf{A}+\mathbf{B})$ will be represented by $\overline{OQ}$. If OQ is extended to a point S chosen so that RS is parallel to PQ it follows from the geometrical similarity of triangles OPQ and ORS that

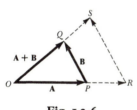

Fig. 1.3-6

$$\frac{RS}{PQ} = \frac{OR}{OP}. \qquad \textbf{1.3-7}$$

But $(OR/OP)=m$, so that the directed line segment $\overline{RS}$ is both parallel to **B** and m times as large. Therefore $\overline{RS}$ represents the vector m**B**. Similarly

$$\frac{OS}{OQ} = \frac{OR}{OP}, \qquad \textbf{1.3-8}$$

so that $\overline{OS}$ represents the vector $m(\mathbf{A}+\mathbf{B})$. Equation 1.3-6 now follows directly from the triangle law of vector addition applied to the large triangle ORS of Fig. 1.3-6.

Now that the operations of vectorial addition, subtraction, and multiplication by a scalar have been shown to conform to all the fundamental rules of algebra, we may bring all the resources of that powerful mathematical tool to bear upon vector equations in which these operations are involved. It must be remembered, however, that only those operations that have been defined may legitimately be employed, and that not every conceivable arrangement of symbols is meaningful. For example, an expression such as $\mathbf{A} - 3\mathbf{B} = 27$ is nonsense because the quantity on the left-hand side of the equality sign is a vector, while that on the right-hand side is a scalar. The two cannot possibly be "equal" in any sense. In mechanics, as in every field of intellectual activity, once a man falls into the habit of writing symbols without troubling himself to formulate the ideas which they represent, his ability to carry out a rational analysis is lost.

Example **1.3-1**

Two vectors, **A** *and* **B,** *are related to a third vector,* **C,** *through the equations*

$$6\mathbf{A} + \mathbf{B} = \mathbf{0},$$
$$-4\mathbf{A} + 2\mathbf{B} = -8\mathbf{C}.$$

Solve for **A** *and* **B.**

Solution: We treat the equations just as in the algebra of ordinary real numbers. Multiply the first equation by 2, the second by -1, and add to get

$$16\mathbf{A} = 8\mathbf{C} \quad \text{or} \quad \mathbf{A} = \frac{1}{2}\,\mathbf{C}.$$

Substitute this result in the first of the given equations to find

$$6\!\left(\frac{1}{2}\mathbf{C}\right) + \mathbf{B} = \mathbf{0} \quad \text{or} \quad \mathbf{B} = -3\mathbf{C}.$$

We conclude that the vector **A** is parallel to **C** and that its magnitude is half as great as the magnitude of **C**. The vector **B** is also parallel to **C** and its magnitude is three times as great as that of **C**. The minus sign before the numeral 3 must be interpreted as meaning that the sense or direction of **B** is opposite to that of **C**. The positive sign, which is understood to appear before the symbol $\frac{1}{2}$, means that the arrowhead on the **A** vector points in the same direction as the arrow associated with the vector **C**.

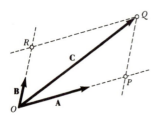

Fig. 1.3-7

Example **1.3-2**

Show that any vector, **C,** *which lies in the plane of the vectors* **A** *and* **B** *can be expressed in the form* $m\mathbf{A} + n\mathbf{B}$.

Solution: The vectors **A** and **B** are shown in Fig. 1.3-7, together with the third vector, **C**. All three have been translated so as to have the same origin O. From point Q, the terminus of $\mathbf{C}$, we construct parallels to $\mathbf{A}$ and $\mathbf{B}$; these intersect the lines of $\mathbf{A}$ and $\mathbf{B}$ at points P and R, respectively, as shown. Now $\overline{OP}$ is a vector that is a suitable multiple of $\mathbf{A}$, say $m\mathbf{A}$, and $\overline{OR}$ is a vector that is a multiple of $\mathbf{B}$, say $n\mathbf{B}$. It follows from the parallelogram law that $\mathbf{C} = m\mathbf{A} + n\mathbf{B}$. The argument assumes that $\mathbf{A}$ and $\mathbf{B}$ are not parallel and, of course, neither of them may be a null vector. The figure is drawn for the case in which m, n exceed unity; the student should sketch the case in which, say, $0 < m < 1$ and the case $m < 0$.

Example 1.3-3

The vector from an origin O to a point P is $\mathbf{s}$ and the vector from P to another point Q is $\mathbf{t}$. What is the vector $\mathbf{r}$ from O to the midpoint of PQ?

Solution: The vectors $\mathbf{s}$ and $\mathbf{t}$ are shown in Fig. 1.3-8, as is the wanted vector $\mathbf{r}$ from the origin to the midpoint, R, of PQ. The vector $\overline{PR}$ has

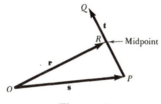

Fig. 1.3-8

the same direction as $\mathbf{t}$ and half of $\mathbf{t}$'s magnitude. From the triangle OPR we have

$$\mathbf{r} = \mathbf{s} + \frac{1}{2}\mathbf{t}.$$

Note that if R, instead of being the midpoint, is *any* point on the line PQ, $\overline{PR}$ will be represented by a vector $m\mathbf{t}$, where m is some number. The vector equation $\mathbf{r} = \mathbf{s} + m\mathbf{t}$ is, therefore, the equation of the line through P in the direction of the vector $\mathbf{t}$. By giving m an appropriate value we can locate any point, R, on that line and to every value of m there corresponds one point on the line.

Example 1.3-4

Show that the line from a vertex of a parallelogram to the center of an opposite side intersects a diagonal of the parallelogram in such a way as to divide it into two parts, one of which is twice as long as the other.

Solution: In Fig. 1.3-9a the vertices of the parallelogram are shown, labeled $OPQR$. The parallelogram is determined by the two non-parallel vectors $\mathbf{A}$ and $\mathbf{B}$, which represent adjacent sides $\overline{OP}$ and $\overline{OR}$, respectively.   Of the other two sides, $\overline{PQ}$ is **B** and $\overline{QR}$ is $-\mathbf{A}$.   It follows that the diagonal $\overline{PR}$ is $\mathbf{B}+(-\mathbf{A})$ or simply $\mathbf{B}-\mathbf{A}$.   The vector from O to the midpoint of PQ is $\mathbf{A}+\frac{1}{2}\mathbf{B}$.   Let us call the intersection of these last two vectors S.   We want to show that the distance from P to S is one-third of the distance from P to R.   To do this we note that, since S is a point on the line PR, the vector $\overline{PS}$ will be some multiple of $\mathbf{B}-\mathbf{A}$, say $m(\mathbf{B}-\mathbf{A})$.   And since S is also a point on the line of the vector $\mathbf{A}+\frac{1}{2}\mathbf{B}$ it follows that $\overline{OS}$ is represented by the vector $n(\mathbf{A}+\frac{1}{2}\mathbf{B})$.   Our task now is to find the value of m.   From the triangle OPS shown in Fig. 1.3-9b we have

$$\mathbf{A}+m(\mathbf{B}-\mathbf{A}) = n\left(\mathbf{A}+\frac{1}{2}\mathbf{B}\right),$$

or

$$\mathbf{A}(1-m-n) = \mathbf{B}\left(\frac{1}{2}n-m\right). \qquad \textbf{1.3-9}$$

At this point we must pause to interpret the last vector equation.   Since **A** and **B** are not parallel, one cannot be equal to a non-zero scalar multiple of the other.   It follows that both scalar multipliers must vanish:

$$1-m-n = 0,$$

$$\frac{1}{2}n-m = 0.$$

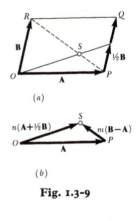

(a)

(b)

Fig. 1.3-9

As a consequence, we have at once $m=\frac{1}{3}$ and $n=\frac{2}{3}$.   The fact that $m=\frac{1}{3}$ means that the vector $\overline{PS}$ is one-third as long as the vector PR, which implies that S divides the line PR into two parts, one of which is one-third of PR and the other of which is two-thirds of PR in length.   That is what we set out to prove.   Notice that the vector equation 1.3-9 yields two ordinary scalar equations in this case.   We also learn from it that $n=\frac{2}{3}$, which shows that the distance OS is two-thirds of the distance from O to the center of the side PQ. It must not be thought that vector algebra is intended to supply proofs of theorems in plane geometry, but the exercise in formulation and interpretation is useful.

Example 1.3-5

The resultant of three forces is a null vector.   Show that the magnitude of each force is proportional to the sine of the angle between the other two.

Solution: The concept of force and its representation by a vector will be discussed in Chapter II; here force is assumed to be a vector quantity. Then if the three forces be denoted **A, B, C** we have

$$\mathbf{A+B+C} = 0.$$

Notice that the forces must lie in the plane of the triangle shown in Fig. 1.3-10. The triangle itself, of course, corresponds to the vector equation just written. Now the magnitude of **A** is A, so that the distance PQ is A, and similarly the distance QR is B and the distance RP is C. From the law of sines of trigonometry,

$$\frac{A}{\sin \alpha} = \frac{B}{\sin \beta} = \frac{C}{\sin \gamma},$$

which is the result wanted. This result, one of the oldest theorems in mechanics, is known as Lamy's Theorem after its discoverer Fr. Bernard Lamy (1645–1715). Although it may be utilized in the analysis of simple equilibrium problems involving three forces, the procedure is not recommended. We shall learn more effective and less restricted ways of handling such analyses.

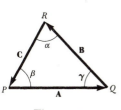

Fig. 1.3-10

I.4 Unit Vectors; Rectangular Cartesian Components

We have seen that the multiplication of a vector, **A**, by a scalar $m > 0$ results in a vector whose magnitude is mA and whose direction is the same as that of **A**. Division of a vector by a scalar may be regarded as multiplication by the reciprocal of the scalar, i.e.,

$$\frac{\mathbf{A}}{m} = \left(\frac{1}{m}\right)\mathbf{A}. \qquad \text{I.4-I}$$

In particular, if the number m is the same as the magnitude of **A**, the vector resulting from the division will have a magnitude of unity. Its direction, of course, will be the same as that of **A**. Such *unit vectors* play an important role in vector algebra. When we come to analyze particular machines and structures, there is usually some symmetry present which makes a particular set of axes convenient. Rectangular cartesian axes $Oxyz$, pictured in Fig. 1.4-1a, constitute the commonest such set. It is customary to denote the unit vector in the x-direction by the symbol **i**, the unit vector in the y-direction by the symbol **j**, and the unit vector in the z-direction by the symbol **k**. The unit vectors **i, j, k** then form what is known as an *orthogonal triad*. Now suppose that we have a vector **A** whose projection on the x-axis is A_x, on the y-axis is A_y, and on the z-axis is A_z. This vector is shown in Fig. 1.4-1b. In Fig. 1.4-1c it is shown shifted so that its origin is at the origin of the cartesian coordinates. We see that the vector **A** may be expressed as the sum of three vectors given by the directed line segments $\overline{OP}, \overline{PQ}, \overline{QR}$. But the vector $\overline{OP}$ is conveniently written $A_x\mathbf{i}$ because the distance OP is A_x and the direction of $\overline{OP}$ is the same as that of the unit vector **i**. Similarly, $\overline{PQ}$ is represented by $A_y\mathbf{j}$ and $\overline{QR}$ by $A_z\mathbf{k}$. Summing up:

$$\mathbf{A} = A_x\mathbf{i} + A_y\mathbf{j} + A_z\mathbf{k}. \qquad \text{I.4-2}$$

Written in this way the vector is said to be expressed as the sum of its *cartesian components* $A_x\mathbf{i}$, $A_y\mathbf{j}$, and $A_z\mathbf{k}$. The numbers A_x, A_y, A_z are known as *scalar* cartesian components of **A**. They may be

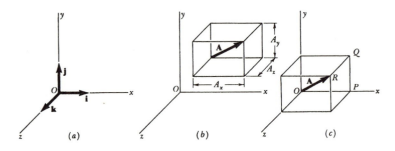

Fig. I.4-I

positive or negative, depending on whether **A** points toward the positive or negative end of the x-, y-, or z-axis. Often, where no confusion will result, A_x, A_y, and A_z are called simply cartesian components of **A**. It would be pedantic always to insist on preserving the distinction between $A_x\mathbf{i}$, the cartesian component in the x-direction, and A_x, the scalar cartesian component associated with the x-direction.

Once a vector has been expressed as the sum of its cartesian components it is a simple matter to compute the magnitude of the vector and the angles it makes with the coordinate axes. We see from Fig. I.4-1c that the distance RP is $(A_y^2 + A_z^2)^{\frac{1}{2}}$ and that OPR is a right triangle. It follows that

$$A = (A_x^2 + A_y^2 + A_z^2)^{\frac{1}{2}}. \qquad \text{I.4-3}$$

The cosine of the angle between OR and the positive x-axis is OP/OR; i.e., it is A_x/A. If we denote the direction cosines of the line OR by the symbols l, m, n we have

$$l = \frac{A_x}{A}, \qquad m = \frac{A_y}{A}, \qquad n = \frac{A_z}{A}. \qquad \text{I.4-4}$$

It follows from the substitution of Eqs. 1.4-4 in Eq. 1.4-3 that $l^2 + m^2 + n^2 = 1$. The separation of a vector into its cartesian components is so common a first step in mechanical analysis that it seems worthwhile to direct attention specifically to the implication of Eqs. 1.4-4. The scalar component of any vector in the direction of any axis is the product of the magnitude of the vector and the cosine of the angle between the vector and the axis.

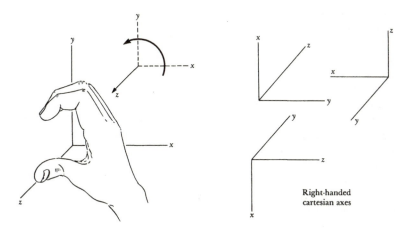

Right-handed
cartesian axes

Fig. 1.4-2   **Fig. 1.4-3**

We see from Eq. 1.4-2 that a vector requires three ordinary real numbers for its complete specification. This implies that, in general, a vector equation will be equivalent to three ordinary scalar equations. For example, the equation

$$\mathbf{A} + 3\mathbf{B} = 2\mathbf{j}$$

implies

$$(A_x\mathbf{i} + A_y\mathbf{j} + A_z\mathbf{k}) + 3(B_x\mathbf{i} + B_y\mathbf{j} + B_z\mathbf{k}) = 2\mathbf{j},$$

$$\mathbf{i}(A_x + 3B_x) + \mathbf{j}(A_y + 3B_y - 2) + \mathbf{k}(A_z + 3B_z) = 0$$

and this can be satisfied only if *all* the parenthetical terms vanish: i.e.,

$$A_x = -3B_x, \qquad A_y = 2 - 3B_y, \qquad A_z = -3B_z.$$

Of course if the vectors lie in one plane, say the *xy*-plane, the *z*-components of all the vectors in the equation will be zero and one of the three equations will simply say that $0 = 0$. In Example 1.3-4 we have seen that a planar vector equation yields two scalar equations.

Before leaving the subject of unit vectors and their relation to the use of cartesian axes, an important convention used in labeling the x-, y-, z-axes should be mentioned. These axes are named in accordance with what is known as the *right-hand rule*. This rule may be stated in a variety of ways. Probably the simplest is that when the fingertips of the right hand are made to pass from the positive end of the first-named axis, x, to the positive end of the second-named axis, y, the thumb of the right hand points toward the positive end of the third-named axis, z. The rule is illustrated in Fig. 1.4-2. Axes named in this way are called *right-handed axes*. (Sometimes we say that the three unit vectors $\mathbf{i}$, $\mathbf{j}$, $\mathbf{k}$ form an orthogonal right-handed triad.) It may be noted that the axes pictured in Fig. 1.4-3 are also examples of right-handed axes. If the symbols x and y were interchanged, however, the axes thus obtained would not be right-handed axes, but left-handed. At the present point there is no essential need to restrict ourselves to right-handed axes. Later, however, when the vector product operation is introduced (Section 1.7) we find that its definition involves a similar right-hand rule. If the same rule is not used for the choice of axes and for the definition of the vector product, sign errors are apt to arise when the results of operations involving the vector product are reduced to cartesian component form.* We therefore always adhere to the use of right-handed cartesian axes.

Example 1.4-1

What is the resultant $\mathbf{R}$ *of the vectors*

$$\mathbf{F}_1 = 2\mathbf{i}+\mathbf{j}-2\mathbf{k}, \qquad \mathbf{F}_2 = 4\mathbf{i}+7\mathbf{j}-4\mathbf{k}, \qquad \mathbf{F}_3 = -8\mathbf{i}-9\mathbf{j}+12\mathbf{k}\,?$$

Solution: These vectors have been expressed in terms of their cartesian components. Since $a\mathbf{i}+b\mathbf{i}=(a+b)\mathbf{i}$, and similarly for the $\mathbf{j}$ and $\mathbf{k}$ components, we need only add the scalar components algebraically:

$$\mathbf{F}_1+\mathbf{F}_2+\mathbf{F}_3 = (2\mathbf{i}+\mathbf{j}-2\mathbf{k})+(4\mathbf{i}+7\mathbf{j}-4\mathbf{k})+(-8\mathbf{i}-9\mathbf{j}+12\mathbf{k}),$$

$$\mathbf{F}_1+\mathbf{F}_2+\mathbf{F}_3 = (2+4-8)\mathbf{i}+(1+7-9)\mathbf{j}+(-2-4+12)\mathbf{k} = -2\mathbf{i}-\mathbf{j}+6\mathbf{k},$$

$$\mathbf{R} = -2\mathbf{i}-\mathbf{j}+6\mathbf{k}.$$

This is the sum or resultant vector. Its magnitude is

$$R = [(-2)^2+(-1)^2+(6)^2]^{1/2} = 6.403.$$

---

* More important theoretically, some characteristics of the vector product are not preserved under such right- to left-handed "reflections"; this has interesting consequences in, among other areas, the theory of turbulent fluid flow.

The direction cosines of the resultant are

$$l = \frac{-2}{6.403} = -0.312, \quad m = \frac{-1}{6.403} = -0.156, \quad n = \frac{6}{6.403} = 0.937$$

so that the resultant vector, **R**, makes an angle of 108.2° with the positive *x*-axis, an angle of 99° with the positive *y*-axis, and an angle of 20.4° with the positive *z*-axis. It may be seen from this example that the efficient way to add and subtract vectors is first to separate them into their rectangular cartesian components.*

Example 1.4-2

What is the unit vector in the direction of $A = 6i - 10j + 15k$?
Solution: The magnitude of **A** is

$$A = (36 + 100 + 225)^{\frac{1}{2}} = 19.$$

The unit vector having the direction of **A** is $(1/A)$**A**. If we use the symbol e_A to denote this unit vector,

$$e_A = \frac{1}{19}(6i - 10j + 15k) = \frac{6}{19}i - \frac{10}{19}j + \frac{15}{19}k.$$

The student should notice that any vector may be written in the form $A = Ae_A$. When the vector **A** of this example is so expressed it is written

$$A = 19\left(\frac{6}{19}i - \frac{10}{19}j + \frac{15}{19}k\right).$$

The numbers $6/19$, $-10/19$, $15/19$ are the direction cosines of the vector **A**.

Example 1.4-3

The origin of the vector **V** *is at the point whose coordinates are* (1, 2, 3) *and its terminus is at the point* (9, −7, 15). *Express this vector as the sum of its cartesian components.*
Solution: The vector is pictured in Fig. 1.4-4. In proceeding from point *A*, the origin of the vector, to point *B*, its terminus, we proceed from a point whose *x*-coordinate is 1 to a point whose *x*-coordinate is 9—that is, 8 units in the positive *x*-direction. Similarly we proceed 9 units in the negative *y*-direction and 12 units in the positive *z*-direction. It follows that $V_x = 8$, $V_y = -9$, $V_z = 12$, and

$$V = 8i - 9j + 12k.$$

---

* Or any other set of orthogonal components, such as those in the polar coordinate systems of the next section.

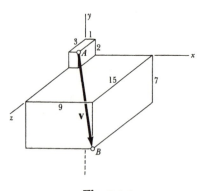

Fig. 1.4-4

Example 1.4-4

Two vectors, **A** *and* **B**, *intersect at O. Find the equation of the bisector of the angle between* **A** *and* **B**.

Solution: Construct the unit vectors e_A and e_B in the directions of **A** and **B**, respectively, as shown in Fig. 1.4-5:

$$e_A = \left(\frac{1}{A}\right)\mathbf{A}, \qquad e_B = \left(\frac{1}{B}\right)\mathbf{B}.$$

Now complete the parallelogram whose sides are e_A and e_B. The vertex of the parallelogram opposite O (point M in Fig. 1.4-5) lies on the bisector of the angle between **A** and **B** because all sides of the parallelogram are of the same unit length. Hence $e_A + e_B$ is a vector whose terminus lies on the bisector. The vector, **r**, to any point on the bisector may therefore be written

$$\mathbf{r} = m\left[\left(\frac{1}{A}\right)\mathbf{A} + \left(\frac{1}{B}\right)\mathbf{B}\right] \qquad \text{1.4-5}$$

where m is simply a scalar multiplier. By assigning numerical values to

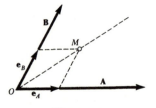

Fig. 1.4-5

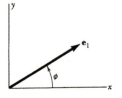

Fig. 1.4-6

m we may have **r** be the vector to any point on the angle bisector, so that Eq. 1.4-5 is described as the equation of the bisector.

Example 1.4-5

Find the cartesian components of a unit vector in the xy-plane making an angle ϕ with the x-axis (Fig. 1.4-6).

Solution: The x-component of the vector, which we may call $\mathbf{e}_1$, is the magnitude of the vector, 1, multiplied by the cosine of the angle between $\mathbf{e}_1$ and the x-axis; the y-component will be 1 multiplied by the cosine of $(\pi/2) - \phi$. We have then

$$\mathbf{e}_1 = (\cos \phi)\mathbf{i} + (\sin \phi)\mathbf{j}.$$

It is easy to verify that this is indeed a vector of unit magnitude.

1.5 Plane and Cylindrical Polar Coordinates

Although rectangular cartesian coordinates form the most commonly encountered coordinate system, a wide variety of special coordinate systems appears in the literature of mechanics. Cylindrical polar coordinates find application when the question under

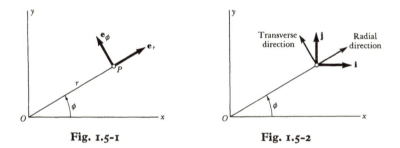

Fig. 1.5-1   Fig. 1.5-2

analysis is symmetrical about some axis. We begin with the case of vectors lying in the xy-plane. Associated with every point P there is a value of the radial distance r measured from the origin or pole, O, and an angle ϕ measured counterclockwise from the positive end of the x-axis, as shown in Fig. 1.5-1. Introduce unit vectors $\mathbf{e}_r$ directed radially outward from P and $\mathbf{e}_\phi$ at right angles to $\mathbf{e}_r$ in the direction of increasing ϕ. Clearly any vector passing through P may be described by means of components in these *radial* and *transverse directions*. As is evident in Fig. 1.5-2, the unit vector in the x-direction will have a component in the radial direction equal to the product of its magnitude, 1, and the cosine of the angle ϕ between the x-axis and the radial direction, together with a component in the transverse direction equal to the magnitude, 1, multiplied by the cosine of the angle $\phi+(\pi/2)$ between the x-axis and the transverse direction. In symbols, since $\cos[\phi+(\pi/2)] = -\sin\phi$,

$$\mathbf{i} = \mathbf{e}_r\cos\phi - \mathbf{e}_\phi\sin\phi. \qquad \textbf{1.5-1a}$$

Similarly,

$$\mathbf{j} = \mathbf{e}_r\sin\phi + \mathbf{e}_\phi\cos\phi. \qquad \textbf{1.5-1b}$$

We may solve these equations to find the cartesian components of $\mathbf{e}_r$ and $\mathbf{e}_\phi$:

$$\mathbf{e}_r = \quad \mathbf{i}\cos\phi + \mathbf{j}\sin\phi,$$
$$\mathbf{e}_\phi = -\mathbf{i}\sin\phi + \mathbf{j}\cos\phi. \qquad \textbf{1.5-2}$$

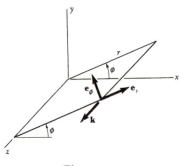

Fig. 1.5-3

Now a vector in the xy-plane may be expressed in terms of its cartesian components, say

$$\mathbf{A} = A_x\mathbf{i} + A_y\mathbf{j};$$

or, alternatively, in terms of its plane polar components,

$$\mathbf{A} = A_r\mathbf{e}_r + A_\phi\mathbf{e}_\phi.$$

Since the vector $\mathbf{A}$ is independent of the coordinate system we use to describe it,

$$A_x\mathbf{i} + A_y\mathbf{j} = A_r\mathbf{e}_r + A_\phi\mathbf{e}_\phi.$$

In view of Eq. 1.5-1 this implies

$$A_x(\mathbf{e}_r\cos\phi - \mathbf{e}_\phi\sin\phi) + A_y(\mathbf{e}_r\sin\phi + \mathbf{e}_\phi\cos\phi) = A_r\mathbf{e}_r + A_\phi\mathbf{e}_\phi,$$
$$\mathbf{e}_r(A_x\cos\phi + A_y\sin\phi - A_r) + \mathbf{e}_\phi(-A_x\sin\phi + A_y\cos\phi - A_\phi) = \mathbf{0}.$$

This relation can be satisfied only if each of the quantities in parentheses vanishes. We may infer that the relations between the plane polar components of a vector and its plane cartesian components are

$$A_r = A_x \cos \phi + A_y \sin \phi,$$
$$A_\phi = -A_x \sin \phi + A_y \cos \phi. \qquad \textbf{1.5-3}$$

These ideas are easily extended to three dimensions by introducing a z-axis through the pole O, at right angles to the $r\phi$-plane. This is the same as the z-axis of the cartesian frame. The unit vectors $\mathbf{e}_r$, $\mathbf{e}_\phi$, $\mathbf{k}$, taken in that order, form a right-handed orthogonal triad, as shown in Fig. 1.5-3. The coordinates (r, ϕ, z) are known as *cylindrical polar coordinates*. The z-direction is, in this context, sometimes called the axial direction.

Example 1.5-1

Express the magnitude of a vector in terms of its cylindrical polar components.

Solution: We have

$$A = (A_x^2 + A_y^2 + A_z^2)^{1/2}.$$

If we solve Eqs. 1.5-3 for A_x and A_y in terms of A_r and A_ϕ we have

$$A_x = A_r \cos \phi - A_\phi \sin \phi,$$
$$A_y = A_r \sin \phi + A_\phi \cos \phi.$$

On substituting these in the expression for A we find that

$$A = (A_r^2 + A_\phi^2 + A_z^2)^{1/2}.$$

This result reflects the fact that cylindrical polar coordinates form an *orthogonal coordinate system*, just as the rectangular cartesian coordinates do.

1.6 The Scalar Product; Orthogonal Projection and Scalar Components

In the last two sections the scalar components of a vector relative to some particular orthogonal coordinate systems were found. In this section, we discuss the problem of finding the (orthogonal) scalar component of a vector in any direction of space. Further, we introduce a most important operation on two vectors—an operation that both generalizes the scalar component concept and permits us to speak about some relations between vectors independently of the choice of a particular coordinate system.

We take as fundamental the theorem which states that, given any point in space and any line not passing through the point, one and only one perpendicular may be drawn from the point to the line. The point of intersection of the given line and the perpendicular is called the *orthogonal projection* of the original point on the line. In order to treat all cases, if the given line passes through the given point, the point itself is taken as its projection.

Project two points orthogonally on the given line. The two points in space determine a line segment; the two projection points also determine a segment on the line of projection. The latter segment is the orthogonal projection of the first. If we now assign directions to the lines involved, we may unambiguously assign an algebraic sign to the lengths of the projections and add the projections of many line segments to determine a resultant projection. We may also define the angle between two directed lines.

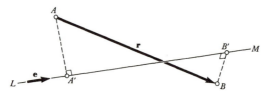

Fig. 1.6-1

Suppose the fixed line LM in space is our line of projection, and we pick a direction on it by assigning a unit vector $\mathbf{e}$ along it (Fig. 1.6-1). Consider the line segment joining any two points A and B in space, and give it direction by taking it as a vector $\mathbf{r}$ from, say, A to B. Project A and B on LM, obtaining points A' and B'; the segment $A'B'$ is the projection of $\overline{AB}$ or of $\mathbf{r}$. If the points A', B' taken in order define a segment having the sense of $\mathbf{e}$, then we attach a positive sign to the length of $A'B'$; if $A'B'$ is opposed to $\mathbf{e}$, we count the length as negative. The signed scalar quantity $A'B'$ is the *orthogonal scalar component* of $\mathbf{r}$ on the directed line LM. Clearly, if we change the assigned direction on LM, i.e., take $-\mathbf{e}$ or M to L as the direction, we change the sign of the scalar projection of $\mathbf{r}$; and, if we change the order of A and B, i.e., use $-\mathbf{r}$ or B to A, we change the sign of the projection on LM.

The next fundamental result is that the sum of the (signed) projections on a fixed line is the projection of the resultant, or sum, of the vectors projected—"the sum of the components is the component of the sum." Consider three points A, B, C and a fixed line LM with direction $\mathbf{e}$ given on it (Fig. 1.6-2). Let the directed line segment $\overline{AC}$ be the vector $\mathbf{R}$; the segment $\overline{CB}$, the vector $\mathbf{p}$;

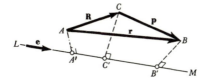

Fig. 1.6-2

and the segment $\overline{AB}$, the vector $\mathbf{r}$ again. Clearly $\mathbf{r} = \mathbf{R} + \mathbf{p}$, by the triangle law definition of vector addition. Let A', B', and C' be the projections of A, B, and C on LM. Our assertion is that the sum of the signed distances $A'C'$ and $C'B'$ equals the signed distance $A'B'$. In Fig. 1.6-2, all three projections are shown as positive; it is clear that the theorem is true in all cases, however, from the very definition of projection. The projection $A'B'$ of $\mathbf{r}$ is, by definition, determined by the projections A' and B' of the points A and B. Wherever the projection C' on LM of point C may be, one starts at the projection of A and ends at the projection of B; symbolically, $A'C' + C'B' = A'B'$, in which ordinary algebraic addition of real numbers is used.

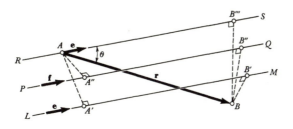

Fig. 1.6-3

From this theorem, it follows that the projection on a fixed line of the vector sum of any finite number of vectors $\mathbf{r}_1$, $\mathbf{r}_2$, $\mathbf{r}_3$, ... $\mathbf{r}_N$, taken in order with one vector starting where the previous one stops, is the algebraic sum of the projections of the separate vectors.

The third fundamental theorem about projections that we need enables us to remove the restriction to projections on a fixed line and hence to define the angle between two lines in space. This theorem states that the projections of a given vector on two parallel lines oriented in the same way are equal. Suppose that we consider the configuration of Fig. 1.6-1 again, but draw another line PQ parallel to LM, assign a unit vector $\mathbf{f}$ on PQ to orient PQ, and project AB on PQ orthogonally obtaining $A''B''$ (Fig. 1.6-3). If $\mathbf{f}=\mathbf{e}$, the theorem asserts that $A''B''=A'B'$. (Of course, if PQ is oriented oppositely to LM, i.e., $\mathbf{f}=-\mathbf{e}$, then the signed projection $A''B''=-A'B'$.) We have already used the idea of this theorem implicitly in defining equality of vectors along different lines and in moving vectors together to define their sum by the triangle or parallelogram law—the idea that a vector may be translated parallel to itself without changing its magnitude or direction or scalar component in any direction.

In particular, draw the line RS through A parallel to LM, and orient RS in the same way as LM (Fig. 1.6-3). A is then its own projection on RS; let B''' be the projection of B, so that $A'B'=A''B''=AB'''$. The two intersecting lines AB and RS determine a plane; in this plane, we may measure the angle θ between AB and RS. We take θ between 0° and 180°, measured between the positive senses of AB and RS, or of $\mathbf{r}$ and $\mathbf{e}$. The orthogonal scalar component of AB in the direction of RS is then the product of the magnitude of AB and the cosine of θ: $AB'''=r\cos\theta$. This is the same as the scalar component of $\mathbf{r}$ on any similarly oriented line parallel to RS, and the angle θ is, by definition, the angle between $\mathbf{r}$ (or $\overline{AB}$) and any of the lines with direction $\mathbf{e}$. The vector component of $\mathbf{r}$ in the direction of $\mathbf{e}$ is simply $r\cos\theta\,\mathbf{e}$.

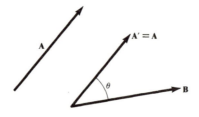

Fig. 1.6-4

These results are independent of the choice of a coordinate system relative to which the points and lines are given coordinate values or direction cosines. It is convenient—more, it is both necessary for a coherent theoretical treatment and useful for computational purposes—to denote the operation of "projection of a vector" in a special way. Indeed, we generalize the projection operation to an operation on any two vectors **A** and **B** (Fig. 1.6-4). Move one, say **A**, parallel to itself, until it coincides with the equal vector **A'** emanating from the initial point of **B**. The two lines along which **A'** and **B** lie determine a plane in which the included angle θ between the directions of **A** = **A'** and **B** may be measured. We define the symbol **A** · **B** by

$$\mathbf{A} \cdot \mathbf{B} = AB \cos \theta, \qquad \text{1.6-1}$$

where A and B are, as usual, the magnitudes of **A** and **B**. The operation indicated by **A** · **B** is known as the *scalar product*, or the "dot" product, of the vectors **A** and **B**. It is a "scalar" because the result of the operation is an ordinary real number; it is a "product" because the operation has some of the algebraic properties of ordinary multiplication.

Let us investigate some of the properties of the scalar product. First, we see that the orthogonal projection or scalar component operation can be represented symbolically by the scalar product. If **B** is a unit vector, its magnitude is unity; then **A** · **B** = $A \cos \theta$, the scalar component of **A** in the direction of **B**. For the configuration of Fig. 1.6-1 or 1.6-3 the scalar component of **r** in the direction LM (or PQ or RS) is **r** · **e**. Second, the scalar product of a vector with itself is the square of the magnitude of the vector:

$$\mathbf{A} \cdot \mathbf{A} = A^2. \qquad \text{1.6-2}$$

This is true since the "included" angle between a vector and itself is zero. Third, the scalar product provides a test for orthogonality or perpendicularity. When $\theta = 90°$, $\cos \theta$ is zero; thus the scalar product of two non-vanishing vectors is zero if, and only if, the vectors are orthogonal (**A** · **B** = 0, **A** ≠ **0**, **B** ≠ **0**). In particular, these results are useful in expressing the relations among the members of the orthogonal unit triad **i**, **j**, **k** along the positive axes of a rectangular cartesian system:

$$\mathbf{i} \cdot \mathbf{i} = \mathbf{j} \cdot \mathbf{j} = \mathbf{k} \cdot \mathbf{k} = 1;$$

$$\mathbf{i} \cdot \mathbf{j} = \mathbf{j} \cdot \mathbf{k} = \mathbf{k} \cdot \mathbf{i} = \mathbf{j} \cdot \mathbf{i} = \mathbf{k} \cdot \mathbf{j} = \mathbf{i} \cdot \mathbf{k} = 0. \qquad \text{1.6-3}$$

Let us turn now to the mathematical properties that lead us to call $\mathbf{A}\cdot\mathbf{B}$ a "product." First of all, the scalar product is *commutative*, as is ordinary multiplication; this follows from the definition, Eq. 1.6-1:

$$\mathbf{A}\cdot\mathbf{B} = AB\cos\theta = \mathbf{B}\cdot\mathbf{A}. \qquad \textbf{1.6-4}$$

Also, multiplication by a scalar is *associative* over the scalar product:

$$k(\mathbf{A}\cdot\mathbf{B}) = (k\mathbf{A})\cdot\mathbf{B} = \mathbf{A}\cdot(k\mathbf{B}). \qquad \textbf{1.6-5}$$

Finally, the scalar product is *distributive over vector addition*:

$$\mathbf{A}\cdot(\mathbf{B}+\mathbf{C}) = \mathbf{A}\cdot\mathbf{B}+\mathbf{A}\cdot\mathbf{C}, \qquad \textbf{1.6-6a}$$

$$(\mathbf{B}+\mathbf{C})\cdot\mathbf{A} = \mathbf{B}\cdot\mathbf{A}+\mathbf{C}\cdot\mathbf{A}. \qquad \textbf{1.6-6b}$$

We shall prove the theorem embodied in Eq. 1.6-6b; Eq. 1.6-6a then follows from the law of commutativity (Eq. 1.6-4).

Equation 1.6-6b follows from the projection theorems and Eq. 1.6-5. Consider any three vectors $\mathbf{A}$, $\mathbf{B}$, and $\mathbf{C}$; bring $\mathbf{B}$ and $\mathbf{C}$ together in order to add them by the triangle law, and draw the line

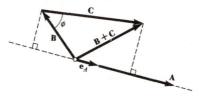

Fig. 1.6-5

through the initial point of $\mathbf{B}$ with the direction of $\mathbf{A}$, i.e., assign the direction $\mathbf{e}_A = \mathbf{A}/|\mathbf{A}|$ to it (Fig. 1.6-5). Since the projection of a sum on a fixed line equals the sum of the projections,

$$(\mathbf{B}+\mathbf{C})\cdot\mathbf{e}_A = \mathbf{B}\cdot\mathbf{e}_A+\mathbf{C}\cdot\mathbf{e}_A.$$

Now multiply by the magnitude of $\mathbf{A}$; by Eq. 1.6-5,

$$A(\mathbf{B}+\mathbf{C})\cdot\mathbf{e}_A = (\mathbf{B}+\mathbf{C})\cdot(A\mathbf{e}_A) = (\mathbf{B}+\mathbf{C})\cdot\mathbf{A},$$

and $A(\mathbf{B}\cdot\mathbf{e}_A+\mathbf{C}\cdot\mathbf{e}_A) = \mathbf{B}\cdot\mathbf{A}+\mathbf{C}\cdot\mathbf{A}$, thus proving Eq. 1.6-6b.

Although, for the reasons given in the last two paragraphs, it is legitimate and suggestive to give the name of scalar *product* to the operation $\mathbf{A}\cdot\mathbf{B}$ on the two vectors $\mathbf{A}$ and $\mathbf{B}$, the reader should be aware that this newly defined operation does not have all the properties of the product of ordinary real numbers. For example, the product, *abc*, of three scalars *a*, *b*, and *c* is a perfectly definite number whereas the group of symbols $(\mathbf{A}\cdot\mathbf{B})\cdot\mathbf{C}$ is meaningless. Further, the law of cancellation does not hold. As we have already noted, whereas $ab=0$ implies either $a=0$ or $b=0$, the equation $\mathbf{A}\cdot\mathbf{B}=0$ allows a third possibility: $\mathbf{A}$ and $\mathbf{B}$ are perpendicular, i.e., make a right angle with one another. Equivalently, whereas the scalar equation $ab=cb$ implies that $a=c$ if $b\neq 0$, the vector equation $\mathbf{A}\cdot\mathbf{B}=\mathbf{C}\cdot\mathbf{B}$, with $\mathbf{B}\neq\mathbf{0}$, implies that either $\mathbf{A}=\mathbf{C}$ or $\mathbf{A}-\mathbf{C}$ is orthogonal to $\mathbf{B}$.

In the development of the theory of mechanics, symbolic operation with the scalar product is often an efficient way of obtaining results generally true without regard to the choice of a particular coordinate frame. In solving particular problems, however, a particular set of coordinates is usually introduced, and we must have a way of reducing the scalar product to a computational form suitable for use. For rectangular cartesian coordinates, this is easily found from the distributive law 1.6-6 and the unit triad relations 1.6-3. Suppose two vectors $\mathbf{A}$ and $\mathbf{B}$ have the representations $\mathbf{A}=A_x\mathbf{i}+A_y\mathbf{j}+A_z\mathbf{k}$, $\mathbf{B}=B_x\mathbf{i}+B_y\mathbf{j}+B_z\mathbf{k}$ in some fixed (x,y,z) system. Then

$$\mathbf{A}\cdot\mathbf{B} = \mathbf{A}\cdot(B_x\mathbf{i}+B_y\mathbf{j}+B_z\mathbf{k})$$

$$= \mathbf{A}\cdot(B_x\mathbf{i})+\mathbf{A}\cdot(B_y\mathbf{j})+\mathbf{A}\cdot(B_z\mathbf{k})$$

$$= A_xB_x(\mathbf{i}\cdot\mathbf{i})+A_yB_x(\mathbf{j}\cdot\mathbf{i})+A_zB_x(\mathbf{k}\cdot\mathbf{i})$$

$$+A_xB_y(\mathbf{i}\cdot\mathbf{j})+A_yB_y(\mathbf{j}\cdot\mathbf{j})+A_zB_y(\mathbf{k}\cdot\mathbf{j})$$

$$+A_xB_z(\mathbf{i}\cdot\mathbf{k})+A_yB_z(\mathbf{j}\cdot\mathbf{k})+A_zB_z(\mathbf{k}\cdot\mathbf{k})$$

or

$$\mathbf{A}\cdot\mathbf{B} = A_xB_x+A_yB_y+A_zB_z. \qquad \text{1.6-7}$$

Finally, we may note that the scalar component and orthogonal projection concept is not the only useful one expressed by the scalar product. We shall see others in the examples of this section and in later sections. In dynamics, the important concept of the power of a force is expressed by the scalar product; if $\mathbf{A}$ is the force on a particle and $\mathbf{B}$ is the instantaneous velocity of the particle, $\mathbf{A}\cdot\mathbf{B}$ is the power, or time-rate of doing work, of the force.

Example **1.6-1**

What is the angle between the vectors $\mathbf{A}=2\mathbf{i}-5\mathbf{j}+6\mathbf{k}$ *and* $\mathbf{B}=4\mathbf{i}-2\mathbf{j}-3\mathbf{k}$ *?*
Solution: Since $\mathbf{A}\cdot\mathbf{B}=AB\cos\theta$, we have

$$\cos\theta = \frac{(\mathbf{A}\cdot\mathbf{B})}{AB}.$$

Here, $\mathbf{A \cdot B} = A_x B_x + A_y B_y + A_z B_z = 8 + 10 - 18 = 0$; therefore, $\cos \theta = 0$, $\theta = 90°$, and $\mathbf{A}$ and $\mathbf{B}$ are perpendicular.

Example 1.6-2

Find the vector or vectors $\mathbf{A} = A_y \mathbf{j} + 4\mathbf{k}$ *making an angle of* 60° *with* $\mathbf{B} = \mathbf{i} + \mathbf{j} + \mathbf{k}$.

Solution: Since $\mathbf{A \cdot B} = AB \cos \theta = AB \cos 60° = \tfrac{1}{2}AB$, we have

$$\mathbf{A \cdot B} = A_y + 4 = \left(\frac{1}{2}\right)(\sqrt{16 + A_y^2})(\sqrt{3}).$$

Upon squaring, we obtain a quadratic equation for the determination of A_y; note that only roots greater than -4 are valid. We find

$$A_y^2 + 8A_y + 16 = 12 + \frac{3}{4} A_y^2,$$

$$A_y^2 + 32A_y + 16 = 0,$$

$$A_y = -16 \pm 4\sqrt{15}.$$

The only root that leads to a solution is $A_y = -16 + 4\sqrt{15}$, and the only vector is

$$\mathbf{A} = -0.508\mathbf{j} + 4\mathbf{k}.$$

Example 1.6-3

Derive the law of cosines of trigonometry.

Solution: Refer to Fig. 1.6-5, where a triangle with sides $B = |\mathbf{B}|$, $C = |\mathbf{C}|$, and $D = |\mathbf{D}| = |\mathbf{B} + \mathbf{C}|$ is shown. Let ϕ be the angle included by the sides of length B and C. We wish to prove $D^2 = B^2 + C^2 - 2BC \cos \phi$.
We have

$$D^2 = (\mathbf{B} + \mathbf{C}) \cdot (\mathbf{B} + \mathbf{C}) = B^2 + 2\mathbf{B} \cdot \mathbf{C} + C^2.$$

Now $\mathbf{B} \cdot \mathbf{C} = BC \cos \theta$, where θ is the angle included between $\mathbf{B}$ and $\mathbf{C}$. The angle ϕ is the supplement of θ, i.e., $\phi = \pi - \theta$; therefore

$$\mathbf{B} \cdot \mathbf{C} = BC \cos (\pi - \phi) = -BC \cos \phi.$$

Thus

$$D^2 = B^2 + C^2 - 2BC \cos \phi.$$

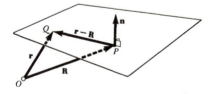

Fig. 1.6-6

Example **1.6-4**

A plane in space is determined by one point in the plane and the normal to the plane. Find the equation of the plane.

Solution: Let **R** be the position vector to the given point, P, in the plane from some fixed origin, O; let **n** be a unit vector normal to the plane; and let **r** be the position vector from O to a general point, Q, in the plane (Fig. 1.6-6). Since **n** is normal to the plane, any vector in the plane is orthogonal to **n**. Thus **r** − **R**, the position of Q relative to P, is perpendicular to **n**, and the equation of the plane is

$$(\mathbf{r} - \mathbf{R}) \cdot \mathbf{n} = 0.$$

1.7 The Vector Product; Moment of a Vector about a Point

A second operation on two vectors, this time producing a vector instead of a scalar and having some attributes of ordinary multiplication, is of fundamental importance for the expression of many

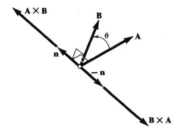

Fig. 1.7-1

physical concepts. Among these may be mentioned the moment of a force, moment of momentum, the force induced by a magnetic field on a moving charge, and the velocity of a point in a rotating rigid body. The *vector product*, or "cross" product, of two vectors is defined, independently of any particular coordinate system, by the relation:

$$\mathbf{A} \times \mathbf{B} = AB \sin \theta \mathbf{n}. \qquad \textbf{1.7-1}$$

Here A and B are the magnitudes of **A** and **B** and θ is the angle included between **A** and **B**, as defined in the previous section. The vector **n** is a unit vector orthogonal to the plane determined by **A** and **B** when they are made to come together by parallel translation. The direction of **n** is determined by the right-hand rule for rotation (Fig. 1.7-1); i.e., if the fingers of one's right hand are made to curl from the direction of **A** to the direction of **B**, the right thumb points in the direction of **n** (see Section 1.4).

Let us investigate the properties of the vector product as we did those of the scalar product. First, the vanishing of the vector product is a test for parallelism of two non-zero vectors. If $\mathbf{A} \times \mathbf{B} = \mathbf{0}$, with $\mathbf{A} \neq \mathbf{0}$, $\mathbf{B} \neq \mathbf{0}$, then $\sin \theta$ must vanish and $\theta = 0°$ or $180°$. Thus **A** and **B** have the same or opposite directions and hence are parallel. This means that the cancellation law of scalar multiplication cannot be extended to the cross product: if $\mathbf{A} \times \mathbf{B} = \mathbf{A} \times \mathbf{C}$, we cannot conclude that $\mathbf{B} = \mathbf{C}$, since the equation is also satisfied if $\mathbf{B} - \mathbf{C}$ is parallel to **A**. The vector product of a vector with itself vanishes.

The vector product operation is not commutative, either; in fact, by the very definition, the inversion of the two "factors" changes the sign of the product:

$$\mathbf{B} \times \mathbf{A} = BA \sin \theta(-\mathbf{n}) = -\mathbf{A} \times \mathbf{B}. \qquad \textbf{1.7-2}$$

The vectors $\mathbf{A} \times \mathbf{B}$ and $\mathbf{B} \times \mathbf{A}$ thus have the same magnitudes but opposite directions (Fig. 1.7-1).

Multiplication of the vector product by a scalar is "associative," i.e.,

$$k(\mathbf{A} \times \mathbf{B}) = (k\mathbf{A}) \times \mathbf{B} = \mathbf{A} \times (k\mathbf{B}). \qquad \textbf{1.7-3}$$

The student should have no difficulty proving this last result for himself.

Meaning can be given to a vector triple product, $\mathbf{A} \times (\mathbf{B} \times \mathbf{C})$; however, the associative law does not hold:

$$\mathbf{A} \times (\mathbf{B} \times \mathbf{C}) \neq (\mathbf{A} \times \mathbf{B}) \times \mathbf{C}.$$

Since $\mathbf{B} \times \mathbf{C}$ is perpendicular to the plane of **B** and **C**, $\mathbf{A} \times (\mathbf{B} \times \mathbf{C})$ must lie in the plane of **B** and **C**. Similarly, $(\mathbf{A} \times \mathbf{B}) \times \mathbf{C}$ must lie in the plane of **A** and **B**. These two planes will not in general be the same, and the two vectors, therefore, will not be equal.

The vector product is distributive over vector addition:

$$\mathbf{A} \times (\mathbf{B} + \mathbf{C}) = \mathbf{A} \times \mathbf{B} + \mathbf{A} \times \mathbf{C},$$

$$(\mathbf{B} + \mathbf{C}) \times \mathbf{A} = \mathbf{B} \times \mathbf{A} + \mathbf{C} \times \mathbf{A}. \qquad \textbf{1.7-4}$$

This will be proved in the next section in Example 1.8-1. One cautionary note is in order: because of the non-commutative nature of the cross product, the order of the symbols must be the same on each side of a vector equation such as 1.7-4.

It may be seen that very few of the attributes of ordinary multiplication are carried over to the vector product. Yet the facts that the distributive law holds as well as that the ordinary product of the two magnitudes appears in the definition are sufficient to justify the use of the word "product" for this operation.

One important concept defined directly in terms of the vector product is that of the *moment of a vector about a point*. Let a vector

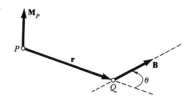

Fig. 1.7-2

B emanate from point Q in space, and let P be any other point in space (Fig. 1.7-2). Let **r** be the position of Q with respect to P. Then the moment vector $\mathbf{M}_P$ of **B** with respect to, or about, P is defined to be

$$\mathbf{M}_P = \mathbf{r} \times \mathbf{B}. \qquad 1.7\text{-}5$$

In statics, the moment with which we shall be most concerned is the moment of a force vector.

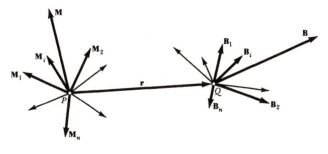

Fig. 1.7-3

There are two basic theorems about moments of vectors that we need. The first is the vector form of *Varignon's theorem*: The sum of the moments of any (finite) number of vectors emanating from one point, Q, about any other point, P, is the moment about point P of the resultant vector. Consider any set of vectors $\mathbf{B}_1, \mathbf{B}_2, \ldots \mathbf{B}_n$ (or simply $\mathbf{B}_i$, $i = 1, 2, \ldots n$) emanating from Q and any other point P

such that $\mathbf{r}$ is the position of Q relative to P (Fig. 1.7-3). Let $\mathbf{M}_1, \ldots \mathbf{M}_n$ be the moments about P of $\mathbf{B}_1, \ldots \mathbf{B}_n$: $\mathbf{M}_i = \mathbf{r} \times \mathbf{B}_i$, $i = 1, 2, \ldots n$. Let $\mathbf{B}$ be the resultant of the $\mathbf{B}_i$:

$$\mathbf{B} = \mathbf{B}_1 + \mathbf{B}_2 + \ldots \mathbf{B}_n = \sum_{i=1}^{n} \mathbf{B}_i,$$

and $\mathbf{M}$ the resultant of the $\mathbf{M}_i$:

$$\mathbf{M} = \mathbf{M}_1 + \mathbf{M}_2 + \ldots \mathbf{M}_n = \sum_{i=1}^{n} \mathbf{M}_i.$$

Varignon's theorem states that $\mathbf{M}$ is the moment of $\mathbf{B}$ about P: $\mathbf{M} = \mathbf{r} \times \mathbf{B}$. The proof follows from the distributive property of the vector product, which we shall prove in Section 1.8:

$$\mathbf{r} \times \mathbf{B} = \mathbf{r} \times \left(\sum_{i=1}^{n} \mathbf{B}_i \right) = \sum_{i=1}^{n} (\mathbf{r} \times \mathbf{B}_i) = \sum_{i=1}^{n} \mathbf{M}_i = \mathbf{M}. \qquad \textbf{1.7-6}$$

For Varignon's theorem as stated, we must have all vectors emanating from Q; that is, the vectors are *bound*, or tied, to the point Q—as

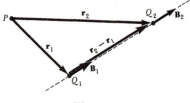

Fig. 1.7-4

opposed to the free vectors of Section 1.2 and succeeding sections. In fact, they need not be bound to Q; they need only have their lines of action concurrent at Q for Varignon's theorem to be true. That is, they need only be *sliding* vectors with concurrent lines of action. The sliding vector is one not free to move anywhere in space for the purposes of the computation to be made, but one which may be "slid" along its line of action (the line through a point on the vector in the direction of the vector itself) without affecting the result of whatever operation is to be performed on it.

In particular, the moment of a vector about a point P is independent of the choice of the point Q from which the vector emanates, provided only that Q be some point on the line of action of the vector.

This is the second fundamental theorem concerning moments. Consider two vectors $\mathbf{B}_1 = \mathbf{B}_2 = \mathbf{B}$ equal in magnitude and direction and having the same line of action, one emanating from Q_1 and the other from Q_2 (Fig. 1.7-4). Let $\mathbf{r}_1$ and $\mathbf{r}_2$ be the positions of Q_1 and Q_2 relative to P. Then the moment of $\mathbf{B}_2$ about P is $\mathbf{r}_2 \times \mathbf{B}_2 = \mathbf{r}_2 \times \mathbf{B}$ and the moment of $\mathbf{B}_1$ is $\mathbf{r}_1 \times \mathbf{B}_1 = \mathbf{r}_1 \times \mathbf{B}$. We have

$$\mathbf{r}_2 \times \mathbf{B}_2 = [\mathbf{r}_1 + (\mathbf{r}_2 - \mathbf{r}_1)] \times \mathbf{B} = \mathbf{r}_1 \times \mathbf{B} + (\mathbf{r}_2 - \mathbf{r}_1) \times \mathbf{B}.$$

But $\mathbf{r}_2 - \mathbf{r}_1$ is the position of Q_2 relative to Q_1 and hence lies along the line $Q_1 Q_2$, i.e., is parallel to $\mathbf{B}$. Therefore, $(\mathbf{r}_2 - \mathbf{r}_1) \times \mathbf{B} = \mathbf{0}$, and

$$\mathbf{r}_2 \times \mathbf{B}_2 = \mathbf{r}_1 \times \mathbf{B}_1.$$

As in the case of the scalar product, we must find a scheme for computing the vector product in a particular coordinate system. This may be done quite simply for right-handed orthogonal systems with the aid of the fundamental definition of the vector product and the distributive law. We introduce, in particular, a right-handed rectangular cartesian system (x, y, z) with unit triad $\mathbf{i}, \mathbf{j}, \mathbf{k}$, as in Section 1.4.* Since the system is right-handed, the unit triad has the following cross-products:

$$\begin{aligned}
\mathbf{i} \times \mathbf{i} &= \mathbf{j} \times \mathbf{j} = \mathbf{k} \times \mathbf{k} = \mathbf{0}, \\
\mathbf{i} \times \mathbf{j} &= -\mathbf{j} \times \mathbf{i} = \mathbf{k}, \\
\mathbf{j} \times \mathbf{k} &= -\mathbf{k} \times \mathbf{j} = \mathbf{i}, \\
\mathbf{k} \times \mathbf{i} &= -\mathbf{i} \times \mathbf{k} = \mathbf{j}.
\end{aligned} \qquad \textbf{1.7-7}$$

The derivation of these follows immediately from the definition of $\mathbf{A} \times \mathbf{B}$. If two vectors $\mathbf{A}$ and $\mathbf{B}$ have the cartesian representations $A_x \mathbf{i} + A_y \mathbf{j} + A_z \mathbf{k}$ and $B_x \mathbf{i} + B_y \mathbf{j} + B_z \mathbf{k}$, then the distributive law and Eqs. 1.7-7 lead to

---

* The fact that we must restrict ourselves to right-handed coordinate systems to preserve the vectorial character of the cross-product is sometimes reflected in the use of the term "pseudo-vector" for moment vectors and cross-product vectors in general. "True" vectors are those which preserve magnitude and direction under transformations of coordinates that include reflections and rotations; the "pseudo-vector" behaves as a vector only under so-called proper rotations that do not change the orientation of the axes. We cannot pursue this topic here, for it is not of concern in elementary rigid-body statics and dynamics. It is of some importance in the formulation of the mechanical theory of the general motions of deformable bodies.

$$\mathbf{A} \times \mathbf{B} = \mathbf{A} \times (B_x\mathbf{i}) + \mathbf{A} \times (B_y\mathbf{j}) + \mathbf{A} \times (B_z\mathbf{k})$$

$$= \quad A_xB_x(\mathbf{i} \times \mathbf{i}) + A_yB_x(\mathbf{j} \times \mathbf{i}) + A_zB_x(\mathbf{k} \times \mathbf{i})$$

$$+ A_xB_y(\mathbf{i} \times \mathbf{j}) + A_yB_y(\mathbf{j} \times \mathbf{j}) + A_zB_y(\mathbf{k} \times \mathbf{j})$$

$$+ A_xB_z(\mathbf{i} \times \mathbf{k}) + A_yB_z(\mathbf{j} \times \mathbf{k}) + A_zB_z(\mathbf{k} \times \mathbf{k}),$$

or

$$\mathbf{A} \times \mathbf{B} = (A_yB_z - A_zB_y)\mathbf{i} + (A_zB_x - A_xB_z)\mathbf{j} + (A_xB_y - A_yB_x)\mathbf{k}. \quad \textbf{1.7-8}$$

If $\mathbf{A} \times \mathbf{B} = \mathbf{C} = C_x\mathbf{i} + C_y\mathbf{j} + C_z\mathbf{k}$, we see that the scalar equivalents to the vector equation $\mathbf{A} \times \mathbf{B} = \mathbf{C}$ are the three equations

$$\begin{aligned} C_x &= A_yB_z - A_zB_y, \\ C_y &= A_zB_x - A_xB_z, \\ C_z &= A_xB_y - A_yB_x. \end{aligned} \qquad \textbf{1.7-9}$$

A convenient way of remembering the sign conventions for the components of the cross product is given by introducing a symbolic determinant. Another way of writing 1.7-9 is by means of two-by-two determinants:

$$C_x = \begin{vmatrix} A_y & A_z \\ B_y & B_z \end{vmatrix},$$

$$C_y = \begin{vmatrix} A_z & A_x \\ B_z & B_x \end{vmatrix} = -\begin{vmatrix} A_x & A_z \\ B_x & B_z \end{vmatrix},$$

$$C_z = \begin{vmatrix} A_x & A_y \\ B_x & B_y \end{vmatrix}.$$

Then $\mathbf{C} = C_x\mathbf{i} + C_y\mathbf{j} + C_z\mathbf{k}$ may be written

$$\mathbf{C} = \mathbf{i}\begin{vmatrix} A_y & A_z \\ B_y & B_z \end{vmatrix} + (-1)\mathbf{j}\begin{vmatrix} A_x & A_z \\ B_x & B_z \end{vmatrix} + \mathbf{k}\begin{vmatrix} A_x & A_y \\ B_x & B_y \end{vmatrix},$$

or

$$\mathbf{C} = \mathbf{A} \times \mathbf{B} = \begin{vmatrix} \mathbf{i} & \mathbf{j} & \mathbf{k} \\ A_x & A_y & A_z \\ B_x & B_y & B_z \end{vmatrix}. \qquad \textbf{1.7-10}$$

The evaluation of the symbolic determinant Eq. 1.7-10 by minors of the first row leads to the extended form (Eq. 1.7-8) of the vector product in cartesian coordinates.

Example **1.7-1**

What are the two unit vectors perpendicular to $\mathbf{A} = 3\mathbf{i} - 4\mathbf{j} + 12\mathbf{k}$ *and* $\mathbf{B} = 5\mathbf{i} + 12\mathbf{j}$ *?*

Solution: $\mathbf{A} \times \mathbf{B}$ is orthogonal to both $\mathbf{A}$ and $\mathbf{B}$; thus the unit vectors $\mathbf{n}_1$, $\mathbf{n}_2$ perpendicular to $\mathbf{A}$ and $\mathbf{B}$ are given by

$$\mathbf{n}_1 = -\mathbf{n}_2 = \frac{\mathbf{A} \times \mathbf{B}}{|\mathbf{A} \times \mathbf{B}|}.$$

Now

$$\mathbf{A} \times \mathbf{B} = \begin{vmatrix} \mathbf{i} & \mathbf{j} & \mathbf{k} \\ 3 & -4 & 12 \\ 5 & 12 & 0 \end{vmatrix} = -144\mathbf{i} + 60\mathbf{j} + 56\mathbf{k},$$

length of $\quad |\mathbf{A} \times \mathbf{B}| = [(-144)^2 + (60)^2 + (56)^2]^{1/2} = 4\sqrt{1717}.$

Therefore,

$$\mathbf{n}_1 = -\mathbf{n}_2 = \frac{\sqrt{1717}}{1717}(-36\mathbf{i} + 15\mathbf{j} + 14\mathbf{k})$$

$$\cong -0.869\mathbf{i} + 0.362\mathbf{j} + 0.338\mathbf{k}.$$

The reader may check that $\mathbf{n}_1 \cdot \mathbf{A} = \mathbf{n}_1 \cdot \mathbf{B} = 0$.

Example **1.7-2**

Find the area of a parallelogram of sides A and B and included angle θ *(Fig.* 1.7-5a*).*

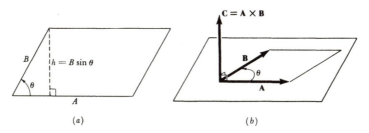

Fig. 1.7-5

Solution: The area is, of course, $Ah = AB \sin \theta$. If we introduce vectors $\mathbf{A}$, $\mathbf{B}$ along the sides of the parallelogram (Fig. 1.7-5b) so that θ is the angle included between their positive directions, the area is given by

$$AB \sin \theta = |\mathbf{A} \times \mathbf{B}|.$$

Thus the magnitude of the cross product of two vectors can be given geometrical interpretation as the area of the parallelogram based on the two vectors as sides. Indeed, the concept of a vectorial representation of area can be developed. This concept is of use in fluid mechanics and other parts of physics where the transport, or flow, of some quantity through a surface is important.

Example 1.7-3

Given a point and a line, find the perpendicular distance from the point to the line.

Solution: Let P be the given point, Q any point on the given line at position $\mathbf{r}$ from P, and $\mathbf{e}$ a unit vector along the given line (Fig. 1.7-6).

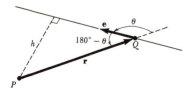

Fig. 1.7-6

Then $|\mathbf{r} \times \mathbf{e}| = r \sin \theta = r \sin (\pi - \theta) = h$, the desired distance. Note that, if $\mathbf{e}$ were reversed in direction, θ would be replaced by $\pi - \theta$ directly in the definition of the vector product.

Example 1.7-4

Use the result of the previous example to compute the perpendicular distance from $(1, 1, 1)$ to the line passing through the points $(-1, 0, 5)$ and $(2, 1, -2)$.

Solution: The point P of Fig. 1.7-6 is the point $(1, 1, 1)$; let us take $(2, 1, -2)$ as the point Q on the line. The vector from P to Q is

$$\mathbf{r} = \mathbf{i} - 3\mathbf{k}.$$

A vector along the line is given by the vector from point Q to point R: $(-1, 0, 5)$. This vector is $-3\mathbf{i} - \mathbf{j} + 7\mathbf{k}$. The unit vector $\mathbf{e}_{QR}$ is thus

$$\mathbf{e}_{QR} = \frac{1}{\sqrt{59}} (-3\mathbf{i} - \mathbf{j} + 7\mathbf{k}).$$

Now

$$\mathbf{r} \times \mathbf{e}_{QR} = \frac{1}{\sqrt{59}} \begin{vmatrix} \mathbf{i} & \mathbf{j} & \mathbf{k} \\ 1 & 0 & -3 \\ -3 & -1 & 7 \end{vmatrix} = \frac{1}{\sqrt{59}} (-3\mathbf{i} + 2\mathbf{j} - \mathbf{k}),$$

and the perpendicular distance is

$$|\mathbf{r} \times \mathbf{e}_{QR}| = \sqrt{\frac{14}{59}} \cong 0.487.$$

Example **1.7-5**

A force of magnitude 10 *lb has line of action passing through* (1, −1, 1) *and* (4, −5, 13) *inches. Find the moment of the force about the point* (0, +1, −1).

Solution: $\mathbf{F} = 10\mathbf{e}_F; \mathbf{e}_F = \dfrac{(4-1)\mathbf{i}+(-5+1)\mathbf{j}+(13-1)\mathbf{k}}{[3^2+(-4)^2+12^2]^{1/2}}.$

The force vector is, therefore,

$$\mathbf{F} = \frac{10}{13}(3\mathbf{i}-4\mathbf{j}+12\mathbf{k}) \quad \text{lb.}$$

The vector for the moment arm is

$$\mathbf{r} = (1-0)\mathbf{i}+(-1-1)\mathbf{j}+(1+1)\mathbf{k}$$
$$= \mathbf{i}-2\mathbf{j}+2\mathbf{k} \quad \text{in.}$$

Hence

$$\mathbf{M} = \mathbf{r}\times\mathbf{F} = \frac{10}{13}\begin{vmatrix} \mathbf{i} & \mathbf{j} & \mathbf{k} \\ 1 & -2 & 2 \\ 3 & -4 & 12 \end{vmatrix} = \frac{10}{13}(-16\mathbf{i}-6\mathbf{j}+2\mathbf{k})$$

$$= -\frac{160}{13}\mathbf{i}-\frac{60}{13}\mathbf{j}+\frac{20}{13}\mathbf{k} \quad \text{lb-in.}$$

The reader should sketch the vectors $\mathbf{r}$, $\mathbf{F}$, and $\mathbf{M}$ to an appropriate scale.

1.8 The Scalar Triple Product; the Moment of a Vector about a Line

Two important extensions of the concept of the product of vectors are needed to complete the discussion of the algebra of vectors in three dimensions. One extension is the *scalar*, or mixed, *triple product* $(\mathbf{A}\times\mathbf{B})\cdot\mathbf{C}$ of three vectors; the other is the *vector triple product* $(\mathbf{A}\times\mathbf{B})\times\mathbf{C}$, treated in the next section. In mechanics, the definition of a scalar "turning moment" or torque about an axis due to a force on a body leads to the expression of that torque as a scalar component of an appropriate moment vector. In terms of the force, the torque may be written in a form given by a scalar triple product. We shall return to this formulation of the moment about an axis after discussing the properties of the scalar triple product for any type of vector.

As the adjective "scalar" implies, $(\mathbf{A}\times\mathbf{B})\cdot\mathbf{C}$ is an ordinary real number. If θ is the angle between $\mathbf{A}$ and $\mathbf{B}$, so that $\mathbf{A}\times\mathbf{B}=AB\sin\theta\mathbf{n}$, and ϕ is the angle between $\mathbf{n}$ and $\mathbf{C}$, then

$$(\mathbf{A}\times\mathbf{B})\cdot\mathbf{C} = ABC\sin\theta\cos\phi. \qquad \textbf{1.8-1}$$

Since $0 \leqq \phi \leqq \pi$, we see that the scalar triple product may be positive, negative, or zero. Assuming that none of the vectors is the zero vector and that **A** is not parallel to **B**($\theta \neq 0, \pi$), we see that the sign of $(\mathbf{A} \times \mathbf{B}) \cdot \mathbf{C}$ is determined by the relative orientation of **A**, **B**, and **C** (Fig. 1.8-1). In Fig. 1.8-1a, **A**, **B**, and **C** are oriented in a right-handed way, with $0 \leqq \phi < \pi/2$, and $(\mathbf{A} \times \mathbf{B}) \cdot \mathbf{C}$ is positive; in Fig. 1.8-1b, $\pi/2 < \phi \leqq \pi$, the three vectors have a left-handed orientation, and $(\mathbf{A} \times \mathbf{B}) \cdot \mathbf{C}$ is negative. When the scalar triple product vanishes with $\phi = \pi/2$, the three vectors are coplanar.

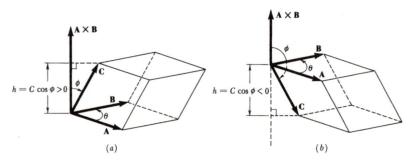

Fig. 1.8-1

A simple geometrical interpretation may be given to the scalar triple product. The magnitude $AB \sin \theta$ of $\mathbf{A} \times \mathbf{B}$ is the area of the parallelogram with sides along **A** and **B** (see Example 1.7-2). The quantity $C \cos \phi$ is the signed altitude of the parallelepiped based on **A** and **B**, with **C** along the slant side. Therefore, $(\mathbf{A} \times \mathbf{B}) \cdot \mathbf{C}$ is the signed volume of the parallelepiped. The two parallelepipeds of Figs. 1.8-1a and 1.8-1b have positive and negative volumes, respectively, according to this sign convention.*

The interpretation of the scalar triple product as a volume leads to the statement of the fundamental property of the scalar triple product. Since the volume does not depend on which face of the parallelepiped is taken as base and on whether the area of the base is pre- or post-multiplied by the altitude, we deduce that the value of $(\mathbf{A} \times \mathbf{B}) \cdot \mathbf{C}$ depends only on the relative, or cyclic, order of the three vectors and not on their absolute order or on the placing of the vector and scalar

---

* If one wishes to keep volume as a positive quantity, then one must either take the volume as $|(\mathbf{A} \times \mathbf{B}) \cdot \mathbf{C}|$ or else choose a right-handed order (**B**, **A**, **C** in Fig. 1.8-1b) from the three.

product operations.   That is, the value will be the same if the "dot" and "cross" are interchanged; and the value will be the same if the cyclic order *ABCAB* is preserved:

$$(\mathbf{A} \times \mathbf{B}) \cdot \mathbf{C} = \mathbf{A} \cdot (\mathbf{B} \times \mathbf{C}) = (\mathbf{B} \times \mathbf{C}) \cdot \mathbf{A} = \mathbf{B} \cdot (\mathbf{C} \times \mathbf{A})$$
$$= (\mathbf{C} \times \mathbf{A}) \cdot \mathbf{B} = \mathbf{C} \cdot (\mathbf{A} \times \mathbf{B}). \quad \textbf{1.8-2}$$

In Fig. 1.8-1, the volume of either parallelepiped is equally well given by the area of the front face based on **A** and **C** multiplied by the altitude given by the projection of **B** on the normal to the front face. To preserve the algebraic sign of the result, we see that we must have $(\mathbf{C} \times \mathbf{A}) \cdot \mathbf{B}$ and not $(\mathbf{A} \times \mathbf{C}) \cdot \mathbf{B}$ for the volume.   This gives the equality of the first and fifth terms in Eq. 1.8-2.   An alternative proof of the relations given in Eqs. 1.8-2 which makes no appeal to geometric intuition is given in Example 1.8-5.   In summary, the foregoing considerations show that the scalar triple product may have either sign, that it vanishes if the three vectors are coplanar, that the dot and cross may be interchanged:

$$(\mathbf{B} \times \mathbf{C}) \cdot \mathbf{A} = \mathbf{B} \cdot (\mathbf{C} \times \mathbf{A}), \quad \textbf{1.8-3a}$$

and that the sign depends on the cyclic order of the vectors:

$$(\mathbf{A} \times \mathbf{B}) \cdot \mathbf{C} = (\mathbf{B} \times \mathbf{C}) \cdot \mathbf{A} \quad \textbf{1.8-3b}$$

but

$$(\mathbf{A} \times \mathbf{B}) \cdot \mathbf{C} = -(\mathbf{B} \times \mathbf{A}) \cdot \mathbf{C}. \quad \textbf{1.8-3c}$$

The last result follows from the non-commutative nature of the vector product.

Using the properties of the scalar triple product, we are now in a position to prove the distributive law for the vector product operation (Eq. 1.7-4).   The proofs of Varignon's theorem and other results of the last section will then be complete.

Example **1.8-1**

Prove the distributive law for the vector product

$$\mathbf{A} \times (\mathbf{B} + \mathbf{C}) = \mathbf{A} \times \mathbf{B} + \mathbf{A} \times \mathbf{C}.$$

Solution: Let $\mathbf{V} = \mathbf{A} \times (\mathbf{B} + \mathbf{C}) - \mathbf{A} \times \mathbf{B} - \mathbf{A} \times \mathbf{C}$; if we can show **V** to be the zero vector, then the distributive law will follow at once.   To compute the magnitude of **V**, we take the dot product of each side of the defining equation with **V**:

$$\mathbf{V} \cdot \mathbf{V} = V^2 = \mathbf{V} \cdot [\mathbf{A} \times (\mathbf{B} + \mathbf{C})] - \mathbf{V} \cdot (\mathbf{A} \times \mathbf{B}) - \mathbf{V} \cdot (\mathbf{A} \times \mathbf{C}).$$

This step is legitimate because the distributive law holds for the scalar product operation (Eq. 1.6-6). In view of Eq. 1.8-3a, the dot and cross operations in each scalar triple product may be interchanged:

$$V^2 = (\mathbf{V} \times \mathbf{A}) \cdot (\mathbf{B} + \mathbf{C}) - (\mathbf{V} \times \mathbf{A}) \cdot \mathbf{B} - (\mathbf{V} \times \mathbf{A}) \cdot \mathbf{C}.$$

Making use of the distributive law for the scalar product once again, we may "factor out" $\mathbf{V} \times \mathbf{A}$ and find

$$V^2 = (\mathbf{V} \times \mathbf{A}) \cdot [(\mathbf{B} + \mathbf{C}) - \mathbf{B} - \mathbf{C}] = 0.$$

Therefore, $\mathbf{V} = 0$, since it has zero magnitude, and Eq. 1.7-4 follows directly.

The computation of the scalar triple product when the vectors are given in cartesian components is straightforward. The symbolic determinant representation of Eq. 1.7-10 for the cross product may be used:

$$\mathbf{D} = (\mathbf{A} \times \mathbf{B}) = \begin{vmatrix} \mathbf{i} & \mathbf{j} & \mathbf{k} \\ A_x & A_y & A_z \\ B_x & B_y & B_z \end{vmatrix}.$$

Now $(\mathbf{A} \times \mathbf{B}) \cdot \mathbf{C} = \mathbf{D} \cdot \mathbf{C}$ is given by

$$(\mathbf{A} \times \mathbf{B}) \cdot \mathbf{C} = \begin{vmatrix} \mathbf{i} & \mathbf{j} & \mathbf{k} \\ A_x & A_y & A_z \\ B_x & B_y & B_z \end{vmatrix} \cdot (C_x \mathbf{i} + C_y \mathbf{j} + C_z \mathbf{k}),$$

or

$$(\mathbf{A} \times \mathbf{B}) \cdot \mathbf{C} = \begin{vmatrix} C_x & C_y & C_z \\ A_x & A_y & A_z \\ B_x & B_y & B_z \end{vmatrix}. \qquad \textbf{1.8-4a}$$

The order properties (Eq. 1.8-2) that we have found for the scalar product are reflected in the row-interchange properties of the determinant, as the student may verify. In particular, the result that $(\mathbf{A} \times \mathbf{B}) \cdot \mathbf{C} = (\mathbf{B} \times \mathbf{C}) \cdot \mathbf{A}$ leads to an alternate form of 1.8-4a that is easier to remember, since the elements of the determinant are in the same order, row by row, as are the vector symbols in the triple product:

$$(\mathbf{A} \times \mathbf{B}) \cdot \mathbf{C} = \begin{vmatrix} A_x & A_y & A_z \\ B_x & B_y & B_z \\ C_x & C_y & C_z \end{vmatrix}. \qquad \textbf{1.8-4b}$$

Let us now consider the concept of the moment of a vector about a line, which was mentioned earlier. We give a rough definition first, make it more precise, then show the relation between the concepts of vector moment about a point and scalar moment about a line. Suppose we are given a line JK (the *moment axis*) in space and a vector **A** emanating from a point R. Drop the perpendicular from R on JK, obtaining point S; the line SR of length h is called the *moment arm* (Fig. 1.8-2). The scalar moment of **A** about the line

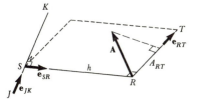

Fig. 1.8-2

JK is defined to be the product of the length h of the moment arm and the component A_{RT} of **A** in a direction RT perpendicular to both JK and SR, i.e., perpendicular to both moment axis and moment arm:

$$M_{JK} = \pm hA_{RT}. \qquad \text{1.8-5}$$

The algebraic sign is determined, if A_{RT} is considered positive, by introducing a right-hand rule for rotations about the moment axis and hence a sign convention for moments.

Let us make this elementary definition of the moment about a line as "moment arm times perpendicular component of the vector" more precise. Choose a direction on JK by assigning a unit vector $\mathbf{e}_{JK}$ (Fig. 1.8-2); direct SR by $\mathbf{e}_{SR}$; and direct the mutual perpendicular RT to JK and SR by $\mathbf{e}_{RT} = \mathbf{e}_{JK} \times \mathbf{e}_{SR}$. The three unit vectors $\mathbf{e}_{JK}$, $\mathbf{e}_{SR}$, $\mathbf{e}_{RT}$ form a right-handed orthogonal triad by construction. Define A_{RT} as the scalar component of **A** in the direction of $\mathbf{e}_{RT}$: $A_{RT} = \mathbf{A} \cdot \mathbf{e}_{RT}$; then, letting A_{RT} be positive or negative, the choice of the plus sign in 1.8-5 will define a signed moment M_{JK} by a right-hand rule on the directions of JK, SR, and RT. Furthermore, if the vector moment $\mathbf{M}_S$ of A about point S is projected on the line JK, we can show that M_{JK} is equal to that projection. That is, M_{JK} is the scalar component of $\mathbf{M}_S$ in the direction JK:

$$M_{JK} = hA_{RT} = \mathbf{M}_S \cdot \mathbf{e}_{JK} = (h\mathbf{e}_{SR} \times \mathbf{A}) \cdot \mathbf{e}_{JK}. \qquad \text{1.8-6}$$

This last is easy to prove, using Eq. 1.8-2; for

$$(h\mathbf{e}_{SR} \times \mathbf{A}) \cdot \mathbf{e}_{JK} = (\mathbf{e}_{JK} \times h\mathbf{e}_{SR}) \cdot \mathbf{A} = h\mathbf{e}_{RT} \cdot \mathbf{A} = hA_{RT}.$$

We are now prepared to generalize. We know, from the previous section, that the moment of **A** about a point, S, is independent of the location of **A** along its line of action.　That is, RS need not be the perpendicular from a point on the line of action PQ of **A** (Fig. 1.8-3)

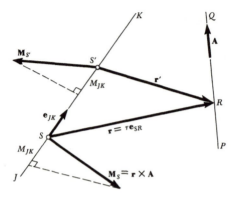

Fig. 1.8-3

to the line JK, but may be any line segment from S to PQ.　It must, however, be directed properly: let $\mathbf{r} = r\mathbf{e}_{SR}$ be the vector *from* point S on the moment axis *to* any point R on the line of action of the vector. Then

$$M_{JK} = (\mathbf{r} \times \mathbf{A}) \cdot \mathbf{e} \qquad\qquad \text{1.8-7}$$

where we set $\mathbf{e} = \mathbf{e}_{JK}$.　We see, in fact, that S may be any point on JK, just as R is any point on PQ.　For, choosing any other point S' on JK, we see that $\mathbf{r}' - \mathbf{r}$ is parallel to $\mathbf{e}$, so that

$$\mathbf{M}_{S'} \cdot \mathbf{e} = (\mathbf{r}' \times \mathbf{A}) \cdot \mathbf{e} = (\mathbf{r} \times \mathbf{A}) \cdot \mathbf{e} + [(\mathbf{r}' - \mathbf{r}) \times \mathbf{A}] \cdot \mathbf{e}$$

$$= M_{JK} + [\mathbf{e} \times (\mathbf{r}' - \mathbf{r})] \cdot \mathbf{A} = M_{JK}. \qquad \text{1.8-8}$$

This is a fundamental result: the moment of a sliding vector about an axis is the scalar component in the direction of the axis of the vector moment about any point on the axis.　In particular, the (x, y, z) components of the moment of a vector about the origin are the moments of that vector about the coordinate axes.

　　The scalar form of Varignon's theorem follows immediately; that is, the moment about an axis of the resultant of a number of con-current vectors is the sum of the moments of the separate vectors about that axis.　This follows from the vectorial form of the theorem (Section 1.7) and the fact that the moment about the axis is simply the projection on the axis of the moment about any point on the axis.　Thus, take the vector equation expressing the vector theorem, $\mathbf{r} \times (\sum \mathbf{A}_i) = \sum (\mathbf{r} \times \mathbf{A}_i)$, and take the scalar product of each side with the appropriate unit vector $\mathbf{e}$.

Example **1.8-2**

Find the scalar triple product $\mathbf{A} \cdot (\mathbf{C} \times \mathbf{L})$ *if*

$$\mathbf{A} = 3\mathbf{i} - 2\mathbf{j} + 5\mathbf{k}, \qquad \mathbf{C} = \mathbf{i} + \mathbf{j} + 2\mathbf{k}, \qquad \mathbf{L} = -\mathbf{i} - \mathbf{j}.$$

Solution: (a) We can compute the cross-product, and then the dot-product:

$$\mathbf{C} \times \mathbf{L} = \begin{vmatrix} \mathbf{i} & \mathbf{j} & \mathbf{k} \\ 1 & 1 & 2 \\ -1 & -1 & 0 \end{vmatrix} = 2\mathbf{i} - 2\mathbf{j},$$

$$\mathbf{A} \cdot (\mathbf{C} \times \mathbf{L}) = (3)(2) + (-2)(-2) + (5)(0) = 10.$$

(b) We can compute the triple product directly:

$$\mathbf{A} \cdot (\mathbf{C} \times \mathbf{L}) = \begin{vmatrix} A_x & A_y & A_z \\ C_x & C_y & C_z \\ L_x & L_y & L_z \end{vmatrix} = \begin{vmatrix} 3 & -2 & 5 \\ 1 & 1 & 2 \\ -1 & -1 & 0 \end{vmatrix}.$$

By any of the standard evaluation methods, the answer is, of course, 10.

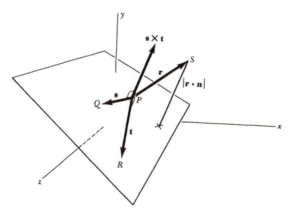

Fig. 1.8-4

Example **1.8-3**

A plane passes through three points, P: $(1, 1, 1)$; Q: $(-1, 2, 3)$; *and* R: $(4, 5, 13)$. *What is the perpendicular distance from the point* S: $(3, 1, -5)$ *to the plane (Fig. 1.8-4)?*

Solution: The perpendicular distance from a point S to a plane PQR is the magnitude of the component of a vector from a point in the plane to S in a direction normal to the plane. With three points given in the plane, two vectors lying in the plane are known. Their cross-product will give a vector normal to the plane. Symbolically, let the directed segment $\overline{PS}$ be **r**, the segment $\overline{PQ}$ be **s**, and the segment $\overline{PR}$ be **t**. Then **s** × **t** is normal to the plane, with **n** = (**s** × **t**)/|**s** × **t**| being a unit normal vector. Then the desired distance is

$$|\mathbf{r}\cdot\mathbf{n}| = \frac{|\mathbf{r}\cdot(\mathbf{s}\times\mathbf{t})|}{|\mathbf{s}\times\mathbf{t}|}.$$

In this example, we have

$$\mathbf{r} = 2\mathbf{i}-6\mathbf{k}, \quad \mathbf{s} = -2\mathbf{i}+\mathbf{j}+2\mathbf{k}, \quad \mathbf{t} = 3\mathbf{i}+4\mathbf{j}+12\mathbf{k};$$

hence

$$\mathbf{s}\times\mathbf{t} = \begin{vmatrix} \mathbf{i} & \mathbf{j} & \mathbf{k} \\ -2 & 1 & 2 \\ 3 & 4 & 12 \end{vmatrix} = 4\mathbf{i}+30\mathbf{j}-11\mathbf{k},$$

$$\mathbf{r}\cdot(\mathbf{s}\times\mathbf{t}) = \begin{vmatrix} 2 & 0 & -6 \\ -2 & 1 & 2 \\ 3 & 4 & 12 \end{vmatrix} = 74,$$

$$|\mathbf{r}\cdot\mathbf{n}| = \frac{74}{\sqrt{16+900+121}} = 2.298.$$

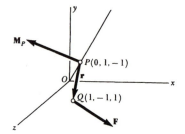

Fig. 1.8-5

Example **1.8-4**

A force vector $\mathbf{F}=(10/13)(3\mathbf{i}-4\mathbf{j}-12\mathbf{k})$ *lb has line of action passing through Q: $(1, -1, 1)$ in. (Fig. 1.8-5). Find its moment about the line joining the origin to P: $(0, 1, -1)$; check the answer by an alternative computation.*

Solution: This is the configuration of Example 1.7-5, wherein we found $\mathbf{r} \times \mathbf{F} = \mathbf{M}_P = -(160/13)\mathbf{i} - (60/13)\mathbf{j} + (20/13)\mathbf{k}$ lb-in. The unit vector $\mathbf{e}$ in the direction OP is $\mathbf{j} - \mathbf{k}$ divided by its magnitude, i.e., $\mathbf{e} = (\sqrt{2}/2)(\mathbf{j} - \mathbf{k})$. Therefore, the moment of $\mathbf{F}$ about OP is

$$M_{OP} = (\mathbf{r} \times \mathbf{F}) \cdot \mathbf{e} = \mathbf{M}_P \cdot \mathbf{e} = -\frac{40}{13}\sqrt{2} \quad \text{lb-in.}$$

Note that, by our sign convention based on the right-hand rule, M_{PO} is the negative of this, i.e., a positive $(40/13)\sqrt{2}$ lb-in.

An alternate computation would be to find the moment of $\mathbf{F}$ about some other point on OP, then project that vector on OP; that is, we also have

$$M_{OP} = \mathbf{M}_O \cdot \mathbf{e} = (\mathbf{r}_{OQ} \times \mathbf{F}) \cdot \mathbf{e}$$

$$= \begin{vmatrix} 0 & \dfrac{\sqrt{2}}{2} & -\dfrac{\sqrt{2}}{2} \\ 1 & -1 & 1 \\ \dfrac{30}{13} & -\dfrac{40}{13} & \dfrac{120}{13} \end{vmatrix} = \frac{5}{13}\sqrt{2} \begin{vmatrix} 0 & 1 & -1 \\ 1 & -1 & 1 \\ 3 & -4 & 12 \end{vmatrix}$$

$$= \frac{5}{13}\sqrt{2} \begin{vmatrix} 0 & 1 & -1 \\ 1 & 0 & 0 \\ 3 & -4 & 12 \end{vmatrix} = \frac{-5}{13}\sqrt{2} \begin{vmatrix} 1 & -1 \\ -4 & 12 \end{vmatrix}$$

$$= -\frac{40}{13}\sqrt{2} \quad \text{lb-in.}$$

Example 1.8-5

Derive relations 1.8-2 without appeal to the geometry of the parallelepipeds of Fig. 1.8-1.

Solution: It is sufficient to show that $(\mathbf{A} \times \mathbf{B}) \cdot \mathbf{C} = (\mathbf{B} \times \mathbf{C}) \cdot \mathbf{A}$; the cyclic ordering property may then be carried one step more to show that these are equal to $(\mathbf{C} \times \mathbf{A}) \cdot \mathbf{B}$, and the commutative property of the scalar product then gives the other three terms of Eq. 1.8-2.

The proof of the formal abstract property is very similar to the arguments given for the triple product representation of the moment about a line, and the reader may wish to compare what follows with that discussion, especially the part dealing with the construction of a right-handed unit triad.

Let $\mathbf{A} = A\mathbf{e}_A$ and $\mathbf{B} = B\mathbf{e}_B$, $\mathbf{e}_A \times \mathbf{e}_B \neq \mathbf{0}$, in the usual way, with $\mathbf{A} \times \mathbf{B} = AB \sin \theta \mathbf{n} = AB\mathbf{e}_A \times \mathbf{e}_B$ defining the unit vector $\mathbf{n}$. Let ϕ be the angle between $\mathbf{C}$ and $\mathbf{n}$, so that $\mathbf{C} \cdot \mathbf{n} = C \cos \phi$. By definition, $(\mathbf{A} \times \mathbf{B}) \cdot \mathbf{C} = ABC \sin \theta \cos \phi$ (Eq. 1.8-1). Write $\mathbf{C} = C \cos \phi \mathbf{n} + \mathbf{D}$; $\mathbf{D}$ must be orthogonal to $\mathbf{n}$ and hence can be expressed as a linear combination of $\mathbf{e}_A$ and $\mathbf{e}_B$. Therefore,

$$\mathbf{C} = C \cos \phi \mathbf{n} + C_A\mathbf{e}_A + C_B\mathbf{e}_B.$$

Note that these are generally oblique components of $\mathbf{C}$, not orthogonal components. Now

$$\mathbf{B} \times \mathbf{C} = BC \cos \phi \mathbf{e}_B \times \mathbf{n} + C_A \mathbf{e}_B \times \mathbf{e}_A,$$

$$(\mathbf{B} \times \mathbf{C}) \cdot \mathbf{A} = ABC \cos \phi (\mathbf{e}_B \times \mathbf{n}) \cdot \mathbf{e}_A + C_A (\mathbf{e}_B \times \mathbf{e}_A) \cdot \mathbf{e}_A.$$

We have reduced the problem to an argument on the unit vectors $\mathbf{e}_A$, $\mathbf{e}_B$, and $\mathbf{n}$. The last term is zero, since the scalar triple product is zero whenever two of the vectors appearing in it are the same. Thus

$$(\mathbf{B} \times \mathbf{C}) \cdot \mathbf{A} = ABC \cos \phi (\mathbf{e}_B \times \mathbf{n}) \cdot \mathbf{e}_A.$$

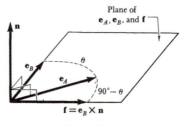

Fig. 1.8-6

Let us consider $(\mathbf{e}_B \times \mathbf{n}) \cdot \mathbf{e}_A$. The cross-product $\mathbf{e}_B \times \mathbf{n} = \mathbf{f}$ is in the plane of $\mathbf{e}_A$ and $\mathbf{e}_B$ and, moreover, is a unit vector. The first statement follows from the orthogonality of $\mathbf{n}$ and $\mathbf{e}_A$, $\mathbf{e}_B$; the second, from the definition of the vector product and the orthogonality of $\mathbf{n}$ and $\mathbf{e}_B$. The three unit vectors $(\mathbf{e}_B, \mathbf{n}, \mathbf{f})$, in that order, form a right-handed orthogonal triad by construction.

Now express $\mathbf{e}_A$ in terms of $\mathbf{e}_B$ and $\mathbf{f}$: since θ is the angle between $\mathbf{A}$ and $\mathbf{B}$, we have (Fig. 1.8-6)

$$\mathbf{e}_A = \cos \theta \mathbf{e}_B \pm \sin \theta \mathbf{f}$$

where we must resolve the ambiguity of sign. It is here that the right-handed nature of the constructed unit vectors is essential. We now have

$$(\mathbf{B} \times \mathbf{C}) \cdot \mathbf{A} = ABC \cos \phi \mathbf{f} \cdot \mathbf{e}_A = \pm ABC \cos \phi \sin \theta,$$

and we must show that the positive sign is the proper choice. But we know

$$\mathbf{e}_A \times \mathbf{e}_B = \sin \theta \mathbf{n} = \pm \sin \theta \mathbf{f} \times \mathbf{e}_B;$$

since the unit vectors $(\mathbf{e}_B, \mathbf{n}, \mathbf{f})$ are a right-handed triad, $\mathbf{f} \times \mathbf{e}_B = \mathbf{n}$ and we must choose the positive sign:

$$(\mathbf{B} \times \mathbf{C}) \cdot \mathbf{A} = ABC \cos \phi \sin \theta = (\mathbf{A} \times \mathbf{B}) \cdot \mathbf{C}.$$

1.9 The Vector Triple Product

The vector triple product, $(\mathbf{A} \times \mathbf{B}) \times \mathbf{C}$, of three vectors, $\mathbf{A}$, $\mathbf{B}$, and $\mathbf{C}$, is a fourth vector, $\mathbf{D}$. In this expression, both the position of the parentheses and the order of the terms must be maintained unaltered for an unambiguous prescription of the result.* As we have seen in Section 1.7, the vector product does not obey the associative law, so that $(\mathbf{A} \times \mathbf{B}) \times \mathbf{C}$ is not equal to $\mathbf{A} \times (\mathbf{B} \times \mathbf{C})$.

If we write

$$\mathbf{D} = (\mathbf{A} \times \mathbf{B}) \times \mathbf{C}, \qquad \text{1.9-1}$$

we observe that $\mathbf{D}$ lies in the plane of $\mathbf{A}$ and $\mathbf{B}$. This follows from the fact that $(\mathbf{A} \times \mathbf{B})$ is perpendicular to that plane. The cross-product with $\mathbf{C}$ provides another ninety-degree rotation, putting the final vector $\mathbf{D}$ back in the plane of $\mathbf{A}$ and $\mathbf{B}$. We know, therefore, that $\mathbf{D}$ may be expressed as the sum of vectors proportional to $\mathbf{A}$ and $\mathbf{B}$

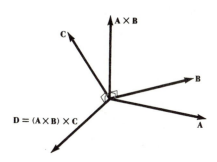

Fig. 1.9-1

(Example 1.3-2): $\mathbf{D} = \alpha\mathbf{A} + \beta\mathbf{B}$. To determine α and β, we follow the pattern of Example 1.8-5 and express $\mathbf{C}$ in terms of oblique components in the directions of $\mathbf{A}$, $\mathbf{B}$, and $\mathbf{A} \times \mathbf{B}$.

Let $\mathbf{A} = A\mathbf{e}_A$, $\mathbf{B} = B\mathbf{e}_B$, $\mathbf{e}_A \times \mathbf{e}_B = \sin\theta\mathbf{n} \neq \mathbf{0}$ in the usual way, and write

$$\mathbf{C} = a\mathbf{e}_A + b\mathbf{e}_B + c\mathbf{n}. \qquad \text{1.9-2}$$

Then

$$\begin{aligned}
\mathbf{D} = (\mathbf{A} \times \mathbf{B}) \times \mathbf{C} &= AB \sin\theta\mathbf{n} \times \mathbf{C} \\
&= AB \sin\theta(a\mathbf{n} \times \mathbf{e}_A + b\mathbf{n} \times \mathbf{e}_B).
\end{aligned}$$

Let $\mathbf{f}$ be the unit vector $\mathbf{e}_B \times \mathbf{n}$, as in Example 1.8-5, so that $(\mathbf{e}_B, \mathbf{n}, \mathbf{f})$ form a right-handed triad; then, as shown in that example, $\mathbf{e}_A = \cos\theta\mathbf{e}_B + \sin\theta\mathbf{f}$ (Fig. 1.9-2).

---

* In the scalar triple product $(\mathbf{A} \times \mathbf{B}) \cdot \mathbf{C}$, the parentheses may be dropped, since there is only one way to give a meaningful interpretation to $\mathbf{A} \times \mathbf{B} \cdot \mathbf{C}$. It is not good practice, however, for one new to the concept to eliminate the parentheses.

Therefore,

$$\mathbf{D} = AB \sin \theta [a \cos \theta (-\mathbf{f}) + a \sin \theta \mathbf{e}_B - b\mathbf{f}]$$
$$= ABa \sin^2 \theta \mathbf{e}_B - AB(a \cos \theta + b) \sin \theta \mathbf{f}.$$

But we want $\mathbf{D}$ in oblique $\mathbf{e}_A$ and $\mathbf{e}_B$ components; therefore, we must replace $\mathbf{f}$. We have $\sin \theta \mathbf{f} = \mathbf{e}_A - \cos \theta \mathbf{e}_B$, so that

$$\mathbf{D} = -AB(a \cos \theta + b)\mathbf{e}_A + AB(a + b \cos \theta)\mathbf{e}_B$$
$$= -B(a \cos \theta + b)\mathbf{A} + A(a + b \cos \theta)\mathbf{B}. \qquad \textbf{1.9-3}$$

What are the coefficients of $\mathbf{A}$ and $\mathbf{B}$ in Eq. 1.9-3? From Eq. 1.9-2, we find

$$\mathbf{A} \cdot \mathbf{C} = A\mathbf{e}_A \cdot \mathbf{C} = A(a + b\mathbf{e}_B \cdot \mathbf{e}_A) = A(a + b \cos \theta),$$
$$\mathbf{B} \cdot \mathbf{C} = B\mathbf{e}_B \cdot \mathbf{C} = B(a\mathbf{e}_B \cdot \mathbf{e}_A + b) = B(a \cos \theta + b).$$

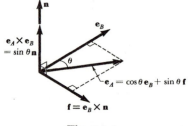

Fig. 1.9-2

Finally, therefore, we obtain the fundamental reduction formula for the vector triple product:

$$(\mathbf{A} \times \mathbf{B}) \times \mathbf{C} = (\mathbf{A} \cdot \mathbf{C})\mathbf{B} - (\mathbf{B} \cdot \mathbf{C})\mathbf{A}. \qquad \textbf{1.9-4}$$

Similarly, the vector triple product obtained by the alternative placing of the parentheses may be written

$$\mathbf{A} \times (\mathbf{B} \times \mathbf{C}) = (\mathbf{A} \cdot \mathbf{C})\mathbf{B} - (\mathbf{A} \cdot \mathbf{B})\mathbf{C}. \qquad \textbf{1.9-5}$$

The non-associative nature of the vector product is apparent here, since the right-hand sides of 1.9-4 and 1.9-5 are different.

The vector triple product and the reduction formula for it are of great use in dynamical theory; in statics proper, we use it only for some considerations of higher-order moments in the next chapter. In the solution of vector equations, the vector triple product expansion is useful, as the first example of Section 1.10 shows.

Example **1.9-1**

Given two non-parallel vectors **A** *and* **B**, *construct three mutually orthogonal vectors* **D**, **E**, *and* **F** *in terms of* **A** *and* **B**.

Solution: Choose **D** equal to any one of the three vectors, say **A**, and **E** equal to **A** × **B**; then **D** and **E** are orthogonal. Furthermore, **D** × **E** = **F** must be orthogonal to both **D** and **E** (Fig. 1.9-3):

$$\mathbf{F} = \mathbf{D} \times \mathbf{E} = \mathbf{A} \times (\mathbf{A} \times \mathbf{B}) = (\mathbf{A} \cdot \mathbf{B})\mathbf{A} - (\mathbf{A} \cdot \mathbf{A})\mathbf{B}$$
$$= (\mathbf{A} \cdot \mathbf{B})\mathbf{A} - A^2\mathbf{B}.$$

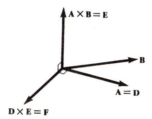

Fig. 1.9-3

Example **1.9-2**

Find the vector triple product (**B** × **C**) × **A** *of*

$$\mathbf{A} = 2\mathbf{i} - 3\mathbf{j}, \qquad \mathbf{B} = \mathbf{k} - \mathbf{i}, \qquad \mathbf{C} = \mathbf{i} + \mathbf{j} + \mathbf{k}.$$

Solution:

(a) $\mathbf{B} \times \mathbf{C} = \mathbf{k} \times \mathbf{C} - \mathbf{i} \times \mathbf{C} = \mathbf{j} - \mathbf{i} - \mathbf{k} + \mathbf{j} = 2\mathbf{j} - \mathbf{i} - \mathbf{k};$

$(\mathbf{B} \times \mathbf{C}) \times \mathbf{A} = (2\mathbf{j} - \mathbf{i} - \mathbf{k}) \times (2\mathbf{i} - 3\mathbf{j}) = -4\mathbf{k} + 3\mathbf{k} - 2\mathbf{j} - 3\mathbf{i} = -3\mathbf{i} - 2\mathbf{j} - \mathbf{k}.$

(b) $(\mathbf{B} \times \mathbf{C}) \times \mathbf{A} = (\mathbf{B} \cdot \mathbf{A})\mathbf{C} - (\mathbf{C} \cdot \mathbf{A})\mathbf{B} = -2\mathbf{C} + \mathbf{B}$

$$= -2\mathbf{i} - 2\mathbf{j} - 2\mathbf{k} + \mathbf{k} - \mathbf{i} = -3\mathbf{i} - 2\mathbf{j} - \mathbf{k}.$$

Example **1.9-3**

Find the vector **D** = (**A** × **B**) × **C** + (**B** × **C**) × **A** + (**C** × **A**) × **B** *in terms of* **A**, **B**, **C**.

Solution: Note that **D** is the sum of all possible triple products (with the parentheses in the same place) formed from **A**, **B**, and **C** in cyclic order. By the reduction formula 1.9-4,

$$(\mathbf{A} \times \mathbf{B}) \times \mathbf{C} = (\mathbf{A} \cdot \mathbf{C})\mathbf{B} - (\mathbf{B} \cdot \mathbf{C})\mathbf{A},$$
$$(\mathbf{B} \times \mathbf{C}) \times \mathbf{A} = (\mathbf{B} \cdot \mathbf{A})\mathbf{C} - (\mathbf{C} \cdot \mathbf{A})\mathbf{B},$$
$$(\mathbf{C} \times \mathbf{A}) \times \mathbf{B} = (\mathbf{C} \cdot \mathbf{B})\mathbf{A} - (\mathbf{A} \cdot \mathbf{B})\mathbf{C}.$$

Summing these and using the commutative nature of the scalar product, we see that $\mathbf{D} = \mathbf{0}$; i.e., for any three vectors $\mathbf{A}$, $\mathbf{B}$, and $\mathbf{C}$,

$$(\mathbf{A} \times \mathbf{B}) \times \mathbf{C} + (\mathbf{B} \times \mathbf{C}) \times \mathbf{A} + (\mathbf{C} \times \mathbf{A}) \times \mathbf{B} = \mathbf{0}.$$

1.10 Vector Equations

The operations defined in preceding sections are intended for use in the solution of equations arising in mechanical analysis and in the development of the theory of mechanics. Such use, to be effective, requires a clear understanding of what the operations mean and of what a vector equation implies. Purely formal manipulation of algebraic relations among vector quantities is not enough for a complete understanding of the geometric or physical model which the quantities are supposed to describe. Furthermore, analogies between the ordinary algebra of scalars and vector algebra may be over-extended if the definitions of the vector operation symbols $+$, $-$, $\cdot$, $\times$ are not carefully kept in mind.

To be meaningful, a vector equation must express an equality between two vectors, not between a vector and a scalar. Expressions such as $7\mathbf{A} - 5\mathbf{B} = 12$ are meaningless. The fact that an ordinary zero symbol is often used for the null vector, as in $7\mathbf{A} - 5\mathbf{B} = 0$, should not lead one to put other scalars in its place. More subtle forms of this error are sometimes encountered: $7\mathbf{A} - 5\mathbf{B} = (\mathbf{A} \times \mathbf{B}) \cdot \mathbf{C}$ is also meaningless. Care in the formulation of equations is an essential first step. Only operations that have been defined should be employed. There is, for example, no operation on vectors corresponding to the division of one number by another, and expressions such as $\mathbf{B}/\mathbf{A}$, $\mathbf{C}^{-1}$, $6\mathbf{i}/(\mathbf{k} + 2\mathbf{j})$ are also meaningless. The reason why it would be fruitless to attempt to define such an operation is, as we have noted, that the ordinary cancellation law of the algebra of scalars does not hold for either the scalar or the vector product of two vectors. That is, neither of the two equations $\mathbf{A} \cdot \mathbf{B} = \mathbf{A} \cdot \mathbf{C}$ and $\mathbf{A} \times \mathbf{B} = \mathbf{A} \times \mathbf{C}$ ensures that $\mathbf{B} = \mathbf{C}$ even when $\mathbf{A} \neq \mathbf{0}$. We shall investigate the complete solution of $\mathbf{A} \times \mathbf{B} = \mathbf{A} \times \mathbf{C}$ in one of the examples of this section.

When a set of vector equations is linear in the vector symbols, with scalar coefficients, they may be solved by the process of successive elimination of unknowns—just as is done in the solution of simultaneous linear equations in scalar unknowns. The technique was demonstrated in Example 1.3-1. Alternatively, a particular coordinate system may be introduced and the set of scalar equations equivalent to the vector equations may be written. In three-dimensional space, there will be three such scalar equations for each vector equation because there can be three linearly independent vectors in the space—for example, **i**, **j**, and $\mathbf{k} = \mathbf{i} \times \mathbf{j}$. Conversely, since there cannot be more than three linearly independent vectors in three dimensions, so that all vectors can be expressed linearly in terms of three such independent ones, there cannot be more than three independent scalar equations corresponding to any set of linear vector algebraic equations. Thus, problems of the *consistency* of sets of four or more vector equations are more complex than the corresponding consistency problems for sets of scalar equations. True understanding of the vector addition, subtraction, and scalar multiplication operations, coupled with the knowledge that the component of a vector in any direction is fixed once its components in three independent directions are given, is essential for the establishment of the independence and consistency of the equations written in any given situation.

When a vector equation embodies any of the product operations on vectors, the geometric relations inherent in the definitions of the scalar and vector products are of use in interpreting and simplifying the equation. Conditions of orthogonality and parallelism can be expressed simply in terms of these operations; orthogonality, for instance, has been useful already in obtaining the vector equation of a plane (Example 1.6-4). The interpretation of the magnitude of the cross-product as an area and of the scalar triple product as a volume is also a useful device.

In the remainder of this text, emphasis is placed on the physical concepts underlying the analysis of mechanical systems in equilibrium. Vector symbolism provides a natural mathematical language to use in expressing these physical concepts. Care taken in its use will help in understanding the physical interrelationships to be described, just as care taken in formulating the physical principles aids in the development of the appropriate equations.

Example 1.10-1

Solve the vector equation $\mathbf{A} \times \mathbf{x} = \mathbf{B}$ *for the unknown vector* **x** *in terms of the known vectors* **A** *and* **B**.

Solution: The vectors **A** and **B**, though "known" or "given," cannot be assigned in a completely arbitrary fashion. The vector **B** must be perpendicular to **A**; if it is not, the equation is self-contradictory and meaningless. Similarly, any solution vector **x** and the vector **B** must be perpendicular. These conclusions follow from the definition of the cross product. They suggest that we look for a particular solution, **x**, which is also perpendicular to **A**, i.e., which satisfies $\mathbf{x} \cdot \mathbf{A} = 0$.

If we take the cross product of each side of the given equation with $\mathbf{A}$, we have

$$(\mathbf{A} \times \mathbf{x}) \times \mathbf{A} = \mathbf{B} \times \mathbf{A}.$$

Making use of the fundamental identity Eq. 1.9-4 for the vector triple product, we may write

$$(\mathbf{A} \cdot \mathbf{A})\mathbf{x} - (\mathbf{x} \cdot \mathbf{A})\mathbf{A} = \mathbf{B} \times \mathbf{A};$$

since $\mathbf{x} \cdot \mathbf{A}$ is to vanish,

$$\mathbf{x} = \frac{\mathbf{B} \times \mathbf{A}}{A^2}.$$

This is certainly a particular solution that satisfies the requirement that $\mathbf{x} \cdot \mathbf{A} = 0$ since it makes $\mathbf{x}$ perpendicular to both $\mathbf{B}$ and $\mathbf{A}$. If we substitute it back into the original equation, the solution is easily checked. Is it, however, the complete solution? Any vector parallel to $\mathbf{A}$—a vector $m\mathbf{A}$, where m is any scalar—may be added to any solution $\mathbf{x}$, since the cross product of $\mathbf{A}$ with such a vector will be the null vector. The complete solution then is

$$\mathbf{x} = \frac{\mathbf{B} \times \mathbf{A}}{A^2} + m\mathbf{A},$$

where m is an arbitrary number.

Example 1.10-2

If $\mathbf{A} \times \mathbf{B} = \mathbf{A} \times \mathbf{C}$, *what are the possible values of* $\mathbf{B}$?

Solution: The equation can readily be reduced to the one discussed in the last example. It is as simple, however, to note that $\mathbf{B} = \mathbf{C}$ is certainly a particular solution and that we may add to this any vector whose vector product with $\mathbf{A}$ is zero: in symbols,

$$\mathbf{B} = \mathbf{C} + m\mathbf{A},$$

where m is again an arbitrary number.

In solving $\mathbf{A} \times \mathbf{B} = \mathbf{C} \times \mathbf{A}$, note that the non-commutative nature of the cross product must be taken into account, the solution now being

$$\mathbf{B} = -\mathbf{C} + m\mathbf{A}.$$

Example 1.10-3

Given the vector $\mathbf{a}$ *and the scalar* m, *solve the equation* $(\mathbf{a} \times \mathbf{x}) + \mathbf{x} + m\mathbf{a} = 0$ *for the vector* $\mathbf{x}$.

Solution: We shall obtain $\mathbf{x}$ by arguing directly from the meanings of the symbols and also by formal algebraic operations. If $\mathbf{a} \times \mathbf{x}$ is not zero, then it must be perpendicular to both $\mathbf{a}$ and $\mathbf{x}$ and hence to the vector $\mathbf{x} + m\mathbf{a}$ in the plane of $\mathbf{a}$ and $\mathbf{x}$. The sum of $(\mathbf{a} \times \mathbf{x})$ and $(\mathbf{x} + m\mathbf{a})$ could not then be zero. Therefore, $\mathbf{a} \times \mathbf{x} = 0$ and $\mathbf{x} = -m\mathbf{a}$.

Formal algebraic solutions can be found in many ways. The simplest is the one corresponding to the argument just given. Take the scalar product of each side of the equation with the vector $\mathbf{a} \times \mathbf{x}$, obtaining

$$(\mathbf{a} \times \mathbf{x}) \cdot (\mathbf{a} \times \mathbf{x}) + (\mathbf{a} \times \mathbf{x}) \cdot \mathbf{x} + (\mathbf{a} \times \mathbf{x}) \cdot m\mathbf{a} = 0.$$

Since a scalar triple product vanishes if two terms are identical, $(\mathbf{a} \times \mathbf{x}) \cdot \mathbf{x} = 0$ and $(\mathbf{a} \times \mathbf{x}) \cdot m\mathbf{a} = m[(\mathbf{a} \times \mathbf{x}) \cdot \mathbf{a}] = 0$. Therefore,

$$(\mathbf{a} \times \mathbf{x}) \cdot (\mathbf{a} \times \mathbf{x}) = |(\mathbf{a} \times \mathbf{x})|^2 = 0,$$

$\mathbf{a} \times \mathbf{x} = 0$, and $\mathbf{x} = -m\mathbf{a}$ as before. Note that this solution is unique.

Exercises

1.2-1: If you have encountered any of the following quantities in your previous experience you should be able to identify them as scalars, vectors, or as neither scalar nor vector: (1) volume, (2) force, (3) displacement (in the sense of change of position), (4) mass, (5) acceleration, (6) stress, (7) temperature, (8) magnetic field intensity.

1.3-1: Solve the vector equations $13\mathbf{A} - 5\mathbf{B} = 49\mathbf{C}$ and $-4\mathbf{A} + \mathbf{B} = -14\mathbf{C}$ to find $\mathbf{A}$ and $\mathbf{B}$ in terms of $\mathbf{C}$.
Ans.: $\mathbf{A} = 3\mathbf{C}$, $\mathbf{B} = -2\mathbf{C}$.

1.3-2: Solve the three vector equations $5\mathbf{A} - \mathbf{B} + 3\mathbf{C} - 3\mathbf{D} = 0$, $10\mathbf{A} + 8\mathbf{B} + 4\mathbf{D} = 0$, $35\mathbf{A} + 14\mathbf{B} - 6\mathbf{C} = 0$ to find $\mathbf{A}$, $\mathbf{B}$, and $\mathbf{C}$ in terms of $\mathbf{D}$. Are the vectors $\mathbf{A}$ and $\mathbf{B}$ necessarily parallel?
Ans.: $\mathbf{A} = (2/5)\mathbf{D}$, $\mathbf{B} = -\mathbf{D}$, $\mathbf{C} = 0$.

1.3-3: Show that if $p\mathbf{A} + q\mathbf{B} + r\mathbf{C} = 0$, where p, q, r are any three non-zero numbers, the vectors $\mathbf{A}$, $\mathbf{B}$, $\mathbf{C}$ must all lie in the same plane.

1.3-4: Show that $|\mathbf{A}| + |\mathbf{B}| \geq |\mathbf{A} + \mathbf{B}|$. Under what circumstances does the equality sign hold?

1.3-5: Show that $||\mathbf{A}| - |\mathbf{B}|| \leq |\mathbf{A} - \mathbf{B}|$. Under what circumstances does the equality sign hold?

1.3-6: The points $PQRSTU$ are the vertices of a regular hexagon. Denote the vector $\overline{PQ}$ by $\mathbf{A}$, the vector $\overline{QR}$ by $\mathbf{B}$, and the vector $\overline{RS}$ by $\mathbf{C}$. Express $\mathbf{C}$ in terms of $\mathbf{A}$ and $\mathbf{B}$.
Ans.: $\mathbf{C} = \mathbf{B} - \mathbf{A}$.

1.3-7: Three vectors, $\mathbf{A}$, $\mathbf{B}$, and $\mathbf{C} = 4\mathbf{A} - 3\mathbf{B}$, have the same origin. Show that their termini lie on a straight line.

1.3-8: Show that if $p\mathbf{A} + q\mathbf{B} + r\mathbf{C} = 0$, where p, q, r are any three numbers such that $p + q + r = 0$ and $\mathbf{A}$, $\mathbf{B}$, $\mathbf{C}$ are vectors with a common origin, the termini of $\mathbf{A}$, $\mathbf{B}$, $\mathbf{C}$ will lie on a straight line. Show that the previous exercise is a special case of this one.

1.3-9: Prove that the diagonals of a parallelogram bisect each other.

1.3-10: The vector $\mathbf{A}$ is six inches long and is directed toward the northeast. The vector $\mathbf{B}$ is three inches long and is directed toward the west.
(a) Express the vector $\mathbf{C}$ which is twelve inches long and is directed to the south in the form $\mathbf{C} = m\mathbf{A} + n\mathbf{B}$.

(b) Do the same for the vector **D** which is one inch long and is directed toward the north.

Ans.: C $= -2\sqrt{2}\mathbf{A}-4\mathbf{B}$ in., **D** $= \dfrac{1}{3}\mathbf{B}+\dfrac{\sqrt{2}}{6}\mathbf{A}$ in.

1.3-11: What vector must be added to the vectors **A** and **B** in order to produce a null resultant? **A** is five inches long and is directed toward the southwest; **B** is eight inches long and is directed toward the north.
Ans.: Magnitude: 5.69 in.; direction: 51.6° south of east.

1.3-12: Prove that the line segment joining the midpoints of two sides of a triangle is parallel to the third side and half as long.

1.3-13: Show that the resultant of the medians of a triangle is a null vector.

1.3-14: A parallelepiped is determined by three vectors **A, B, C** having a common origin. Express the body diagonal of the parallelepiped in terms of **A, B,** and **C**.
Ans.: A+B+C.

1.3-15: (Desargues' theorem) Two triangles *ABC* and *DEF* are so positioned in space that the lines *AD, BE, CF* intersect at *O*. Then the lines *AB* and *DE* intersect, as do the lines *AC* and *DF* and the lines *BC* and *EF*. Show that these points of intersection are collinear.

1.4-1: What is the resultant of the forces $\mathbf{F}_1 = 6\mathbf{i}+8\mathbf{j}$ lb, $\mathbf{F}_2 = \mathbf{i}-2\mathbf{j}+2\mathbf{k}$ lb, $\mathbf{F}_3 = 4\mathbf{i}+4\mathbf{j}-7\mathbf{k}$ lb? What is its magnitude and what angle does it make with the *y*-axis? Sketch this vector.
Ans.: R $= 11\mathbf{i}+10\mathbf{j}-5\mathbf{k}$ lb, $|\mathbf{R}| = 15.7$ lb, $\beta = 50.4°$.

1.4-2: What vector must be added to the vectors $\mathbf{F}_1=24\mathbf{i}-45\mathbf{j}+68\mathbf{k}$ and $\mathbf{F}_2=21\mathbf{i}-40\mathbf{j}-72\mathbf{k}$ in order to produce a null resultant vector? What is the magnitude of this so-called *equilibrating* vector and what are its direction cosines?
Ans.: $-45\mathbf{i}+85\mathbf{j}+4\mathbf{k}$; 96.3; $-0.467, 0.883, 0.042$.

1.4-3: What is the unit vector in the direction of the vector $\mathbf{A}=6\mathbf{i}-10\mathbf{j}+15\mathbf{k}$?
Ans.: $\mathbf{e}_A = \dfrac{1}{19}(6\mathbf{i}-10\mathbf{j}+15\mathbf{k})$.

1.4-4: What is the unit vector in the direction opposite to that of the vector $\mathbf{A}=8\mathbf{i}+9\mathbf{j}-12\mathbf{k}$?
Ans.: $\dfrac{-1}{17}(8\mathbf{i}+9\mathbf{j}-12\mathbf{k})$.

1.4-5: What is the vector equation of the straight line through the termini of the vectors $\mathbf{A}=13\mathbf{i}+14\mathbf{j}+34\mathbf{k}$ and $\mathbf{B}=24\mathbf{i}-45\mathbf{j}+68\mathbf{k}$? **A** and **B** emanate from the origin of coordinates.
Ans.: $(13+11m)\mathbf{i}+(14-59m)\mathbf{j}+34(1+m)\mathbf{k}$.

1.4-6: Show that the bisector of an angle of a triangle intersects the opposite side dividing it into two segments, the ratio of whose lengths is the same as the ratio of the lengths of the other two sides.

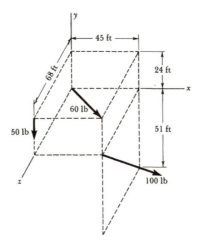

Exer. 1.4-7

1.4-7: Find the resultant of the three vectors shown. What is its magnitude? What are the direction cosines of its line of action?

Ans.: $31.8\mathbf{i} - 93.1\mathbf{j} - 32\mathbf{k}$ lb; 103.5 lb; 0.307, -0.900, -0.309.

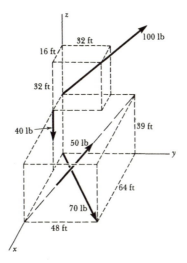

Exer. 1.4-8

1.4-8: Find the resultant of the four vectors shown.   What is its magnitude ? What are the direction cosines of its line of action ?
Ans.: $53.4\mathbf{i}+135.6\mathbf{j}+48.6\mathbf{k}$ lb; 153.6 lb; 0.348, 0.883, 0.316.

1.4-9: A vector of magnitude 10 is directed along the line from the point P: $(13, -5, 0)$ to the point Q: $(0, 9, 34)$.   What are the cartesian components of this vector ?
Ans.: $-3.33\mathbf{i}+3.59\mathbf{j}+8.72\mathbf{k}$.

1.4-10: What is the equation of the line in the plane formed by the origin O and the points P and Q of the previous exercise bisecting the angle POQ ?
Ans.: $\mathbf{r} = m(0.933\mathbf{i}-0.103\mathbf{j}+0.967\mathbf{k})$.

1.4-11: Consider the planar vector $\mathbf{A}=3\mathbf{i}+4\mathbf{j}$.   What will the cartesian components of this vector be if the x-, y-axes are rotated counterclockwise through an angle of 120 degrees ?
Ans.: $A_{x'} = 1.96$, $A_{y'} = -4.60$.

1.4-12: Show that the three angle bisectors of a triangle intersect at a common point.

1.4-13: Find the cartesian components of a unit vector making equal angles with the axes and directed from the origin into the first octant. What is the angle between the vector and any one of the axes ?

Ans.: $\mathbf{e} = \dfrac{1}{\sqrt{3}}(\mathbf{i}+\mathbf{j}+\mathbf{k})$; $54.7°$.

1.4-14: Taking an x-axis directed toward the east and a y-axis directed northward, solve Exercise 1.3-11 by expressing all vectors in cartesian form.

1.4-15: Show that the sum of the squares of the diagonals of a parallelogram is equal to the sum of the squares of the sides.

1.5-1: If $\mathbf{e}_r=0.6\mathbf{i}+0.8\mathbf{j}$, express the vector $\mathbf{A}=4\mathbf{i}+3\mathbf{j}-12\mathbf{k}$ in terms of its cylindrical polar components.
Ans.: $\mathbf{A} = 4.8\mathbf{e}_r-1.4\mathbf{e}_\phi-12\mathbf{k}$.

1.5-2: Suppose a set of rectangular cartesian coordinates (x, y, z) and a set of cylindrical polar coordinates (r, ϕ, z) are given, the z-direction of both being the same.   The angle between the x-direction and the r-direction is $30°$.   A vector $\mathbf{A}=10\mathbf{e}_r+7.5\mathbf{k}$ is given.   Find the magnitude of $\mathbf{A}$ and its direction cosines with respect to the (x, y, z) axes.
Ans.: $|\mathbf{A}| = 12.5$; $(\cos\alpha, \cos\beta, \cos\gamma) = (0.693, 0.4, 0.6)$.

1.5-3: If a set of cylindrical polar coordinates (r, ϕ, z) is based on the vector $\mathbf{A}=3\mathbf{i}+4\mathbf{j}+12\mathbf{k}$, i.e., if $\mathbf{A}=A_r\mathbf{e}_r+12\mathbf{k}$, what are the polar components in that set of the vector $\mathbf{B} = -4\mathbf{i}+3\mathbf{j}-12\mathbf{k}$?
Ans.: $B_r = 0$, $B_\phi = 5$, $B_z = -12$.

1.5-4: If, in the same set of cylindrical polar coordinates, three vectors $\mathbf{A}=10\mathbf{e}_r-5\mathbf{k}$, $\mathbf{B}=3\mathbf{e}_r+2\mathbf{e}_\phi+7\mathbf{k}$, $\mathbf{C}=-2\mathbf{e}_r+2\mathbf{e}_\phi-2\mathbf{k}$ are given, what is the magnitude of $\mathbf{A}+\mathbf{B}+\mathbf{C}$?
Ans.: 11.7.

1.5-5: At any point in the vicinity of a long, magnetized, straight rod the magnetic field exerts a radial force on a charged particle, the magnitude of the attraction varying inversely as the square of the distance of the particle from the rod and being unity when that distance is unity. Write an expression for the force exerted on the particle by the field as a function of the coordinates of the point at which the charged particle is located. Do this first using cylindrical, then cartesian coordinates.

Ans.: $F = -\dfrac{1}{r^2}\, e_r = -\dfrac{x i + y j}{[x^2 + y^2]^{3/2}}$.

1.5-6: The velocity of a point in a rigid body rotating about a fixed z-axis is proportional to the distance of the point from the axis of rotation and is directed at right angles to that axis and to the shortest line joining the point to the axis. Denote the constant of proportionality by the symbol ω and express the velocity vector of any point in the body in cylindrical polar components.

Ans.: $v = \omega r e_\phi = \omega(-y i + x j)$.

1.6-1: Find the scalar component of the vector $A = 3i - 4j + 12k$ in the direction of:
(a) the line through the origin making equal angles with the coordinate axes;
(b) the line from $(-2, 1, -6)$ to $(-5, 5, -18)$;
(c) the line from $(1, 1, 1)$ to $(-1, 2, -3)$.

Ans.: (c) $-58\sqrt{21}/21$.

1.6-2: Find the angle between $A = 6i - 2j + 5k$ and $B = 2i + k$.

Ans.: arccos $(17\sqrt{13}/65)$.

1.6-3: Determine the value or values of x that make $A = xi - 2j + 4k$ orthogonal to $B = 3i + 2j - 2k$.

1.6-4: Are there any vectors in the yz-plane perpendicular to the line making equal angles with the coordinate axes? If so, find the unit vector or vectors that are.

Ans.: $\pm\dfrac{\sqrt{2}}{2}\,(j - k)$.

1.6-5: The position vector r to any point in a plane passing through r_0 and having unit normal vector n satisfies the relation $(r - r_0) \cdot n = 0$. Suppose three points with coordinates $(1, 1, 1)$, $(2, -1, 5)$, and $(-1, -1, 0)$ are given. Find a unit normal vector n perpendicular to the plane through the three points.

Ans.: $n = \pm(0.735i - 0.515j - 0.441k)$.

1.6-6: If $A = 3i + 4j + 12k$, $B = -4i + 3j$, use the scalar products $C \cdot A$ and $C \cdot B$ to show that all vectors $C = C_x i + C_y j + C_z k$ perpendicular to both A and B can be written in the form

$$C = \pm\frac{|C|}{65}\,(-36i - 48j + 25k).$$

1.6-7: What is the angle between the line making equal angles with the positive (xyz) coordinate directions and the line in the xz-plane making equal angles with the negative coordinate directions ?

Ans.: arccos $(-\sqrt{6}/3) \cong 144.7°$.

1.6-8: A cone with vertex at the origin has semi-vertical angle $45°$ and axis along the x-axis.   Find the position vector **r** from the origin to any point in the conical surface.

Ans.: $\mathbf{r} = \sqrt{y^2+z^2}\,\mathbf{i}+y\mathbf{j}+z\mathbf{k}$.

1.6-9: Two vectors $\mathbf{A}=6\mathbf{e}_r-7\mathbf{k}$ and $\mathbf{B}=5\mathbf{e}_r-2\mathbf{e}_\phi+4\mathbf{k}$ are given in the same set of cylindrical polar coordinates.   What is the angle between them ?

Ans.: $88.1°$.

1.6-10: For each pair of vectors **A** and **B** that follow, find (i) the scalar component of **B** in the direction of **A** and (ii) the angle θ between **A** and **B**.

(a) $\mathbf{A} = \mathbf{i}-\mathbf{j}-\mathbf{k}, \quad \mathbf{B} = 5\mathbf{i}+6\mathbf{j}$;

(b) $\mathbf{A} = \mathbf{i}-\mathbf{j}-2\mathbf{k}, \quad \mathbf{B} = \mathbf{i}-3\mathbf{j}+2\mathbf{k}$;

(c) $\mathbf{A} = 3\mathbf{i}-4\mathbf{j}+12\mathbf{k}, \quad \mathbf{B} = -4\mathbf{i}-9\mathbf{j}-\mathbf{k}$;

(d) $\mathbf{A} = -\mathbf{i}-\mathbf{j}+2\mathbf{k}, \quad \mathbf{B} = -3\mathbf{i}-9\mathbf{j}+2\mathbf{k}$.

Ans.: (a) $\theta = 94.25°$; (c) $\theta = 84.65°$.

1.7-1: Find the vector product $\mathbf{A} \times \mathbf{B}$ of:

(a) $\mathbf{A} = 3\mathbf{i}-2\mathbf{j}+5\mathbf{k}, \mathbf{B} = \mathbf{i}-\mathbf{j}-\mathbf{k}$;

(b) $\mathbf{A} = \mathbf{i}-\mathbf{j}+\mathbf{k}, \mathbf{B} = -6\mathbf{i}+6\mathbf{j}-6\mathbf{k}$;

(c) $\mathbf{A} = \mathbf{i}-\mathbf{j}-\mathbf{k}, \mathbf{B} = 3\mathbf{i}-2\mathbf{j}+5\mathbf{k}$;

(d) $\mathbf{A} = \mathbf{i}+2\mathbf{j}+\mathbf{k}, \mathbf{B}$ the vector from $(1, 1, 1)$ to $(3, 0, 0)$.

Sketch the vectors **A**, **B**, and $\mathbf{A} \times \mathbf{B}$.

Ans.: (a) $7\mathbf{i}+8\mathbf{j}-\mathbf{k}$; (d) $-\mathbf{i}+3\mathbf{j}-5\mathbf{k}$.

1.7-2: Find the moment of the force $\mathbf{F} = 3\mathbf{i}-2\mathbf{j}+6\mathbf{k}$ newtons acting at $\mathbf{r}=2\mathbf{i}-2\mathbf{j}$ meters about the point $\mathbf{R}=7\mathbf{k}$ meters.

Ans.: $-26\mathbf{i}-33\mathbf{j}+2\mathbf{k}$ newton-meters.

1.7-3: Given two non-parallel vectors **A** and **B**, construct an orthogonal triad $\mathbf{e}_1, \mathbf{e}_2, \mathbf{e}_3$ of unit vectors.

Ans.: $\mathbf{e}_1 = \mathbf{A}/A, \mathbf{e}_2 = \mathbf{A} \times \mathbf{B}/|\mathbf{A} \times \mathbf{B}|, \mathbf{e}_3 = \mathbf{e}_1 \times \mathbf{e}_2$ is one such set.

1.7-4: Solve Exercise 1.6-5 using the cross product.

1.7-5: Suppose that, in addition to the force **F** of Exercise 1.7-2, we have a second force $-\mathbf{F}$ acting at $3\mathbf{i}+3\mathbf{k}$ meters.   Find the sum of the moment vectors of **F** and $-\mathbf{F}$ about $\mathbf{R}=7\mathbf{k}$.   What is the sum of the moment vectors of **F** and $-\mathbf{F}$ about the origin ?   Generalize to any vectors $(\mathbf{A}, -\mathbf{A})$ acting at points $\mathbf{r}_1, \mathbf{r}_2$.

Ans.: $-18\mathbf{i}-3\mathbf{j}+8\mathbf{k}$ newton-meters for $(\mathbf{F}, -\mathbf{F})$ about any point of space.

1.7-6: Derive the result of Exercise 1.6-6 using the vector product.

1.7-7: Derive the result of Exercise 1.6-7 using the vector product.

1.7-8: Derive the result of Exercise 1.6-9 using the vector product.

1.7-9: For each pair of vectors **A** and **B** of Exercise 1.6-10, find the vector product $\mathbf{B} \times \mathbf{A}$ and the sine of the angle between **A** and **B**.

1.7-10: Suppose that two lines in space are given, one passing through the origin and making equal angles with the axes and the other passing through the points (0, 0, 1) and (3, 2, 0). Find the length of the common perpendicular to these two lines.

Ans.: $\sqrt{26}/26$.

1.8-1: Find the moment of the force $\mathbf{F}$ of Exercise 1.7-2 about a line through $\mathbf{R} = 7\mathbf{k}$ meters making equal angles with the axes.

Ans.: $-19\sqrt{3}$ newton-meters.

1.8-2: A parallelepiped has face $ABCD$ in the xz-plane, with the 10 in. edge AB along the z-axis and the 8 in. edge AD at 60° to AB in the first quadrant of the xz-plane (draw a picture). The opposite face, $A'B'C'D'$, is so situated that corner A', 6 in. away from A, lies on the line through the origin A and the point $3\mathbf{i} + 12\mathbf{j} + 4\mathbf{k}$ in. Find the volume of the parallelepiped.

Ans.: 383.7 in³.

1.8-3: A line with direction $\mathbf{e}_1$ passes through point $\mathbf{r}_1$; a line with direction $\mathbf{e}_2$ passes through point $\mathbf{r}_2$. What is the length of the common perpendicular to the two lines?

Ans.: $|(\mathbf{r}_2 - \mathbf{r}_1) \cdot (\mathbf{e}_1 \times \mathbf{e}_2)|/|\mathbf{e}_1 \times \mathbf{e}_2|$.

1.8-4: Check the answer to Exercise 1.7-10 using the result of Exercise 1.8-3.

1.8-5: In each of the following parts, a vector $\mathbf{B}$ and a point $\mathbf{r}$ on its line of action are given. An axis through the origin is specified in direction by the vector $\mathbf{A}$ in each case. Find the moment of $\mathbf{B}$ about that axis.

(a) $\mathbf{A} = 3\mathbf{i} - 2\mathbf{j} + 5\mathbf{k}$, $\mathbf{B} = \mathbf{i} - \mathbf{j} - \mathbf{k}$, $\mathbf{r} = \mathbf{i} + \mathbf{j} + \mathbf{k}$;
(b) $\mathbf{A} = \mathbf{i} - \mathbf{j} + \mathbf{k}$, $\mathbf{B} = -6\mathbf{i} + 6\mathbf{j} - 6\mathbf{k}$, $\mathbf{r} = -2\mathbf{j} + 3\mathbf{k}$;
(c) $\mathbf{A} = \mathbf{i} - \mathbf{j} - \mathbf{k}$, $\mathbf{B} = 3\mathbf{i} - 2\mathbf{j} + 5\mathbf{k}$, $\mathbf{r} = -2\mathbf{i} + 4\mathbf{j} + 4\mathbf{k}$;
(d) $\mathbf{A} = \mathbf{i} + 2\mathbf{j} + \mathbf{k}$, $\mathbf{B} = -2\mathbf{i} - 4\mathbf{j} - 2\mathbf{k}$, $\mathbf{r} = \mathbf{i} + \mathbf{j} + \mathbf{k}$.

Ans.: (a) $-7\sqrt{38}/19$; (b) 0; (c) $14\sqrt{3}/3$; (d) 0.

1.9-1: Find the vector product $\mathbf{B} \times (\mathbf{C} \times \mathbf{A})$ of $\mathbf{A} = \mathbf{i} + \mathbf{j}$, $\mathbf{B} = 7\mathbf{i} - 2\mathbf{k}$, $\mathbf{C} = 6\mathbf{i} - \mathbf{j} - 2\mathbf{k}$ both by carrying out the indicated products and by using the fundamental identity for the reduction of such products.

Ans.: $-4\mathbf{i} - 53\mathbf{j} - 14\mathbf{k}$.

1.9-2: Show that the component of $\mathbf{A} \times (\mathbf{A} \times \mathbf{B})$ in the direction of $\mathbf{B}$ is never positive and hence that $(\mathbf{A} \cdot \mathbf{B})^2 \leqq A^2 B^2$. (Hint: use both the reduction identity for the vector triple product and the cross-dot interchange identity for the scalar triple product.)

1.9-3: Show that, if $\mathbf{A} \neq \mathbf{0}$, $\mathbf{B} \neq \mathbf{0}$, the vanishing of $\mathbf{A} \times [\mathbf{B} \times (\mathbf{A} \times \mathbf{B})]$ implies that either $\mathbf{A}$ is parallel to $\mathbf{B}$ or $\mathbf{A}$ is perpendicular to $\mathbf{B}$.

1.9-4: Given the three vectors $\mathbf{A} = 2\mathbf{i} - 5\mathbf{j} + 6\mathbf{k}$, $\mathbf{B} = \mathbf{i} - \mathbf{j} - \mathbf{k}$, $\mathbf{C} = \mathbf{i} + \mathbf{j} + 4\mathbf{k}$, compute all possible different vector triple products that can be formed from these. Do not include both products if two differ by only a factor of -1.

1.10-1: Prove that, if **A** and **B** are any two vectors and **e** is a unit vector at right angles to the plane of **A** and **B**, then the expression $(\mathbf{A} \times \mathbf{B}) \times \mathbf{C} = [(\mathbf{A} \times \mathbf{B}) \cdot \mathbf{e}][\mathbf{e} \times \mathbf{C}]$ is true for any vector **C** whatsoever; i.e., the expression is an identity, whatever **C** may be.

1.10-2: Given any two vectors, **A** and **B**, show that $|\mathbf{A} \times \mathbf{B}|^2 = A^2 B^2 - (\mathbf{A} \cdot \mathbf{B})^2$. Derive the result of Exercise 1.9-2 from this. What is the trigonometrical interpretation of this identity when **A** and **B** are unit vectors which include an angle θ?

1.10-3: What is the general solution of the equation $\mathbf{a} \cdot \mathbf{x} = \mathbf{a} \cdot \mathbf{b}$?

Ans.: $\mathbf{x} = \mathbf{b} + \mathbf{a} \times \mathbf{c}$, where **c** is an arbitrary vector.

1.10-4: Find a vector **x** that satisfies the equations $\mathbf{a} \times \mathbf{x} = \mathbf{b}$ and $\mathbf{a} \cdot \mathbf{x} = b$, where $\mathbf{a} \cdot \mathbf{b} = 0$.

Ans.: $\mathbf{x} = \dfrac{1}{a^2}(b\mathbf{a} + \mathbf{b} \times \mathbf{a})$.

1.10-5: Solve the simultaneous equations
$$x + y + z = \mathbf{a},$$
$$x - y + z = \mathbf{b},$$
$$x + y - z = \mathbf{c},$$
for the unknown vectors **x**, **y**, **z**, given the constant vectors **a**, **b**, and **c**.

1.10-6: (a) Solve the simultaneous equations
$$w + x + y + z = \mathbf{a},$$
$$w + x - y + z = \mathbf{b},$$
$$w + x + y - z = \mathbf{c},$$
$$w - x + y + z = \mathbf{d},$$
for the unknown vectors **w**, **x**, **y**, **z**, given the constant vectors **a**, **b**, **c**, and **d**.

Ans.: $\mathbf{x} = (\mathbf{a} - \mathbf{d})/2$, $\mathbf{w} = (-\mathbf{a} + \mathbf{b} + \mathbf{c} + \mathbf{d})/2$.

(b) Suppose $2\mathbf{a} - 3\mathbf{b} + \mathbf{c} - \mathbf{d} = 0$ is the relation expressing the linear dependence of **a**, **b**, **c**, and **d**. What are **w**, **x**, **y**, and **z** in terms of **a**, **b**, and **c**? What is the relation among **w**, **x**, **y**, **z** expressing the linear dependence of these four vectors?

Ans.: $\mathbf{w} - \mathbf{x} - 5\mathbf{y} + 3\mathbf{z} = 0$.

1.10-7: Given four vectors **a**, **b**, **c**, and **d**, solve the equation $\mathbf{x} \times \mathbf{a} + (\mathbf{x} \cdot \mathbf{b})\mathbf{c} = \mathbf{d}$ for the unknown vector **x**. It may be assumed that $\mathbf{a} \cdot \mathbf{b}$ and $\mathbf{a} \cdot \mathbf{c}$ are not equal to zero.

Ans.: $\mathbf{x} = \dfrac{1}{(\mathbf{a} \cdot \mathbf{b})}\left[\dfrac{(\mathbf{d} \cdot \mathbf{a})}{(\mathbf{c} \cdot \mathbf{a})}(\mathbf{c} \times \mathbf{b} + \mathbf{a}) - \mathbf{d} \times \mathbf{b}\right]$.

CHAPTER **II**

Principles of Statics

2.1 Newton's Laws of Motion; Equilibrium of a Particle

Mechanics is based upon Newton's laws of motion which, in modern language, take the form:

(1) A particle acted on by forces whose resultant always vanishes will move with constant velocity.

(2) A particle acted on by forces whose resultant is not zero will move in such a way that the time rate of change of its momentum will at any instant be proportional to the resultant force.

(3) If one particle, A, exerts a force on a second particle, B, then B exerts a collinear force of equal magnitude and opposite direction on A.

The study of the motion of material systems has been developed by systematic generalization and extension of these laws. Here we shall say only enough to indicate the general pattern of this development; detailed proofs are left to dynamical studies.*

---

* See L. E. Goodman and W. H. Warner, *Dynamics* (Belmont, Calif: Wadsworth Publishing Company, Inc., 1963), Chapters I and II.

First, we note that the laws relate to the motion of "particles." The particle plays the same role in mechanics that the point plays in geometry. An actual object may be treated as a particle if, insofar as its motion is concerned, it can be regarded with sufficient accuracy as a moving point. The laws of motion, therefore, presuppose a space in which the particle moves. To that space certain properties— homogeneity, isotropy, and continuity—must be attributed so that changes in position may be measured. The familiar Euclidean three-dimensional space of elementary geometry has the properties wanted and is the one used in newtonian mechanics. Time is introduced as a scalar variable taking on the values of the real-number continuum. Regarding position vectors as functions of this scalar variable, we can define the rate of change of the position vector, called the *velocity* vector. This definition requires the development of a vector calculus, with velocity as the derivative of position. *Acceleration* is also a vector quantity, the time derivative of velocity. The interrelationships between the position, velocity, and acceleration vectors is that subject of study known as *kinematics*. Finally, a time scale is introduced by our comparing the motions we study to a standard observable motion, say the rotation of the earth.

Further attributes are given to particles by introducing the concepts of "force," related to a measure of the mutual influence of particles on one another's motion, and "mass," a scalar property of each particle which does not change with time or location. Mass may be defined by considering a hypothetical experiment in which two particles are supposed to move free from all influences except their mutual interaction. Since force is a measure of this interaction, the third law assures us that the two particles will move under the action of forces of equal magnitude. The accelerations of the two particles will, however, usually have different magnitudes; one particle will, in general, be more sluggish or inert than the other. To account for this property of inertia, a measure of it called *mass* is assigned to each particle, the more sluggish particle being said to have the larger mass. To be precise, the ratio of the masses of the two particles is taken as the reciprocal of the ratio of the magnitudes of their accelerations at any instant. Once a standard body or mass is selected, all other masses are determined by comparison with it, by means of the experiment described or by means of one shown to be equivalent. The *momentum* vector of a particle then is taken as the product of the scalar mass and the vector velocity, and the second law states that the derivative of this is the proper measure of the change of motion of the particle due to other particles.

This derivative of the momentum is the product of mass and acceleration.

Force is defined through the second law as the derivative of the momentum vector. The second law requires only a proportionality, but by proper choice of units of measurement—so-called *dynamically consistent units*—the constant of proportionality may be taken as unity. Since the derivative of a vector is a vector, force is a vector quantity. Apparently we are then introducing an unnecessary name for a previously defined vector, the product of mass and acceleration. However, force has a definite utility as a separate concept. As a name for the "mechanical interaction" (i.e., effect on motion) it is subject to the third law. The third law is of great use in statics. It makes possible the simplification of the analysis of complex systems by permitting the examination of parts of the system without introducing completely new and unknown "interactions" (i.e., forces) at each step. Through it, we may extend newtonian mechanics from particles to objects of finite size. The vectorial character of force as defined by the second law enables us to combine, by vectorial addition, the forces arising from a number of sources; and, finally, we have learned to recognize a variety of interactions and write appropriate mathematical expressions for them as force vectors to use in the equations of motion. In the next few sections, we shall discuss some of the important types of forces and their representations.

We now come to the fundamental concept underlying statics: that of equilibrium. The first law of motion states that, when the resultant interaction effect on a particle vanishes—when the vector sum of the forces on it is the zero vector—the particle moves with constant velocity. First of all, this law asserts the existence of at least one preferred coordinate frame relative to which the constant velocity is measured. Such a frame of reference is called *inertial*, *newtonian*, or *galilean*. In practice, of course, we cannot apply the test of removing an object from all influence. What we do is use a convenient frame of reference and check the predictions of the laws of motion against experiment. Accurate agreement between prediction and measurement in a very wide range of physical and engineering situations is the basis for the enormous power and scope of the methods of newtonian mechanics. For most engineering problems axes fixed in the earth, regarded as stationary, form a satisfactory inertial frame. More complex problems, such as the motion of a gyrocompass, require that we not ignore the rotation of the earth and, indeed, problems of celestial mechanics require an astronomical frame of reference fixed with respect to the so-called "fixed" stars.

The existence of other inertial frames, granted the existence of one such frame, is not a matter of immediate concern.   The important assumption is that one exists, relative to which a particle free of mechanical influence moves at constant velocity.   Since velocity is a vector, this means the particle moves at constant speed, or magnitude of velocity, in a straight line—a fixed direction.   The particle is said to be in *equilibrium*.   There are two apparent aspects to such an equilibrium motion: the vanishing of the resultant force and hence of the acceleration, and the steady motion of the particle.   Which is the proper basis for generalization of the concept of equilibrium to more complex systems?   It can be shown that, if a particle is moving with constant velocity relative to an inertial frame of reference, then there is another inertial frame relative to which the particle has zero velocity; i.e., it is at *rest*.   It is this concept of equilibrium that is taken as fundamental for generalization to systems of particles. A system is in equilibrium if no part of the system has any acceleration and if the system is at rest in some inertial frame of reference.

Statics, then, is the study of material systems in equilibrium. If we regard our material system as a particle, we conclude directly from the laws of motion that it is in equilibrium if and only if the resultant force on the particle vanishes.   We must see how this principle is to be generalized as we consider more and more complex models of systems.   In the remainder of this chapter, we shall be concerned with two things.   First, how do we describe particular forces and work with them mathematically?   Second, how do we describe systems and develop the principles of equilibrium for them?

2.2 Force and Moment Vectors; the Couple

Force, as defined by the second law of motion, is a vector quantity. The separate effects on a particle's motion due to any number of other particles are represented by force vectors $\mathbf{F}_1, \mathbf{F}_2, \ldots \mathbf{F}_n$ acting on the particle; the vector nature of force then implies that these may be replaced by a single vector $\mathbf{F}$, the *resultant force*, given by the sum of the separate force vectors:

$$\mathbf{F} = \mathbf{F}_1 + \mathbf{F}_2 + \ldots \mathbf{F}_n = \sum_{i=1}^{n} \mathbf{F}_i. \qquad \textbf{2.2-1}$$

In statics, we are concerned with the case in which $\mathbf{F} = 0$.

The force vectors are *bound vectors* in the sense in which they are used at present; they have a definite point of application.   To say that another force is completely equivalent to a given force means more than the simple statement that the vectors representing the forces have the same magnitude and direction, i.e. are equal vectors in the sense of Chapter I. They must also be given the same point of application in general. That is, though force is a vector quantity, not all of its properties are represented by its vector alone. This "bound vector" concept of force will be relaxed later when we consider distributed mass systems, moments, and the definition of mechanical equivalence, or equipollence, of force sets.

In the course of scientific progress large classes of inter-actions between particles and between finite bodies have come to be recognized and have found mathematical expression as force vectors. Among those commonly encountered in mechanics and engineering, contact forces, gravitational forces, electromagnetic forces, and viscous drag forces deserve special mention. Whenever two objects are in contact, forces of the first type mentioned arise. Such forces may be considered as distributed over an area or idealized as con-centrated forces. In technology, cases of common occurrence are those of two massive objects connected by a light cable or by a light spring; the forces representing the action of the cable or spring are usually idealized as concentrated contact loads. Gravitational forces, including the weight force on an object near the earth, are introduced as a consequence of Newton's law of universal gravitation. Coulomb's law for the force on electrically charged particles takes the same inverse square form as the gravitational force. Viscous drag forces, depend-ent on the velocity of a body relative to a fluid, and frictional forces at a rough contact are also common representations of interaction effects. As we proceed through the developments of the next few sections, we shall see how we represent the force vectors of some of the more important of these.

A unit of magnitude is required for the force vector. In statics, we need only be sure that all forces are measured in the same units, the unit itself being unimportant since we shall set the resultant force equal to zero. In dynamics a set of dynamically consistent units is needed for force, mass, length, and time (and hence acceleration) if we are to write $\mathbf{F} = m\mathbf{a}$. A complete discussion of the various metric and English, "scientific" and "engineering," absolute and gravita-tional systems of units would be out of place in material concerned only with statics.[*]   In technology, as in everyday speech, it is custom-ary in English-speaking countries to use the pound-force as the unit

---

[*] See *Dynamics*, Section 2.3.

of force and to refer to it simply as the *pound*. This is the force
necessary to give a standard of mass (the so-called engineering unit
of mass or "slug") an acceleration of one foot per second per second.
Alternatively, the unit of mass may be taken as the pound-mass
avoirdupois and the unit of force as that force necessary to give one
pound-mass avoirdupois an acceleration of one foot per second per
second. This unit of force, called the *poundal*, is not much used in
engineering. One pound-force is the same as 32.174 poundals.
Multiples of the pound-force are the kilopound, or "kip," and the
ton. In metric units, the unit of force, known as the *newton*, is that
force necessary to give a mass of one kilogram an acceleration of one
meter per second per second. One pound is the same as 4.44 newtons.
Where a smaller unit of force is wanted, the *dyne*, which is 10^{-5}
newtons, is employed. Continental European and South American
engineers usually employ the kilogram-force as a unit; one kilogram-
force is the same as 9.811 newtons. Whatever unit is chosen, it
should be associated with the magnitude of the force vector or with
its scalar components and not with the unit direction vector.

The necessary and sufficient condition for particle equilibrium
is the vanishing of the resultant force **F** on the particle. From this,
a secondary condition follows: if a particle is in equilibrium, the
resultant moment of the forces on the particle must vanish. In
Chapter I, we defined the moment of any vector with respect to an
arbitrarily chosen base point. We also proved Varignon's theorem,
that the resultant of the moment vectors of the separate forces is the
moment of the resultant of the forces. It is a trivial consequence of
the force equilibrium condition, then, to say that the moment must
also be zero if the particle is in equilibrium. The vanishing of the
moment about one point is not, however, a sufficient condition for
particle equilibrium. That is, if we know that the resultant moment
M_P of the forces on a particle about a point P vanishes, we cannot
conclude that the force vanishes and hence that the particle is in
equilibrium. For a particle, the establishment of equilibrium conditions in terms of moments is not an important task; for extended
systems, we shall want to develop moment conditions necessary for
equilibrium as well as the ordinary force equilibrium equations that
are necessary. Now we point out only that we shall need moments
of force vectors; that we shall need the interpretation of scalar components of moments as moments about axes; that moment vectors
have units: pound-feet, kilogram-force-meters, newton-meters, pound-
inches, etc.; and that moment vectors should be considered in general as
bound vectors, with point of application at the base point for moments.

Varignon's theorem treats of the moments of vectors with concurrent lines of action. The forces on extended systems of bodies will not have concurrent lines of action in general. We need a means of extending the resultant moment concept to such forces. The formalism by which this may be done is based on consideration of a special force system, called a *couple*. A couple is a pair of force vectors equal in magnitude, opposite in direction, and on parallel lines of action. That is, if two forces $\mathbf{F}_1$ and $\mathbf{F}_2$ act on a body so that the vectors representing the forces are negatives of one another ($\mathbf{F}_2 = -\mathbf{F}_1 = -\mathbf{F}$) but the lines of action of the forces are not the same (and hence are parallel, since the line of action of a force has, by

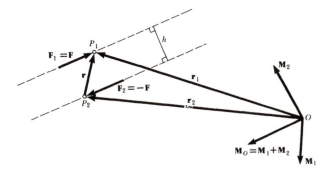

Fig. 2.2-1

definition, the same direction as the force), then we call the pair of force vectors $(\mathbf{F}_1, \mathbf{F}_2) = (\mathbf{F}, -\mathbf{F})$ a couple. The plane containing the parallel lines of action is called the *plane of the couple*.

Let us consider the formal force-and-moment resultant properties of a couple, even though we are not yet prepared to discuss the physical significance of these. Let $(\mathbf{F}_1, \mathbf{F}_2) = (\mathbf{F}, -\mathbf{F})$ be a couple, with P_1 and P_2 the points of application of $\mathbf{F}_1$ and $\mathbf{F}_2$, respectively (Fig. 2.2-1). Choose any point O of space and let $\mathbf{r}_1$ be the position of P_1 relative to O, $\mathbf{r}_2$ the position of P_2 relative to O, and $\mathbf{r} = \mathbf{r}_1 - \mathbf{r}_2$ the position of P_1 relative to P_2. The vector sum of the two forces is zero:

$$\mathbf{F}_1 + \mathbf{F}_2 = \mathbf{F} + (-\mathbf{F}) = \mathbf{0}. \qquad \textbf{2.2-2}$$

It can be shown (see *Dynamics*, Section 4.3) that, if a body is subject only to a couple, its mass center will have no acceleration. The body will, however, rotate about its mass center: to see this, think of the couple as consisting of two oppositely directed forces applied at the ends of a bar and directed at right angles to the bar. A measure of the rotational effect of the forces is given by their resultant moment. Let us take moments about an arbitrary point O for the couple shown in Fig. 2.2-1. The moment $\mathbf{M}_1$ about O of $\mathbf{F}_1$ is given by $\mathbf{r}_1 \times \mathbf{F}_1 = \mathbf{r}_1 \times \mathbf{F}$; the moment $\mathbf{M}_2$ about O of $\mathbf{F}_2$ is given by $\mathbf{r}_2 \times \mathbf{F}_2 = \mathbf{r}_2 \times (-\mathbf{F})$. The formal moment resultant, $\mathbf{M}_O$, of the two moments is their vector sum:

$$\mathbf{M}_O = \mathbf{M}_1 + \mathbf{M}_2 = \mathbf{r}_1 \times \mathbf{F} + \mathbf{r}_2 \times (-\mathbf{F}) = (\mathbf{r}_1 - \mathbf{r}_2) \times \mathbf{F},$$

or

$$\mathbf{M}_O = \mathbf{r} \times \mathbf{F}. \qquad \text{2.2-3}$$

Examining this result, we see that the resultant moment of the couple is *independent of the choice of the base point for moments.* It depends only on the relative placement of the lines of action of the forces and the force vector $\mathbf{F}$ itself. The moment of a couple is its only characteristic that is important in many applications, and the vector representing the moment of the couple is often the only information about the couple that is needed. We shall speak of a couple vector $\mathbf{C}$ applied to a body and understand by this that a moment $\mathbf{C}$ is applied to the body.

The couple vector is a *free* vector, since it does not depend on the choice of base point for the moment computation. For this reason, a couple is often called a "pure moment" or "pure torque." Since the moment computation is independent of where along its line of action the force vector is placed, we may use *any* relative position vector $\mathbf{r}$ from the line of action of $\mathbf{F}_2$ to the line of $\mathbf{F}_1 = \mathbf{F}$ to find $\mathbf{C} = \mathbf{r} \times \mathbf{F}$. In particular, the magnitude of the couple vector is Fh, where F is the magnitude of either force constituting the couple and h is the perpendicular distance between the lines of action of the forces. The direction of $\mathbf{C}$ must be determined by the usual right-hand rule; one must be careful only to take $\mathbf{r}$ *from* $\mathbf{F}_2$ *to* $\mathbf{F}_1$, once one has chosen which of the two force vectors is to be $\mathbf{F}_1 = \mathbf{F}$. The direction of $\mathbf{C}$ is, of course, perpendicular to the plane of the couple.

It is well to reiterate that $\mathbf{C}$ is a moment vector and must not be added to force vectors in equations. Indeed, a couple has zero force resultant. If a couple is acting, however, it must be added into all vectorial moment equations, whatever base point has been chosen.

Given a couple vector $\mathbf{C}$, all representations of it in terms of forces in any plane perpendicular to the direction of $\mathbf{C}$ are considered "equivalent." That is, take forces of half the magnitude of our original $\mathbf{F}_1$ and $\mathbf{F}_2$ but on lines of action a distance $2h$ apart; this couple will also have moment $\mathbf{C}$.   Take a pair of forces whose vectors are the same as those of $\mathbf{F}_1$ and $\mathbf{F}_2$, but in a plane parallel to the original plane of the couple; this couple will also have moment $\mathbf{C}$. This does not mean that these force systems are equivalent in all their physical effects on the motion of general systems, but only that their force and moment sums are the same.   We shall examine later in just what sense such a mathematical equivalence has significance as a physical equivalence.

Example 2.2-1

A particle in equilibrium is subjected to four forces, $\mathbf{F}_i$, $i = 1, 2, 3, 4$. Three of the forces are $\mathbf{F}_1 = 2\mathbf{i} - 5\mathbf{j} + 6\mathbf{k}$ lb, $\mathbf{F}_2 = \mathbf{i} + 3\mathbf{j} - 7\mathbf{k}$ lb, $\mathbf{F}_3 = 2\mathbf{i} - 2\mathbf{j} - 3\mathbf{k}$ lb.   What is $\mathbf{F}_4$?

Solution: The resultant of all forces must vanish since the particle is in equilibrium:

$$\sum_{i=1}^{4} \mathbf{F}_i = \mathbf{F}_1 + \mathbf{F}_2 + \mathbf{F}_3 + \mathbf{F}_4 = 0.$$

Therefore, $\mathbf{F}_4$ is the negative of the resultant of the other three:

$$\begin{aligned}\mathbf{F}_4 &= -(\mathbf{F}_1 + \mathbf{F}_2 + \mathbf{F}_3) \\ &= -(5\mathbf{i} - 4\mathbf{j} - 4\mathbf{k}) \\ &= -5\mathbf{i} + 4\mathbf{j} + 4\mathbf{k} \quad \text{lb.}\end{aligned}$$

Example 2.2-2

A couple consists of forces $\mathbf{F}_1 = 2\mathbf{i} - 3\mathbf{j}$ newtons, acting at $-5\mathbf{j} + 2\mathbf{k}$ meters, and $\mathbf{F}_2 = -2\mathbf{i} + 3\mathbf{j}$ newtons, acting at $\mathbf{i} - 2\mathbf{j}$ meters.   What is the couple vector $\mathbf{C}$?

Solution: Let $\mathbf{F}_1 = \mathbf{F}$; then $\mathbf{C} = \mathbf{r} \times \mathbf{F}$, where $\mathbf{r}$ is a vector from $\mathbf{F}_2$ to $\mathbf{F}_1$.   Therefore,

$$\mathbf{r} = (-5\mathbf{j} + 2\mathbf{k}) - (\mathbf{i} - 2\mathbf{j}) = -\mathbf{i} - 3\mathbf{j} + 2\mathbf{k},$$

$$\mathbf{C} = \mathbf{r} \times \mathbf{F} = \begin{vmatrix} \mathbf{i} & \mathbf{j} & \mathbf{k} \\ -1 & -3 & 2 \\ 2 & -3 & 0 \end{vmatrix} = 6\mathbf{i} + 4\mathbf{j} + 9\mathbf{k} \quad \text{newton-meters.}$$

Example 2.2-3

Can the couple of the previous example be replaced by an equivalent couple $(-\mathbf{P}, \mathbf{P})$ acting at points $(1, 1, 1)$ and $(3, -2, -1)$ meters respectively?

Solution: Let $\mathbf{R}$ be the vector from $(1, 1, 1)$ to $(3, -2, -1)$: $\mathbf{R} = 2\mathbf{i} - 3\mathbf{j} - 2\mathbf{k}$.   The question asked is: can $\mathbf{P}$ be found such that $\mathbf{R} \times \mathbf{P} = \mathbf{C} = 6\mathbf{i} + 4\mathbf{j} + 9\mathbf{k}$?   The answer is no; for $\mathbf{C}$ would have to be perpendicular to $\mathbf{R}$, and the given $\mathbf{C}$ and $\mathbf{R}$ are not:

$$\mathbf{C} \cdot \mathbf{R} = 12 - 12 - 18 = -18 \neq 0.$$

Example **2.2-4**

Find the resultant of the three forces shown in Fig. 2.2-2 *as well as the total moment of these forces about the point O.*

Solution: We first express the various forces in cartesian form:

$$\mathbf{F}_1 = 10(\cos 45°\mathbf{i} + \cos 65°\mathbf{j} + \cos 55.5°\mathbf{k}) = 7.07\mathbf{i} + 4.23\mathbf{j} + 5.66\mathbf{k} \quad \text{lb;}$$

$$\mathbf{F}_2 = 15\left[-\left(\frac{15}{19}\right)\mathbf{i} + \left(\frac{6}{19}\right)\mathbf{j} - \left(\frac{10}{19}\right)\mathbf{k}\right] = -11.84\mathbf{i} + 4.74\mathbf{j} - 7.89\mathbf{k} \quad \text{lb;}$$

$$\mathbf{F}_3 = 20\left(\frac{15}{17}\mathbf{j} - \frac{8}{17}\mathbf{k}\right) = 17.65\mathbf{j} - 9.41\mathbf{k} \quad \text{lb.}$$

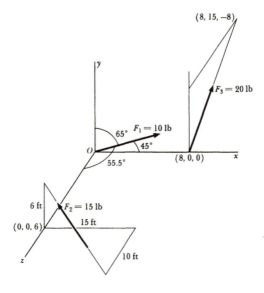

Fig. 2.2-2

We find the resultant simply by adding components:

$$\mathbf{F} = \mathbf{F}_1 + \mathbf{F}_2 + \mathbf{F}_3 = -4.77\mathbf{i} + 26.62\mathbf{j} - 11.64\mathbf{k} \quad \text{lb.}$$

The magnitude of the resultant is

$$[(-4.77)^2 + (26.62)^2 + (-11.64)^2]^{1/2} = 29.44 \quad \text{lb.}$$

The angles that the resultant force makes with the positive directions of the x, y, z axes are

$$\theta_x = \arccos\left(\frac{-4.77}{29.44}\right) = \arccos(-0.162) = 99.3°,$$

$$\theta_y = \arccos\left(\frac{26.62}{29.44}\right) = \arccos(0.904) = 25.3°,$$

$$\theta_z = \arccos\left(\frac{-11.64}{29.44}\right) = \arccos(-0.395) = 113.3°.$$

Since the cartesian form determines the magnitude and direction angles of a vector, we shall not in future compute these quantities unless they are of particular interest.

We turn now to the moment computation. It is clear that $\mathbf{F}_1$ has zero moment about O, since it passes through O. The force $\mathbf{F}_2$ passes through a point whose coordinates are (0, 6, 6) ft and the force $\mathbf{F}_3$ passes through the point whose coordinates are (8, 0, 0) ft. Therefore

$$\mathbf{M}_O = (6\mathbf{j}+6\mathbf{k})\times\mathbf{F}_2+8\mathbf{i}\times\mathbf{F}_3$$

$$= \begin{vmatrix} \mathbf{i} & \mathbf{j} & \mathbf{k} \\ 0 & 6 & 6 \\ -11.84 & 4.74 & -7.89 \end{vmatrix} + \begin{vmatrix} \mathbf{i} & \mathbf{j} & \mathbf{k} \\ 8 & 0 & 0 \\ 0 & 17.65 & -9.41 \end{vmatrix}$$

$$\cong (-76\mathbf{i}-71\mathbf{j}+71\mathbf{k})+(75\mathbf{j}+141\mathbf{k}) = -76\mathbf{i}+4\mathbf{j}+212\mathbf{k} \quad \text{lb-ft.}$$

This means that the forces can be regarded as producing a torque of magnitude 225 lb-ft about an axis through O whose direction cosines are $-0.34, 0.02, 0.94$.

Example 2.2-5

Given a force, say $\mathbf{F} = 21\mathbf{i}+40\mathbf{j}+72\mathbf{k}$ *lb, and a direction, say the direction of the unit vector* $\mathbf{e} = 0.800\mathbf{i}+0.424\mathbf{j}+0.424\mathbf{k}$, *show how to express* $\mathbf{F}$ *as the sum of two components,* $\mathbf{F}_1$ *parallel to* $\mathbf{e}$, *and* $\mathbf{F}_2$ *perpendicular to* $\mathbf{e}$.

Solution: We first find the scalar component of $\mathbf{F}$ in the given direction. This is $\mathbf{F}\cdot\mathbf{e} = 64.3$ lb. Then $\mathbf{F}_1 = (\mathbf{F}\cdot\mathbf{e})\mathbf{e} = 51.4\mathbf{i}+27.3\mathbf{j}+27.3\mathbf{k}$ lb. The other component, $\mathbf{F}_2$, is most easily found when we observe that $\mathbf{F}_1+\mathbf{F}_2 = \mathbf{F}$ so that

$$\mathbf{F}_2 = \mathbf{F}-\mathbf{F}_1 = \mathbf{F}-(\mathbf{F}\cdot\mathbf{e})\mathbf{e} = -30.4\mathbf{i}+12.7\mathbf{j}+44.7\mathbf{k} \quad \text{lb.}$$

We can check the fact that $\mathbf{F}_2$ is at right angles to $\mathbf{F}_1$ by observing that $\mathbf{F}_1\cdot\mathbf{F}_2 = 0$.

Example 2.2-6

A pair of vertical forces comprising a couple act at points A and B of the wrench shown in Fig. 2.2-3a. Replace these by a pair of horizontal forces at points C and D. The replacement forces are to produce the same moment about any axis as would be produced by the original set.

Solution: The forces at A and B comprise a couple whose magnitude is $(10)(14) = 140$ lb-in. and whose direction is at right angles to the (vertical) plane containing the forces of the couple. If we choose (x, y, z) axes as shown, this plane is the xy-plane and the moment vector is $-140\mathbf{k}$, as shown. The replacement forces at C and D must form a couple with the same moment vector. Since these forces are to be horizontal, the distance

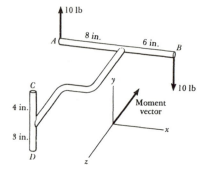

Fig. 2.2-3a

between them will be 7 in. and the forces must therefore be of magnitude 20 lb. They must produce a clockwise moment as viewed from the positive end of the z-axis. The necessary forces are shown in Fig. 2.2-3b.

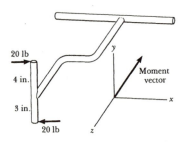

Fig. 2.2-3b

The sets of forces shown in Figs. 2.2-3a and b each have zero resultant force vector and the same moment vector. It follows that the moment produced by the first set of forces about any point will be the same as the moment produced by the second set of forces about any point.

Example 2.2-7

A cube of metal in a turret lathe is being drilled simultaneously along lines at right angles to two adjacent faces and along a body diagonal, as shown in Fig. 2.2-4. Assuming that each drill exerts a torque of 50 lb-in., what is the magnitude of the resultant torque about the center of the cube and what is the direction of the axis about which this torque acts?

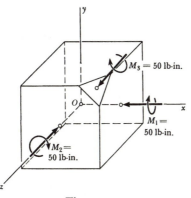

Fig. 2.2-4

Solution: The couples acting about the x and z axes are

$$\mathbf{M}_1 = -50\mathbf{i}, \qquad \mathbf{M}_2 = -50\mathbf{k} \quad \text{lb-in.}$$

The couple exerted by the twist drill acting along the body diagonal is

$$\mathbf{M}_3 = \left(\frac{-50}{\sqrt{3}}\right)(\mathbf{i}+\mathbf{j}+\mathbf{k}) \quad \text{lb-in.}$$

Adding these we have

$$\mathbf{M}_O = \left(\frac{-50}{\sqrt{3}}\right)[(1+\sqrt{3})(\mathbf{i}+\mathbf{k})+\mathbf{j}]$$

$$= -115(0.684\mathbf{i}+0.251\mathbf{j}+0.684\mathbf{k}) \quad \text{lb-in.}$$

The resultant moment vector has been put into the standard form in which the quantity in parentheses is a vector of unit magnitude. We conclude that the magnitude of the resultant torque is 115 lb-in. and that it acts about an axis whose direction cosines are 0.684, 0.251, 0.684. This axis is pictured in Fig. 2.2-5 as the directed line segment $\overline{OA}$. The minus sign in the expression for $\mathbf{M}_O$ tells us that $\mathbf{M}_O$ will be clockwise as viewed from A looking toward O, as it is pictured in the figure. This follows from the right-hand rule which is used to give the sense of a moment vector; when the thumb of the right hand points along the directed line segment $\overline{OA}$ in its positive direction, from O to A, the fingers of the right hand curl round the axis in a counterclockwise direction, as viewed from A. We conclude that a positive sign on the magnitude of $\mathbf{M}_O$ would be associated with a counterclockwise moment as viewed from A. A negative sign must therefore mean a clockwise moment as viewed from A. In the present case, of course, the sense of the moment vector is fairly obvious from inspection of Fig. 2.2-4.   Notice that if we were to write

$$\mathbf{M}_O = 115(-0.684\mathbf{i} - 0.251\mathbf{j} - 0.684\mathbf{k})$$

we should come to exactly the same conclusion.   The axis would then be the extension of the one shown in Fig. 2.2-5 into the region on the other side of the origin, which is the same as saying it would be the directed line segment $\overline{AO}$ rather than $\overline{OA}$.

   In the statement of the example we are asked for the resultant torque about the *center* of the cube.   Since we dealt only with pure couples, why do we need to mention any particular point?   The answer lies in the fact

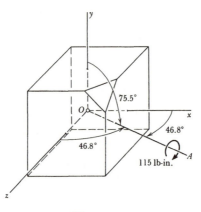

Fig. 2.2-5

that in reality any practical drill will exert an axial force as well as a torque. In Fig. 2.2-4 these forces would be directed along the x-, z-, and body-diagonal axes toward the origin O at the center of the cube.   Since, for the purpose of this example, we have not wished to include these forces, the statement has been worded in such a way that their presence would not affect the answer.

2.3 The Mechanical System and the Free-Body Diagram

   At this point we introduce the concept of a *mechanical system*, an idea that is of fundamental importance in all mechanical analysis. A mechanical system is an identified set of material objects.   It may be a steel beam, the gas enclosed in the cylinder of an internal combustion engine, or the armature of an electric motor—mechanics admits the most diverse material systems imaginable.   Once the system has been selected, however, there are two classes of objects in the universe: those that are part of the system and those that are not. The objects that are not part of the system exert forces on the system wherever they come in contact with it (and even, in the case of electromagnetic and gravitational effects, where there is no actual contact). These forces exerted on the system by something outside the system are known as *external forces*.   Forces exerted on one part of the system by another part of the system are known as *internal forces*.   The distinction is fundamental because, as we shall see, it is the external forces that are critical in the study of mechanical systems in equilibrium.

The foregoing definition of a system is the same as that which is employed in thermodynamics.*   There, however, the "identified set of material objects" is usually a fluid.   The boundary of the prescribed material often changes shape.   This boundary, or some other boundary chosen to contain the material system, is said to enclose a *control volume*; a distinction is sometimes made between so-called *closed systems*, in which no material enters or leaves the control volume, and so-called *open systems*, for which such a mass change is possible.   In statics we deal for the most part with solids. These fall naturally into the class of closed systems.

The first step in mechanical analysis, then, is the isolation of a system.   This should be self-evident, perhaps, but lack of clear identification of the system is a common source of confusion and error.   To aid in this identification a second step is made, which involves drawing a picture of the system and of all the external forces exerted on it.   This picture is known as a *free-body diagram*.   It is important that the student actually draw the free-body diagram and not merely imagine it.   Teaching experience shows the folly of elaborate calculations based on a false concept of the physical system under investigation.   In the free-body diagram the external forces are represented by a combination of arrow for direction and symbol for magnitude—either a literal symbol such as T or F, or a positive number.   If coordinate axes are to be used, as they generally are when we come to grips with particular problems, the positive directions of these axes must appear upon the figure.   Then, in writing the scalar equations of equilibrium (or of motion in general), a force component is entered as a positive quantity if its arrow in the free-body diagram points in the positive direction.   If the arrow points

---

* See, e.g., J. H. Keenan, *Thermodynamics* (New York: John Wiley & Sons, Inc., 1949), p. 1, or N. A. Hall and W. E. Ibele, *Engineering Thermodynamics* (Englewood Cliffs, N.J.: Prentice-Hall, Inc., 1960), p. 4ff.

in the negative direction, the force component is entered as a negative quantity. Clearly, in the case of an unknown force which is itself the object of investigation, we must begin by assuming a direction for the arrow associated with the unknown force. Should the direction assumed be incorrect, the equations of equilibrium will subsequently indicate this because their solution will give a negative numerical value to the symbol representing the magnitude of the unknown force.

The mathematical model by means of which we choose to describe the behavior of a particular piece of engineering equipment may be as simple or as elaborate as is felt necessary for the purposes at hand. The particle model discussed in the first section of this chapter is useful as the representation of a pendulum bob or a piston or even the earth itself in some problems. In most examples dealt with in this text, however, we shall want to treat beams and shafts, airplanes, framed structures, and other products of engineering design which cannot be regarded as single particles. More will be said about models of extended systems later in this section, and in Section 2.5 we shall consider in detail the rigid-body model and the equilibrium conditions for it.

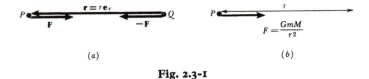

$$\mathbf{r} = r\mathbf{e}_r$$

(a) $F = \dfrac{GmM}{r^2}$ (b)

Fig. 2.3-1

Let us consider some of the standard representations of interactions between systems by force vectors. The first force we consider is the gravitational force on an isolated particle. Suppose a particle P has mass m and is subjected to the gravitational attraction of another particle Q of mass M (Fig. 2.3-1a). Newton's *law of universal gravitation* asserts that the force on P due to Q has magnitude proportional to the product of the masses of the two particles and inversely proportional to the square of the distance between them. The direction of the force on P is from P toward Q. The constant of proportionality is the *universal gravitational constant* G, the same for any two bodies; the numerical value of G depends on the units of measurement used for mass, force, and distance. Let $\mathbf{r}$ be the position of P relative to Q; then $\mathbf{e}_r = \mathbf{r}/r$ is the unit vector from Q to P. The force function is then

$$\mathbf{F} = -\frac{GmM}{r^2}\,\mathbf{e}_r = -\frac{GmM}{r^3}\,\mathbf{r}. \qquad \text{2.3-1}$$

The force on Q due to P is, by the third law of motion or by the gravitational force law itself, the negative of $\mathbf{F}$. On a free-body diagram of P as our isolated system, the force $\mathbf{F}$ is shown as in Fig. 2.3-1b: an arrow from P to Q giving the direction of the force and a letter F representing the magnitude of the force. Since the form of the magnitude function, GmM/r^2, is known, we can also insert that as shown.

By assumption, every particle of matter in the universe exerts a gravitational attraction on every other. In most engineering problems, all gravitational attractions are neglected except that of the earth. In most problems, also, the anomalies of gravitation due to the non-uniformity of mass distribution in the earth, its non-spherical shape, and its rotation may be neglected. That is, we may treat the attraction of the earth on bodies exterior to it as though the earth were a particle of mass equal to the mass of the earth placed at the center of the earth. A particle of mass m in free fall near the surface of the earth has an acceleration $g = 32.2$ ft/sec^2 (a number accurate enough for most engineering purposes) and is subject to a force of magnitude GmM/R^2, where M is the mass of the earth and R is its mean radius. The second law of motion, $\mathbf{F} = m\mathbf{a}$, applied to the particle leads to $GmM/R^2 = mg$, or $g = GM/R^2$. The force $GmM/R^2 = mg$ is the *weight* force on the particle at the earth's surface. For a particle near the earth's surface, the gravitational force may be considered as a weight force of constant magnitude $W = mg$ directed toward the earth's center. Suppose the particle is at altitude h above the surface, or a distance $r = R + h$ from the center; its weight at that altitude has magnitude w given by

$$w = \frac{GmM}{r^2} = \frac{GmM}{(R+h)^2} = \frac{GmM}{R^2\left(1+\dfrac{h}{R}\right)^2} = \frac{mg}{\left(1+\dfrac{h}{R}\right)^2}. \qquad \text{2.3-2}$$

For small h/R ratios, i.e., small distances from the earth's surface, we can expand the denominator in a Taylor's series:

$$w = W\left(1+\frac{h}{R}\right)^{-2} \cong W\left[1 - 2\left(\frac{h}{R}\right) + 3\left(\frac{h}{R}\right)^2 - \cdots\right]. \qquad \text{2.3-3}$$

For $h/R \ll 1$, $w \cong W$, the weight at the surface. On a free-body diagram, the weight force is shown as an arrow pointing vertically downward and the constant magnitude is denoted by the symbol W (Fig. 2.3-2).

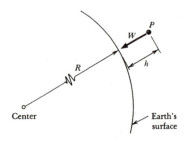

Fig. 2.3-2

In connection with gravitational forces, we may comment on the use of the words "light" and "heavy" in mechanics problems. The use of the word "light," even in dynamics problems, almost always means that the mass of the body may be neglected and the body may be considered as in equilibrium with no weight force acting on it. Physically, this usually means that the other loads or forces acting on the body are of so much greater magnitude than the weight force that the latter may be neglected in the analysis. "Heavy" does not have such a usual meaning; it always means that the weight force on that body must be considered if the body is part of the system. Sometimes it may mean that other weights may be neglected by comparison with it.

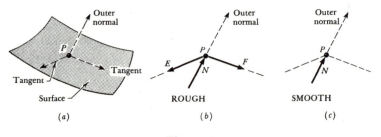

Fig. 2.3-3

Next we consider contact forces. Suppose a particle P, which can be fixed or moving, is in contact with a given surface, which can also be fixed or moving (Fig. 2.3-3a). When we isolate the particle, how is the effect of the surface on the particle represented as a contact force? Such a condition on the position of a point—in this case, that the position coordinates of the point must satisfy the equation of the surface—is known as a *constraint condition*, and the force exerted by the surface is a *constraint force* or *constraint reaction*. Here the constraint is *one-sided*; that is, the material that exerts the force on the particle lies on one side of the surface, and the constraint can be broken by lifting the particle. The surface may, in general, exert a force of arbitrary magnitude in an arbitrary direction in space, subject to a condition to be discussed shortly. The force is ordinarily represented in *normal* and *tangential* components.   An *outer normal* direction away from the surface into space is supposed known, and a component of magnitude N (the *normal force*) acting on the particle (Figs. 2.3-3b, c) is drawn in that direction, as shown by the arrow. Once that direction is assumed, the one-sided nature of the constraint is established by the inequality $N \geq 0$.   That is, it is assumed that an inward force toward the surface cannot be exerted on the particle by the material surface.   The other components of the force are given by the *tangential forces* of magnitudes E and F (Fig. 2.3-3b), which are drawn in orthogonal directions in the tangent plane to the surface at P.   Such tangential forces are also called *frictional forces*; since the vectors **E** and **F** are orthogonal, the total friction force has magnitude $[E^2 + F^2]^{1/2}$.   If non-zero frictional forces can be exerted on the particle, the surface is called *rough*; a *smooth* surface exerts only a normal force (Fig. 2.3-3c).   In our first statical analyses, we shall assume smooth contacts for the most part.   The laws of friction will be studied in Chapter V.

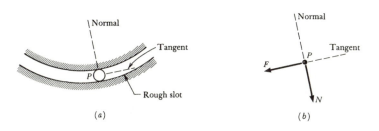

(a)

(b)

Fig. 2.3-4

A *two-sided constraint* is exemplified by a particle constrained to move between two surfaces—a pin moving in a slot, for example. The representation of the contact force from both surfaces by a single vector of unknown magnitude and direction can be given by the division into normal and frictional components, with the normal force now unrestricted by the $N \geq 0$ condition.   But what does $N < 0$

mean? What is the "negative magnitude" of a vector? As mentioned earlier, if the solution of a problem gives $N < 0$, it simply means that the wrong direction for the vector **N** has been assumed originally.

Two-dimensional counterparts of the general surface contact problem are given by particles constrained to move in a given plane in contact with a curve or in a groove. Forces perpendicular to the plane are neglected and only forces in the plane are shown. Figure 2.3-4a shows a pin in a rough plane slot; Fig. 2.3-4b is a free-body diagram of the pin showing the normal and frictional reactions on it.

The force exerted by a cord or cable on a particle is a "one-sided" type of constraint force, because our basic assumption about "cords" is that they can transmit only a tensile force in a direction tangent to the cord. That is, the cable must be either taut or slack;

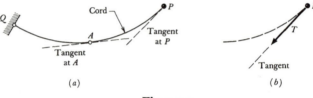

(a)   (b)

Fig. 2.3-5

when it is slack it transmits zero force, and when it is taut it transmits a tensile force. Consider our particle P at one end of a cord (Fig. 2.3-5a), the other end of which is at Q. When we isolate P (Fig. 2.3-5b), the effect of the cord is represented by a force of magnitude T and direction tangent to the cord and pointing from P toward the cord. If $T > 0$, the cord is taut; if $T = 0$, the cord is slack; and T can never be less than zero.

The fact that the cord does not exert a reactive force normal to itself is expressed by the statement that the cord is assumed to be *perfectly flexible* and has no *stiffness in shear*—that is, it does not transmit shear forces. We usually make two further idealizing assumptions about cords. We may assume the cord to be *inextensible*; i.e., neglect any stretching of the cord and assume its length to be constant. We may also assume the cord to be *light*; i.e., neglect its mass so that it is always in equilibrium. If a cable in equilibrium has a shape like that in Fig. 2.3-5a, some other force, such as the weight of the cable, must be acting on the cable. The light, inextensible cord is straight when it is taut and the force exerted by it is directed from P to the other end Q of the cord (Figs. 2.3-6a, b).

Fig. 2.3-6

A stiff rod, AB, with the particle P attached to end B while end A is attached to a foundation in some way (Fig. 2.3-7a) is a "two-sided" constraint in the sense that the force component tangent to the rod may be in either direction: tensile, as in Fig. 2.3-6b, or compressive, as N is shown to be in the free-body diagram (Fig. 2.3-7b) of the particle. A force normal to the rod—the shearing force Q of Fig. 2.3-7b—is also generally exerted. The assumptions of perfect flexibility, inextensibility, and lightness made for the cord have counterparts here. Usually in elementary work inextensibility

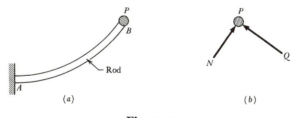

Fig. 2.3-7

and *no* flexibility in shear are assumed to hold; these are expressed by saying that AB is *rigid*, i.e., that any two points in the bar are always the same distance apart. If we assume also that the rigid bar is *light*, again neglecting its mass, we can replace N and Q by a single force having line of action AB (Figs. 2.3-8a, b); the reason for this will be given in the next section, where such "two-force" members will be discussed.

The elastic spring is the next fundamental element we wish to discuss. An elastic body is, roughly, one that deforms under load in such a way that it recovers its initial shape at once when the loads are removed. The only such body we shall consider here is the *ideal linear spring.* The ideal spring is considered to be massless and hence always in equilibrium. When no force is applied to the spring, it has a natural or undeformed length, denoted by l_0 (Fig.

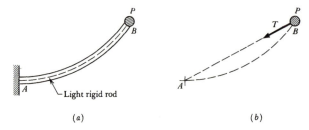

(a) (b)

Fig. 2.3-8

2.3-9a). Suppose one end is fixed to a foundation point and a force is applied at the other end; the line of action of the force is the line joining the ends of the spring (Fig. 2.3-9). Under the action of **F**, the spring changes length to a final length l. The extension, or deformation, of the spring is $e = l - l_0$, positive when **F** is tensile

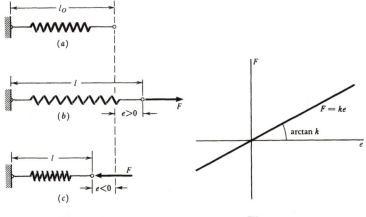

Fig. 2.3-9 **Fig. 2.3-10**

(Fig. 2.3-9b) and negative when **F** is compressive (Fig. 2.3-9c). Thus, if we draw the force always as a tensile force as in Fig. 2.3-9b, then a negative magnitude F will correspond to a compressive force and a negative extension e.

If F is plotted against e (Fig. 2.3-10), we shall assume that the graph is a straight line (linear elasticity) with slope k. The number k is known as the *spring constant*, with dimensions force/length; k and the unstretched length l_0 are the properties of the ideal linear spring which, together with the geometry of the spring, are needed to specify the spring force. The adjective "linear" in "ideal linear spring" refers, by the way, to the fact that the spring is considered to lie along a straight line and transmit a force along that line, not to the fact that the force-deformation graph is linear. Later we shall consider the ideal *torsional* spring which transmits a pure torque, and "linear" is to be contrasted with "torsional."

An analytical expression can be written for the spring force on a particle as a function of the relative position of the ends of the spring. If the particle P is attached to one end of the spring while the other end is at Q (Fig. 2.3-11a), with $\mathbf{r}$ the position of P relative to Q, then

$$(a) \qquad\qquad\qquad (b)$$

Fig. 2.3-11

the force on P due to the spring is represented on the free-body diagram (Fig. 2.3-11b) by an arrow pointing toward Q and a magnitude $F = ke$ (supposing the spring to be in tension). The vector $\mathbf{F}$ has the opposite direction to $\mathbf{r}$ and has magnitude equal to the product of the spring constant k and the extension $e = l - l_0$, where l is the magnitude of $\mathbf{r} = l\mathbf{e}_r$. Therefore, the force on the particle is

$$\mathbf{F} = -k(l-l_0)\mathbf{e}_r = -k(\mathbf{r}-\mathbf{r}_0) \qquad\qquad \textbf{2.3-4}$$

where $\mathbf{r}_0 = l_0\mathbf{e}_r$. It is easy to see that the compressive case ($e < 0$) is also properly represented by this. Since the force *on the particle* always "points back" to a position where the spring is unstretched, the spring force is sometimes called a *restoring force*.

Before considering the isolation of systems containing more than a single particle, let us work a few examples utilizing the force functions considered so far.

Example **2.3-1**

A small ball weighing 5 lb rests on a smooth incline of slope angle 20°; it is held in position by a light inextensible cord which is fastened at its other end at a point such that the taut cord makes an angle of 30° with the vertical (*Fig.* 2.3-12). *Draw the free-body diagram of the particle; introduce a coordinate system and write the force vectors in components.*

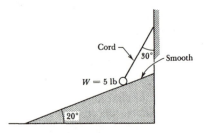

Cord — 30° — Smooth

$W = 5$ lb

20°

Fig. 2.3-12

Solution: Isolating the ball and considering it as a particle, we see that three forces act on it: the weight force, the cord tension, and the slope reaction. Taking the vertical plane through the ball and the other end of the cord as the *xy*-plane, with positive *x* horizontal to the right and positive *y* vertically upward (Fig. 2.3-13), we see that the weight force is

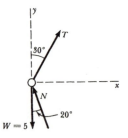

Fig. 2.3-13

known in magnitude and direction, while the cord tension and the normal reaction of the smooth incline are known only in direction. The three forces, with all pertinent information, are shown in the complete free-body diagram 2.3-13. The analytical expressions for the forces are

$$\mathbf{W} = -5\mathbf{j}, \quad \mathbf{T} = T(\sin 30°\mathbf{i} + \cos 30°\mathbf{j}), \quad \mathbf{N} = N(-\sin 20°\mathbf{i} + \cos 20°\mathbf{j}).$$

Since the unit of W is the pound, T and N should be given in that unit also.

Example 2.3-2

Three identical springs of constant k and unstretched length a are attached to a particle of weight W. The other ends of the springs are attached to a horizontal ceiling at the vertices of an equilateral triangle of side b (Fig. 2.3-14). The particle is found to be in equilibrium at a distance h below the ceiling. What is the free-body diagram of the particle? What is the force in each spring?

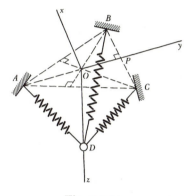

Fig. 2.3-14

Solution: Four forces act on the isolated particle: three spring forces and the weight force. Because of the symmetry of *both* geometrical *and* spring-force characteristic conditions, we may assume that all (tensile) spring forces have the same magnitude and that the particle will be on the vertical line through the centroid of the equilateral triangle. The free-body diagram is given in Fig. 2.3-15, where axes have been chosen as in Fig. 2.3-14: positive z vertically downward, positive y along an altitude of the triangle, and positive x parallel to a side so that (x, y, z) form a right-handed system. There remains only to compute the magnitude F of each spring force. The locations of the points A, B, and C relative to the axes are given in Fig. 2.3-15, from the properties of an equilateral triangle. The lengths of all three springs are given by

$$\left[h^2+\left(\frac{b\sqrt{3}}{3}\right)^2\right]^{\frac{1}{2}} = \left[h^2+\left(\frac{b\sqrt{3}}{6}\right)^2+\left(\frac{b}{2}\right)^2\right]^{\frac{1}{2}} = \left[h^2+\frac{b^2}{3}\right]^{\frac{1}{2}}.$$

Therefore,

$$F = k\left\{\left[h^2+\frac{b^2}{3}\right]^{\frac{1}{2}}-a\right\}.$$

The relation between the distance h and the weight W of the particle can now be found from the equilibrium equation in the vertical direction.

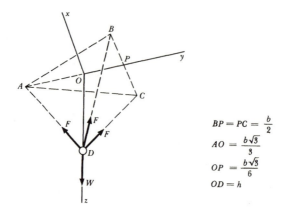

$$BP = PC = \frac{b}{2}$$

$$AO = \frac{b\sqrt{3}}{3}$$

$$OP = \frac{b\sqrt{3}}{6}$$

$$OD = h$$

Fig. 2.3-15

Mathematical models of most systems require more than one particle. We cannot describe the motion of every point in a complex body, nor do we wish to; we settle for some description of average motion characteristics that enables us to say something meaningful about the system. Moreover, we may not be able to write down appropriate expressions for the force interactions in all cases. It is for such systems that the multiparticle and continuous models are made, and where the third law of motion becomes important for developing the governing principles and equations.

Let us start with a system that we model as two particles: the earth and the moon, perhaps. Each particle has its own free-body diagram as a separate isolated system; we take, as a first free-body

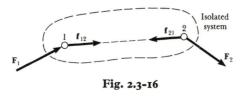

Fig. 2.3-16

diagram of the two-particle system, the two diagrams together. Let the internal force on the first particle due to the second particle be denoted by $\mathbf{f}_{12}$ and the force on the second due to the first, by $\mathbf{f}_{21}$. Let the resultant external force on the first particle (due to all other bodies except the second) be $\mathbf{F}_1$, and the corresponding force on the second particle be $\mathbf{F}_2$ (Fig. 2.3-16). The third law of motion tells us that $\mathbf{f}_{12} = -\mathbf{f}_{21}$ and that the forces are collinear. Because of the third law, we shall find that we can express the principles of motion and equilibrium in terms of the external forces only. This means that, on a free-body diagram, we need only show the external forces; for the two-particle system of Fig. 2.3-16, a second free-body diagram of this type would therefore show only $\mathbf{F}_1$ and $\mathbf{F}_2$.

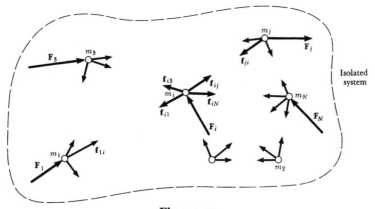

Fig. 2.3-17

The general discrete multiparticle system has a free-body diagram, including internal forces, of the schematic type of Fig. 2.3-17; if internal forces need not be considered, then they need not be shown on the diagram. The system here consists of N particles. We speak of the typical element as the i-th particle where i has the numerical range (i.e., may take on any of the values) 1, 2, 3,... N. The i-th particle has mass m_i and position $\mathbf{r}_i$. The resultant external force on the i-th particle is $\mathbf{F}_i$. If another particle, say number j, with mass m_j is located at $\mathbf{r}_j$, then the relative position of the i-th with respect to the j-th is $\mathbf{r}_i - \mathbf{r}_j$, which we denote by $\mathbf{r}_{ij}$. The internal force on the i-th due to the j-th is $\mathbf{f}_{ij}$; by the third law, $\mathbf{f}_{ji} = -\mathbf{f}_{ij}$. The resultant internal force $\mathbf{f}_i$ on the i-th particle is the sum of the $\mathbf{f}_{ij}$ over all other particles $j, j \neq i$:

$$\mathbf{f}_i = \sum_{\substack{j=1 \\ j \neq i}}^{N} \mathbf{f}_{ij}, \quad i = 1, 2,... N. \qquad \text{2.3-5}$$

No force functions other than those similar to the types we have described already need be introduced for such a multiparticle system.

The important steps are (1) the isolation of a system with a clear statement, through the use of the free-body diagram, of just what is included in the system, and (2) a clear identification of the external, and sometimes the internal, forces acting on the system.

Most often, the many-particle model we make is that of a *continuous system*, or continuum. We ignore the atomic or molecular constitution of matter and say that we may consider the mass of a body to be given by a density function ρ of position in the body such that the total mass of any part of the body, however small or large, is given by the integral of the mass density function over the volume occupied by that part: $\int_V \rho \, dV$. Mass densities based on unit area or unit length may also be defined where appropriate.

Continuous systems are of two basic types: fluid and solid. The motions of deformable continuous systems are governed by partial differential equations of motion which are too complex to consider at a first encounter with mechanics. Indeed, even the analysis of the statics of deformable solids is complex enough not to warrant treatment at the beginning of our study of mechanics; fluid statics is somewhat simpler and will be considered in Chapter IV. The continuous model we treat first in statics and dynamics is the rigid solid.

A *rigid body* is one in which every pair of points is considered to be at a constant distance apart at all times. Such a model is an excellent first approximation to the behavior of real solids for the analysis of many engineering problems. Some of the reasons for its mathematical usefulness and simplicity in theoretical work will be brought out later both in statics and in dynamics. It must not be forgotten, however, that its utility as a concept is based on its practicality as well as on theory. Our interest in the rigid body is a natural result of the fact that, for many well-designed engineering products, deformations are small and do not affect the mechanical analysis. The student should appreciate, however, that model systems consisting of rigid parts will not serve for all purposes. The rigid model would not enable us to determine internal stresses in most cases and would be completely inappropriate for the purposes, say, of fluid mechanics. The principles of statics apply to any of the mathematical models; it is only for the rigid body, however, that we may base our analysis in many cases on the principles of statics alone without the need for supplementary physical hypotheses.

Besides the forces due to the connections we have discussed already, which may equally well connect points on two extended bodies, there are a number of others that may be mentioned. Some of the commoner ones are shown in Fig. 2.3.18. The symbolic representation of the connection is shown in the left-hand column. The free-body diagram of the attached member, in the vicinity of the connection, is shown in the second column. It should be noted

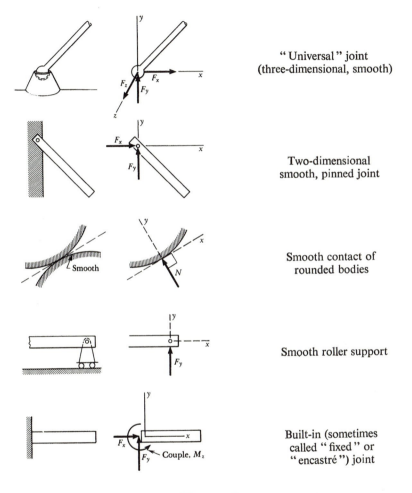

"Universal" joint
(three-dimensional, smooth)

Two-dimensional
smooth, pinned joint

Smooth contact of
rounded bodies

Smooth roller support

Built-in (sometimes
called "fixed" or
"encastré") joint

Fig. 2.3-18

that at the built-in joint the forces shown are due to a distribution over the end face of the horizontal member; these forces give rise to a moment about the z-axis as well as to forces in the x- and y-directions. Also, if the connections given are rough instead of smooth, couples as well as the forces shown should be considered as applied to the members.

Example 2.3-3

A 150-lb man stands on a light ladder that is supported on a rough floor and on the top edge of a smooth vertical wall. Draw the free-body diagram of the ladder (Fig. 2.3-19).

Solution: The free-body diagram of the ladder is shown in Fig. 2.3-19b. Students often have difficulty deciding on the proper direction for N_2. Should it be vertical, horizontal, or at right angles to the ladder? The question is easily answered: We draw an enlarged view of the contact

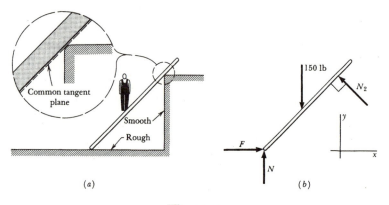

Fig. 2.3-19

region, as shown in Fig. 2.3-19a, with all sharp corners rounded, as they are in reality. Then the direction of the common tangent plane at the point of contact is easily seen to be along the line of the ladder.

The senses of F, N and N_2 have been assumed. In the present example it is intuitively evident that these senses are correctly drawn. We should therefore expect the analysis to show that F, N, and N_2 are positive quantities. The 150-lb downward force is that which the man exerts on the ladder. If we were to draw the free-body diagram of the man, an upward force equal and opposite to this downward force would appear on the diagram as the force exerted on the new system (the man) by the ladder. There would also appear on the free-body diagram of the man a downward force of 150 lb exerted on him by the gravitational attraction of the earth.

Example 2.3-4

Suppose that the smooth wall of the previous example is high enough so that the ladder cannot reach its upper edge and must, instead, rest against the side of the wall (Fig. 2.3-20a).   Draw the free-body diagram of the ladder.

Solution: The free-body diagram for the ladder is shown in Fig. 2.3-20b.   Note that the force N_2 exerted by the wall on the ladder is now

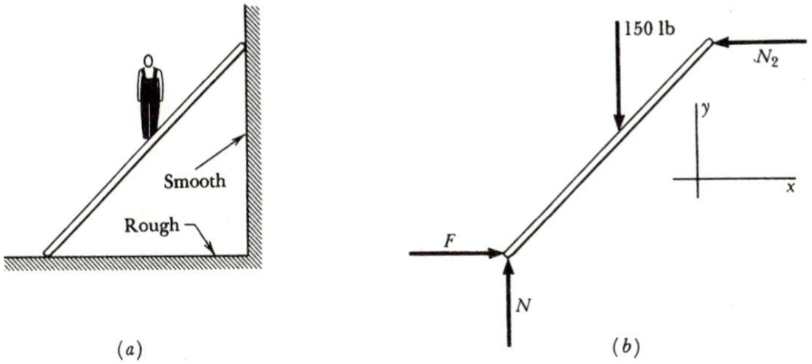

Smooth

Rough

150 lb

N_2

y

x

F

N

(a)   (b)

Fig. 2.3-20

horizontal.   The student should sketch a magnified view of the neighborhood of the contact to assure himself that the tangent plane at the point of contact is now vertical.   Of course the numerical values of F, N, and N_2 will be different in this case than in that of Fig. 2.3-19.

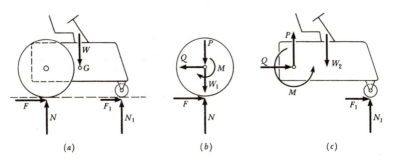

W

G

F F_1

N N_1

P

Q M

W_1

F

N

P

Q W_2

M

F_1

N_1

(a)   (b)   (c)

Fig. 2.3-21

Example 2.3-5

Draw the free-body diagram (a) of a light farm tractor, (b) of its rear wheels and axle, and (c) of its chassis and front wheels.

Solution: The free-body diagram of the tractor is shown in Fig. 2.3-21a. The downward force W is the weight of the entire tractor.   This force is actually the resultant of a distributed force, acting on every particle of the system. As we shall see, and as is probably well known to the student through his study of elementary physics, this distributed force may be replaced by a concentrated one acting through the mass center. The point G must therefore be the mass center of the entire tractor. Since the ground is rough, we have, in general, both a normal and a frictional force exerted at the ground contact. The free-body diagram of the rear wheel alone is shown in Fig. 2.3-21b. The symbol W_1 denotes the weight of the rear wheels and axle. The forces exerted by the remainder of the tractor on the rear axle now appear on the diagram, since these are now external forces. They have a resultant whose vertical and horizontal components are denoted P and Q respectively and they exert a torque about the rear axle of magnitude M. This is the so-called driving torque. Finally, in Fig. 2.3-21c, the free-body diagram of the forward part of the chassis and wheels is shown. Note that the moment and force components P, Q, and M appear again, *now reversed in direction*. If the forward part of the vehicle exerts a downward force, P, on the rear part, the rear must, in view of Newton's third law, exert an upward force, P, on the forward part. It does not matter that the directions of P, Q, and M are unknown. What is essential is that the assumptions made in drawing the free-body diagrams must not embody any implicit violation of the third law. The attentive student will have noticed that Figs. 2.3-21a, b, and c are not independent. In fact, Fig. 2.3-21a may be regarded as the result of combining (b) and (c). It follows that, once the equations of equilibrium corresponding to any two of (a), (b), and (c) have been written down, no new information can be extracted by writing the equations of equilibrium for the third free-body diagram.

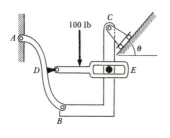

Fig. 2.3-22

In engineering practice a designer would probably omit the force F_1 from the start, on the grounds that it must be negligibly small compared with F. No driving torque is supplied to the front axle and, since the front wheels are light, only a very small torque is needed to keep the front wheels in motion. Strictly speaking, $F_1 = 0$ only if the mass of the front wheels is negligible.

Example 2.3-6

Draw the free-body diagrams for members AB, BC, and DE of the plane framework shown in Fig. 2.3-22. Neglect the weights of the members and all friction.

Solution: The free-body diagrams are shown in Fig. 2.3-23. The only point that requires comment is associated with the direction of the force exerted on bar *BC* at *C*. This force must be directed upward and

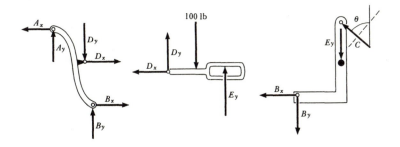

Fig. 2.3-23

to the left rather than downward to the right. If, on analysis, it should
appear that the force C is negative, the rollers would, in reality, lift off
the base. A support such as that at C is an example of a one-sided constraint. At A, on the other hand, the forces of constraint may turn out
to be either positive or negative without affecting the applicability of the
analysis.

Example 2.3-7

*The tailgate of a truck is a uniform rectangle ABCD weighing 50 lb.
It is hinged at A and B and supported at an angle θ to the horizontal by a
cable connected to the tailgate at D and to the truck at a point E vertically
above B. Draw the free-body diagram of the tailgate.*

Solution: The free-body diagram is shown in Fig. 2.3-24. The hinges
at A and B can each supply a force with any direction: these forces are

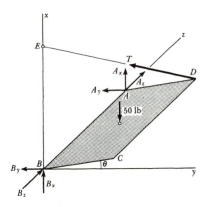

Fig. 2.3-24

most conveniently represented by their cartesian components. The force exerted by the cable must be directed along the line from D to E. Note that A_y and B_y have both been assumed positive in the negative y-direction by the choice of arrows for those components. The 50-lb weight force is directed vertically downward through the mass center.

2.4 Equilibrium of Systems

An isolated particle is in equilibrium if, and only if, the resultant force on the particle is zero:

$$\mathbf{F} = \sum_{i=1}^{n} \mathbf{F}_i = \mathbf{0}. \qquad \text{2.4-1}$$

The vector equation of equilibrium 2.4-1 is equivalent to three scalar *equations of equilibrium*. We may, therefore, solve Eq. 2.4-1 for, at most, three independent scalar unknowns.

The equilibrium condition for a particle is a necessary and sufficient condition: if the particle is in equilibrium, the resultant force necessarily vanishes; and it is sufficient that the resultant force vanish in order that the particle be in equilibrium. This follows from the first law of motion and the meaning of the term equilibrium. As a consequence, if the particle is in equilibrium, the moment of the resultant force about any point, and hence about any axis, in space is zero. The fact that the resultant, $\mathbf{F}$, has no moment about some particular point, however, does not ensure that $\mathbf{F} = \mathbf{0}$; the point might be on the line of action of $\mathbf{F}$.

Now consider an isolated multiparticle system like that of Fig. 2.3-17, repeated here as Fig. 2.4-1. A necessary condition for the

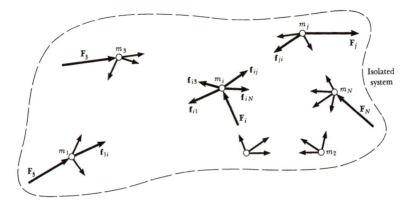

Fig. 2.4-1

equilibrium of such a system is the vanishing of the resultant external force, i.e., of the formal vector sum of the external forces on each particle. We assume that the system is in equilibrium. We recall that this means for a system that there is an inertial frame of reference with respect to which every particle is at rest. Since each particle is individually in equilibrium, Eq. 2.4-1 holds for each one. For the i-th particle, the resultant force is the sum of the resultant external force, $\mathbf{F}_i$, and the resultant internal force, $\mathbf{f}_i$; the latter, in turn, is the sum of the internal forces, $\mathbf{f}_{ij}$, on the i-th particle (as given by Eq. 2.3-5):

$$\mathbf{f}_i = \sum_{\substack{j=1 \\ j \neq i}}^{N} \mathbf{f}_{ij}.$$

The restriction that $j \neq i$ may be removed by defining the symbol $\mathbf{f}_{ii}$ to be the zero vector $\mathbf{0}$. Then the equilibrium equations for the separate particles may be written

$$\mathbf{F}_i + \sum_{j=1}^{N} \mathbf{f}_{ij} = \mathbf{0}, \quad i = 1, 2, 3, \ldots N. \qquad \text{2.4-2}$$

These N vector equations are equivalent to $3N$ scalar equations. We may now eliminate the internal forces by adding the N equations vectorially:

$$\sum_{i=1}^{N} \mathbf{F}_i + \sum_{i=1}^{N} \sum_{j=1}^{N} \mathbf{f}_{ij} = \mathbf{0}. \qquad \text{2.4-3}$$

The first term is the resultant, $\mathbf{F}$, of the external forces only. The second term is the resultant internal force, and we assert that this vanishes. For, in the terms where i and j are the same, $\mathbf{f}_{ii} = \mathbf{0}$ by definition; and where $i \neq j$, the terms vanish in pairs. Consider $\mathbf{f}_{ij}$, the force exerted on particle i by particle j; it may be paired off in the double summation with the force $\mathbf{f}_{ji}$ exerted on particle j by particle i. In view of Newton's third law, these are negatives of one another and therefore the sum of them is zero. Therefore, all the non-zero terms in the double sum may be rearranged so that the sum of each pair vanishes. We have shown the desired result:

> If a system of particles is in equilibrium, the resultant external force is zero:

$$\mathbf{F} = \mathbf{0}. \qquad \text{2.4-4}$$

The same principle may be extended to continuous systems immediately, providing we assume that the internal forces in a continuous system obey the third law of motion and are of such magnitude as to keep accelerations finite. While $\mathbf{F} = \mathbf{0}$ for any system which is in equilibrium, the fact that $\mathbf{F} = \mathbf{0}$ does not assure that the system will, in fact, be in equilibrium. If we imagine a system consisting of two particles connected by a spring, and if we stretch the spring, release the particles, and suppose that all effects other than the spring may be ignored, the external force on the system particles-and-spring is zero. The individual particles are certainly not in equilibrium. For a system consisting of either particle alone, there is an external force exerted by the spring.

Another necessary condition for the equilibrium of a multiparticle or a continuous mechanical system is the vanishing of the resultant moment of the external forces about any point whatever. To prove this, we suppose the system of Fig. 2.4-1 is in equilibrium with the i-th particle at rest at position $\mathbf{r}_i$ relative to an arbitrary point O fixed in the inertial frame. Since each particle is in equilibrium, the resultant moment about O of the forces on the particle is zero; the N moment equations are

$$\mathbf{r}_i \times \mathbf{F}_i + \mathbf{r}_i \times \left(\sum_{j=1}^{N} \mathbf{f}_{ij} \right) = \mathbf{0}, \quad i = 1, 2, 3, \ldots N. \qquad \textbf{2.4-5}$$

Add the N equations, as was done with the force equations 2.4-2; then we have

$$\sum_{i=1}^{N} \mathbf{r}_i \times \mathbf{F}_i + \sum_{i=1}^{N} \left\{ \mathbf{r}_i \times \left(\sum_{j=1}^{N} \mathbf{f}_{ij} \right) \right\} = \mathbf{0}. \qquad \textbf{2.4-6}$$

The first sum is the sum of the moment vectors of the external forces; we define this to be the *resultant external moment* about O, $\mathbf{M}_O$. The second sum is the resultant internal moment, and this vanishes as does the resultant internal force. Rearrange the summation:

$$\sum_{i=1}^{N} \left\{ \mathbf{r}_i \times \left(\sum_{j=1}^{N} \mathbf{f}_{ij} \right) \right\} = \sum_{i=1}^{N} \sum_{j=1}^{N} (\mathbf{r}_i \times \mathbf{f}_{ij}).$$

Each term with $i=j$ is zero since $\mathbf{f}_{ii} = \mathbf{0}$. To each term with $i \neq j$, add the corresponding term with i and j interchanged:

$$\mathbf{r}_i \times \mathbf{f}_{ij} + \mathbf{r}_j \times \mathbf{f}_{ji}.$$

Since $\mathbf{f}_{ji} = -\mathbf{f}_{ij}$, we have

$$\mathbf{r}_i \times \mathbf{f}_{ij} + \mathbf{r}_j \times \mathbf{f}_{ji} = \mathbf{r}_i \times \mathbf{f}_{ij} + \mathbf{r}_j \times (-\mathbf{f}_{ij}) = (\mathbf{r}_i - \mathbf{r}_j) \times \mathbf{f}_{ij}.$$

But $\mathbf{r}_i - \mathbf{r}_j = \mathbf{r}_{ij}$, the position vector of m_i relative to m_j, and this is parallel to $\mathbf{f}_{ij}$. Therefore, the non-zero terms in the internal moment sum also vanish by pairs. The extension of the result to continuous systems again follows from the hypotheses on internal forces, and we have the result:

If a system is in equilibrium, the resultant moment of the external forces only about any point fixed in an inertial frame must be zero:

$$\mathbf{M}_O = \mathbf{0}. \qquad\qquad 2.4\text{-}7$$

The equations of equilibrium 2.4-4 and 2.4-7 in vector or equivalent scalar form are the equations with which we shall work. There are no other independent equations involving external forces only that can be written, although we shall look at some other equations equivalent to these. The force equation and the moment equation with respect to one point are the only independent ones; adding another moment equation with respect to a second point may be useful but will give no information not contained in or implied by $\mathbf{F} = \mathbf{0}$ and $\mathbf{M}_O = \mathbf{0}$. Higher-order moments $[\mathbf{r} \times (\mathbf{r} \times \mathbf{F})]$ do not help in three-dimensional space, since any vector can be expressed in components in three independent directions only, say the directions of $\mathbf{r}$, $\mathbf{F}$, and $\mathbf{M} = \mathbf{r} \times \mathbf{F}$.

We shall consider two examples of the application of the equilibrium equations before turning to a more detailed study of the equilibrium of a rigid body. These examples introduce the useful concepts of the *two-force* and *three-force* members.

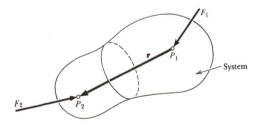

Fig. 2.4-2

Example 2.4-1

Show that if a system is in equilibrium with but two external forces acting on it, the forces must be equal in magnitude, be opposite in direction, and have the same line of action which passes through their points of application.

Solution: Suppose the system is isolated and its free-body diagram drawn with but two external forces acting on it, $\mathbf{F}_1$ at P_1 and $\mathbf{F}_2$ at P_2 (Fig. 2.4-2). Since the system is given in equilibrium, Eq. 2.4-4 holds:

$\mathbf{F}_1 + \mathbf{F}_2 = \mathbf{0}$, or $\mathbf{F}_1 = -\mathbf{F}_2 = \mathbf{F}$. The magnitudes F_1 and F_2 are equal; the directions are opposed. The two forces must, therefore, constitute at most a couple (Fig. 2.4-3). The resultant external moment about any

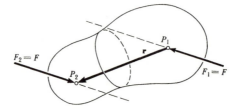

Fig. 2.4-3

point must vanish also (Eq. 2.4-7). Take P_1 as the base point for moments and $\mathbf{r}$ as the vector from P_1 to P_2; then $\mathbf{F}_1$ has no moment about P_1, and the moment equilibrium equation is

$$\mathbf{M}_{P_1} = \mathbf{r} \times \mathbf{F}_2 = \mathbf{0}.$$

By hypothesis, $\mathbf{F}_2 \neq \mathbf{0}$; therefore either $\mathbf{r} = \mathbf{0}$ or $\mathbf{r}$ and $\mathbf{F}_2$ are parallel. If the latter is true, the line of action of $\mathbf{F}_2$, and hence of $\mathbf{F}_1$, is the same as the line of $\mathbf{r}$, i.e., the line $P_1 P_2$ (Fig. 2.4-4a). If $\mathbf{r} = \mathbf{0}$, P_1 and P_2 coincide, the two forces are applied at the same point, and, since they are negatives of one another, they necessarily have the same line of action (Fig. 2.4-4b).

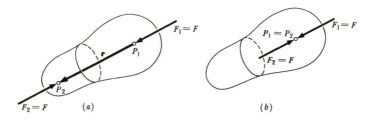

Fig. 2.4-4

Bodies subject to just two external forces are known as *two-force members*. If a two-force member is in equilibrium, we know what the line of action of the forces on it must be and need not solve for the direction. In splitting up complex systems, when we remove a two-force member, we need only represent the equivalent mechanical action by appropriate force vectors in the known direction with (usually) unknown magnitudes. We have already encountered some simple two-force members in the last section: the ideal spring, the light cord, and the light rigid rod loaded at its

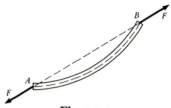

Fig. 2.4-5

ends.   For instance, the light rod of Fig. 2.3-8 will have a free-body diagram as in Fig. 2.4-5.

Example 2.4-2

If a system is in equilibrium under three external forces, show that the force vectors must be coplanar and concurrent or coplanar and parallel.

Solution: Let Fig. 2.4-6 be the free-body diagram of the isolated system

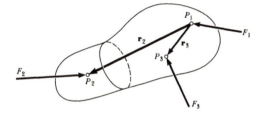

Fig. 2.4-6

in equilibrium. Take moments about P_1, with $\mathbf{r}_2$ and $\mathbf{r}_3$ the position vectors of P_2 and P_3:

$$\mathbf{r}_2 \times \mathbf{F}_2 + \mathbf{r}_3 \times \mathbf{F}_3 = \mathbf{0}.$$

Since $\mathbf{r}_2 \times \mathbf{F}_2 = -\mathbf{r}_3 \times \mathbf{F}_3$, the line normal to the plane of $\mathbf{r}_2$ and $\mathbf{F}_2$ is the same as the line normal to the plane of $\mathbf{r}_3$ and $\mathbf{F}_3$.   Both these planes pass

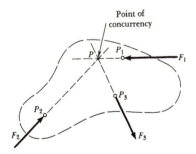

Fig. 2.4-7

through P_1 and so are in fact the same plane: $\mathbf{F}_2$ and $\mathbf{F}_3$ lie in the same plane. Since $\mathbf{F}_1 + \mathbf{F}_2 + \mathbf{F}_3 = \mathbf{0}$, $\mathbf{F}_1 = -(\mathbf{F}_2 + \mathbf{F}_3)$ has direction that is parallel to the plane of the other two.   Since $\mathbf{F}_1$ has line of action through P_1, it must lie in the plane of the other two.   The three force vectors are coplanar, in the plane of points P_1, P_2, and P_3.   Let us look at that plane (Fig. 2.4-7), which is a section of the body shown in Fig. 2.4-6.   Either $\mathbf{F}_1$ and $\mathbf{F}_2$ are parallel or they are not.   If they are not, since they are in the same plane their lines of action must intersect at some point P.   Using P as a base point for moments, then, only $\mathbf{F}_3$ could have a moment.   Since this must vanish for equilibrium, the line of action of $\mathbf{F}_3$ must also pass through P and the three forces have lines of action that intersect at P, i.e., they are concurrent.   If $\mathbf{F}_1$ is parallel to $\mathbf{F}_2$, then $\mathbf{F}_3$ is also parallel to these since $\mathbf{F}_3 = -(\mathbf{F}_1 + \mathbf{F}_2)$.   The three parallel lines of action could be said to be concurrent, also, if we admit a point at infinity as the point of concurrency.

Bodies subjected to three external forces are called *three-force members*; the knowledge that the forces must be concurrent and coplanar for equilibrium is often helpful in drawing free-body diagrams and writing appropriate moment equations.

2.5 The Rigid Body: Constraints and Equilibrium

The deformations that occur in well-designed structures under load are usually small in comparison with the dimensions of the body when it is not loaded.   For many engineering purposes, as we have remarked, we take a rigid body as the appropriate model of the system or of part of the system.   The ideal rigid body is characterized by internal constraints on its motion, which keep every pair of points at a constant distance apart throughout the motion.   As we shall see, this assumption has important consequences in the mathematical theory of mechanics.   But the simplifications in the mathematics that arise from the hypothesis of rigidity are not the primary justification for introducing the concept of the rigid body.   The fact that the engineering applications of rigid-body motion analysis have proved to be useful and most satisfactory for many purposes provides that justification.

The basic simplification that the rigid body introduces into dynamical theory is one of sufficiency.   That is, the equations of motion—and, for our immediate purposes, of equilibrium in particular—in terms of the external forces alone are sufficient for the determination of the motion of a rigid body.   The analysis of motion and equilibrium of deformable bodies requires additional hypotheses about the nature of the internal forces, i.e., requires assumptions about the material properties of the body.   The rigidity hypothesis allows us to ignore the nature of the material of which the rigid body is formed.

We shall not prove that the six independent scalar equations of equilibrium in terms of external forces and couples are sufficient to ensure rigid-body equilibrium. That is a task for dynamics,* following a discussion of the kinematics of a rigid body and the development of the full equations of motion. The general argument of the proof may be described briefly, however. It consists in showing that by the assignation of values to six independent scalar position coordinates a rigid body can be completely restrained from moving.

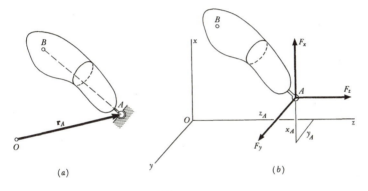

Fig. 2.5-1

Six equations would be needed to determine these position coordinates if they were unknown; the equilibrium equations are special forms of the general equations of motion sufficient for the task.

How can we fix the position of a rigid body? First of all, we can fix the position $\mathbf{r}_A$ of some point A in the body. This is equivalent to fixing the values of three scalar quantities, say the cartesian components (x_A, y_A, z_A) of the position of A. By doing this, we have prevented translational motion of the body as a whole; the only motion still permitted for the body is a rotational motion about A as a "pivot point." That is, the rigidity of the body requires that any other point, B, in the body must move on a spherical surface about A, the radius of the sphere on which B moves being the distance AB. The smooth ball-and-socket joint (Fig. 2.5-1a) of the last section

---

* See *Dynamics*, Section 6.8.

will constrain point A in the desired fashion. On a free-body diagram of the body, the connection at A is represented as shown in Fig. 2.5-1b in terms of the cartesian components (F_x, F_y, F_z) of the constraint force **F**.

Let us now continue further to fix the location of the body. Let us also prescribe the position $\mathbf{r}_B$ of a second point B. We cannot, however, prescribe the three components of $\mathbf{r}_B$ independently, for B must be a constant distance from A if the body is rigid. Only two new independent scalar quantities have been fixed. Equivalently, what we have done is to fix the direction from A to B, i.e., fixed the direction of the axis AB in space. A way of doing this is to introduce a *smooth sleeve bearing* at B with the axis of the bearing

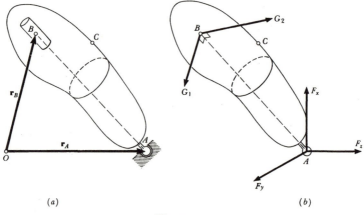

(a)   (b)

Fig. 2.5-2

along AB. The bearing provides a reaction force **G** normal to the line AB, with two independent components G_1, G_2, when the body is isolated (Fig. 2.5-2).

With these two constraints, the only way the body can move is by rotation about the axis AB. Now fix the body entirely by requiring a plane ABC in the body to be fixed in space. Fixing the position $\mathbf{r}_C$ of C does not require three additional independent scalars, but only one, to be prescribed; for the coordinates of C are subject to the internal constraints that the distances from A and B to C must be constant. By placing a smooth roller at C between parallel planes, such a constraint is obtained; when the body is isolated, the constraint is represented by a force **H** normal to the planes (Fig. 2.5-3).

Once three such points are fixed by the assignation of values to six independent generalized position coordinates, the position of any other point in the body is fixed by the rigidity condition and the geometry of the body.   There are, of course, other means of providing the necessary geometrical constraints than by the three particular ones mentioned above; in the examples and exercises, others will be used.   The important point is that six independent kinematical quantities determine general rigid-body motion, and that there are exactly six independent equations of motion in terms of external forces and couples.   It is on this basis that we may argue that the equilibrium equations in terms of external loads are both necessary and sufficient for rigid-body equilibrium.

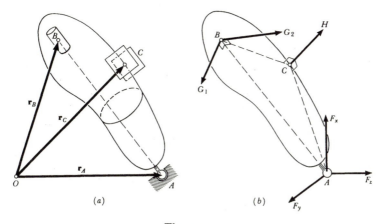

Fig. 2.5-3

One consequence of the fact that only formal vector sums of external forces and their moments are needed for the determination of rigid-body motion and equilibrium is the so-called *principle of transmissibility* of force.   As we have seen in Chapter I, any vector may be considered as a "free" vector for purposes of vector addition if that is the only principle of combination of vectors we wish to use; and any vector may be considered as a "sliding" vector if we also wish to preserve the moment of the vector about a point.   That is, the vector may be considered to act at any point along its line of action and it will still have the same moment about any point of space.   Since only the sums of force vectors and of their moments are of concern in the equations governing rigid bodies, any external force *acting on a rigid body* may be considered to be a "sliding" vector.

The mechanical effects of the force on the rigid body are independent of the point of application of the force on its line of action. This transmissibility of force in its motional effects is limited to its effect on a single rigid body, and does not hold for non-rigid systems, or indeed for the submembers of a complex rigid system such as a truss. "Transmissibility" of external forces may, of course, be used for computing the moments needed in the over-all equations that necessarily govern any system in equilibrium. But we must not forget that the forces have effects on bodies that are not rigid other than those given by their vectorial resultant and moment.

2.6 Equipollent Sets of Forces; Reduction of Sets of Forces

Since only external forces and their moments enter into the determination of the state of motion or of equilibrium of a rigid body, different sets* of forces may be compared on the basis of their effects on a rigid body. We are therefore led to define equivalent or *equipollent* sets of forces in the following way: two sets of forces and couples are said to be equipollent if (1) the vector sum of the forces of the first set is the same as the vector sum of the forces of the second set and if (2) the vector sum of the moments of the forces and couples of the first set about any point in space is the same as the vector sum of the moments of the forces and couples of the second set about the same point. The vector sum of the forces is termed the *resultant force* and the vector sum of the moments about any point is called the *resultant moment* about that point.

In the foregoing definition of equipollence the moments of the two sets are to be compared at every point of space. This presents obvious difficulties. The need for comparison at every point may be removed by a fundamental theorem which relates the resultant force and the resultant moment about any particular point to the resultant moment about any other particular point. Let the forces of the set be denoted $\mathbf{F}_i$, with $i = 1, 2, \ldots m$, and the couples, $\mathbf{C}_j$, with $j = 1, 2, \ldots n$. Suppose that $P_1, P_2, \ldots P_m$ are points on the lines of action of $\mathbf{F}_1, \mathbf{F}_2, \ldots \mathbf{F}_m$, respectively, and that the directed line segments from some point O to points $P_1, P_2, \ldots P_m$ are called $\mathbf{r}_1, \mathbf{r}_2, \ldots$

---

* What we are calling a "force set," meaning a particular collection of forces and couples applied to a mechanical system, is sometimes called a "force system." We have selected the present terminology so that there will be no confusion between the mechanical *system* under discussion and any particular *set* of forces exerted on that system.

$\mathbf{r}_m$, respectively. Then for this force and couple set the resultant moment about O is

$$\mathbf{M}_O = \sum_{j=1}^{n} \mathbf{C}_j + \sum_{i=1}^{m} \mathbf{r}_i \times \mathbf{F}_i. \qquad \text{2.6-1}$$

Now let P be some other point, with the relative position of P and O given by the directed line segment $\overline{PO}$ or by the vector $\mathbf{r}_{PO}$ (see Fig. 2.6-1). Then the directed line segment from P to P_1 is the

Fig. 2.6-1

vector sum of the directed line segment from P to O and the directed line segment from O to P_1. The moment of the force and couple set about P is given by the expression

$$\mathbf{M}_P = \sum_{j=1}^{n} \mathbf{C}_j + \sum_{i=1}^{m} (\mathbf{r}_{PO} + \mathbf{r}_i) \times \mathbf{F}_i. \qquad \text{2.6-2}$$

In this equation $\mathbf{r}_{PO}$, being the same in every term, may be factored out of the summation. Then

$$\mathbf{M}_P = \sum_{j=1}^{n} \mathbf{C}_j + \mathbf{r}_{PO} \times \left(\sum_{i=1}^{m} \mathbf{F}_i \right) + \sum_{i=1}^{m} \mathbf{r}_i \times \mathbf{F}_i. \qquad \text{2.6-3}$$

In view of Eq. 2.6-1, the first and third terms on the right-hand side of Eq. 2.6-3 are equal to $\mathbf{M}_O$. If we denote the resultant force by $\mathbf{F}$, i.e., take

$$\mathbf{F} = \sum_{i=1}^{m} \mathbf{F}_i, \qquad \text{2.6-4}$$

we have

$$\mathbf{M}_P = \mathbf{M}_O + \mathbf{r}_{PO} \times \mathbf{F}. \qquad \text{2.6-5}$$

This is a basic result that is used in dynamical theory as well as in statics. Once the resultant force, $\mathbf{F}$, of a force set and the resultant moment, $\mathbf{M}_O$, are known for any point O, the moment about any other point, P, is determined. We can now appreciate why there are no more than six independent scalar force and moment equations of equilibrium for a given system; if the resultant force and the resultant moment about any one point vanish, the moment about every point must vanish. We are also in a position to rephrase the definition of equipollent force sets. Let $\mathbf{F}_1, \mathbf{F}_2, \ldots \mathbf{F}_m$ and $\mathbf{C}_1, \mathbf{C}_2, \ldots \mathbf{C}_n$ be a set of forces and couples. Their resultant force is denoted $\mathbf{F}$ and their resultant moment about some point O is denoted $\mathbf{M}_O$. Let $\mathbf{F}_1', \mathbf{F}_2', \ldots \mathbf{F}_k'$ and $\mathbf{C}_1', \mathbf{C}_2', \ldots \mathbf{C}_l'$ be a second set of forces and couples having resultant force $\mathbf{F}'$ and resultant moment about O, $\mathbf{M}_O'$. The two sets of forces and couples are equipollent if and only if

$$\mathbf{F}' = \mathbf{F} \qquad \text{and} \qquad \mathbf{M}_O' = \mathbf{M}_O. \qquad \text{2.6-6}$$

It is meaningless to speak of *the* equipollent force set. Corresponding to any given set of forces and couples there will be many possible equipollent sets. It is natural to ask, however, whether there is a simplest possible force set which can be regarded as typifying all sets equipollent to it. Given a set of forces $\mathbf{F}_1, \mathbf{F}_2, \ldots \mathbf{F}_m$ and couples $\mathbf{C}_1, \mathbf{C}_2, \ldots \mathbf{C}_n$ having a resultant force $\mathbf{F}$ and moment $\mathbf{M}_O$ we can find an equipollent set consisting of a single force $\mathbf{F}'$ and a single couple $\mathbf{C}'$. Furthermore, we may choose any point as the point of application of the force $\mathbf{F}'$. Say that we select the point P whose position relative to O is given by the vector $\mathbf{r}$. Then if we choose

$$\mathbf{F}' = \mathbf{F} = \mathbf{F}_1 + \mathbf{F}_2 + \ldots \mathbf{F}_m$$

and

$$\mathbf{C}' = \mathbf{M}_O - \mathbf{r} \times \mathbf{F}, \qquad \text{2.6-7}$$

both of Eqs. 2.6-6 will be satisfied. The single force $\mathbf{F}'$ applied at the point with position vector $\mathbf{r}$ together with the single couple $\mathbf{C}'$ will be equipollent to the force and couple set $\mathbf{F}_1, \mathbf{F}_2, \ldots \mathbf{F}_m$ and $\mathbf{C}_1, \mathbf{C}_2, \ldots \mathbf{C}_n$. The force and couple set $\mathbf{F}'$, $\mathbf{C}'$ is called a *resultant force set* and the process of replacing any set of forces by an equipollent resultant force set is described as *reduction* of the force set.

Needless to say, corresponding to any force set having resultant force $\mathbf{F}$ and moment $\mathbf{M}_O$ there will be any number of equipollent resultant force sets, one for each choice of $\mathbf{r}$ which locates the line of action of $\mathbf{F}$. Is it possible to choose $\mathbf{r}$ so that the couple $\mathbf{C}'$ is zero?

In view of the second of Eqs. 2.6-7, this condition requires that $\mathbf{r}$ be chosen so as to satisfy the vector equation

$$\mathbf{r} \times \mathbf{F} = \mathbf{M}_O. \qquad \text{2.6-8}$$

This vector equation has been discussed previously (see Example 1.10-1). If it is to be a meaningful equation the vector $\mathbf{M}_O$ must be at right angles to the vector $\mathbf{F}$. But both $\mathbf{F}$ and $\mathbf{M}_O$ are given properties of the original force system. It follows that it is possible to choose $\mathbf{r}$ so that $\mathbf{C}'$ vanishes only if $\mathbf{F}$ and $\mathbf{M}_O$ happen to be orthogonal. It is, however, always possible to choose $\mathbf{r}$ (i.e., to locate the point of application of $\mathbf{F}$) so that the vector $\mathbf{C}'$ is parallel to the vector $\mathbf{F}'$. To see how this is done, write $\mathbf{C}' = l\mathbf{F}$ in Eqs. 2.6-7 and rearrange the terms. Then

$$\mathbf{r} \times \mathbf{F} = (\mathbf{M}_O - l\mathbf{F}). \qquad \text{2.6-9}$$

This equation contains two unknowns, l and $\mathbf{r}$. If we take the scalar product of each side with $\mathbf{F}$, the left-hand side will vanish because $(\mathbf{r} \times \mathbf{F}) \cdot \mathbf{F} = 0$: the scalar triple product represents the volume of a parallelepiped whose edges are the three vectors, and it therefore vanishes whenever two of those vectors are identical. We have, then,

$$0 = \mathbf{M}_O \cdot \mathbf{F} - lF^2, \quad \text{and} \quad l = \frac{\mathbf{M}_O \cdot \mathbf{F}}{F^2}. \qquad \text{2.6-10}$$

This determines the scalar unknown l which has the dimensions of a length. If we take the vector product of each side of Eq. 2.6-9 with $\mathbf{F}$ we have

$$(\mathbf{r} \times \mathbf{F}) \times \mathbf{F} = \mathbf{M}_O \times \mathbf{F} - l\mathbf{F} \times \mathbf{F}.$$

The last term is zero. The left-hand side may be expanded by means of Eq. 1.9-4:

$$(\mathbf{r} \cdot \mathbf{F})\mathbf{F} - F^2\mathbf{r} = \mathbf{M}_O \times \mathbf{F}.$$

This equation is satisfied by taking (note the interchange of $\mathbf{F}$ and $\mathbf{M}_O$ in the cross-product)

$$\mathbf{r} = \frac{\mathbf{F} \times \mathbf{M}_O}{F^2} \qquad \text{2.6-11}$$

which makes $\mathbf{r}$ perpendicular to $\mathbf{F}$ so that $\mathbf{r} \cdot \mathbf{F} = 0$. Equations 2.6-10 and 11 determine l and $\mathbf{r}$. The student will have noted that the solution of Eq. 2.6-9 for $\mathbf{r}$ parallels Example 1.10-1. It was there shown, and is obvious here, that we may add to $\mathbf{r}$, as determined by Eq. 2.6-11, any vector parallel to $\mathbf{F}$ without changing the value of

$\mathbf{r} \times \mathbf{F}$. Physically this simply means that, having found a particular point on the line of action of the resultant, we can shift the force along its line of action without affecting its equipollent nature. The particular resultant force set in which $\mathbf{C}'$ and $\mathbf{F}'$ are parallel vectors is known technically as a *wrench*. It may be visualized as a thrust combined with a twist about the line of action of the thrust. The discovery that every set of forces and couples is equipollent to a wrench is due to Poinsot.[*]

The exceptional case mentioned in the previous paragraph, in which the given $\mathbf{M}_O$ and $\mathbf{F}$ are orthogonal, is actually of considerable practical importance. It occurs whenever the given set consists of (a) forces all of which lie in the same plane, or (b) forces that are parallel. In the former case we may think of the plane in which all the forces lie as the xy-plane and take the point O as the origin of coordinates in this plane. Then the resultant force set consists of a single force $\mathbf{F}'$ acting at a point whose coordinates are $(x, y, 0)$. Equations 2.6-6 then require that

$$\mathbf{F}' = \sum_{i=1}^{m} \mathbf{F}_i = F_x\mathbf{i} + F_y\mathbf{j}$$

and

$$\mathbf{M}'_O = (x\mathbf{i} + y\mathbf{j}) \times (F_x\mathbf{i} + F_y\mathbf{j}) = M_O\mathbf{k}.$$

Note that the forces, lying in the xy-plane, can produce a moment of magnitude M_O only about the z-axis. The coordinates of the point of application are given by any values of x and y which satisfy this last requirement, that is, which satisfy the equation

$$xF_y - yF_x = M_O. \qquad\qquad \textbf{2.6-12}$$

The latter case, (b), is of interest in connection with the idea of a center of mass, to be developed in the next section. Suppose that the given set of forces consists of two parallel forces $F_1\mathbf{e}$, $F_2\mathbf{e}$ applied at points P_1, P_2 whose positions relative to O are given by the vectors $\mathbf{r}_1$, $\mathbf{r}_2$, as shown in Fig. 2.6-2. These two forces may be replaced by a single force acting at a point on the line joining P_1 and P_2. In order to satisfy the first of Eqs. 2.6-6 we must take this force to be $(F_1 + F_2)\mathbf{e}$, and in order to satisfy the second of Eqs. 2.6-6 we must locate the force at a point whose position vector $\mathbf{r}$ satisfies the relation

$$\mathbf{r} \times (F_1 + F_2)\mathbf{e} = \mathbf{r}_1 \times F_1\mathbf{e} + \mathbf{r}_2 \times F_2\mathbf{e}.$$

---

[*] L. Poinsot, *Éléments de statique*, 1806. The term "couple" appears also to have originated with Poinsot.

Take $\mathbf{r} = \mathbf{r}_1 + x(\mathbf{r}_2 - \mathbf{r}_1)$; this locates the point of application of the single equivalent force at a point on the line* joining P_1 and P_2, the precise position to be determined by the unknown x. Inserting this trial value of $\mathbf{r}$ in the preceding expression we find

$$(F_1 + F_2)\mathbf{r}_1 \times \mathbf{e} + x(F_1 + F_2)(\mathbf{r}_2 - \mathbf{r}_1) \times \mathbf{e} = F_1\mathbf{r}_1 \times \mathbf{e} + F_2\mathbf{r}_2 \times \mathbf{e},$$

or

$$[F_2 - x(F_1 + F_2)](\mathbf{r}_1 - \mathbf{r}_2) \times \mathbf{e} = \mathbf{0}.$$

This is satisfied whatever $\mathbf{e}$ and $\mathbf{r}_1 - \mathbf{r}_2$ may be if we choose

$$x = \frac{F_2}{F_1 + F_2}.$$ 2.6-13

Fig. 2.6-2

This elementary result has interesting consequences. We see that the location of the equipollent force is between the points P_1 and P_2 if F_1 and F_2 have the same sign, i.e., if the forces $F_1\mathbf{e}$, $F_2\mathbf{e}$ are in the same direction. Otherwise the point of application lies outside the line segment P_1P_2. Of course the case $F_1 = -F_2$ makes the given force set a couple; we tacitly exclude this case from the discussion. It is important to note that the location of the resultant force given by Eq. 2.6-13 is unchanged if we increase the magnitudes of both the given forces by any common multiple; that is, x depends only on the ratio F_1/F_2. The location is also unchanged if the

---

* Indeed, the equation for $\mathbf{r}$ is the vector equation of the line through P_1 and P_2; see Example 1.3-3, and compare the equation here with the equation $\mathbf{r} = \mathbf{s} + m\mathbf{t}$ occurring in that example.

directions of the given forces are altered, provided they remain parallel. This follows from the fact that **e** does not enter Eq. 2.6-13. If now we consider the reduction of three or more parallel forces applied at points $P_1, P_2, P_3, \ldots$, we see that this may be accomplished by taking the resultant force for any two of them, then combining it with the third force, and so on. Ultimately we arrive at a single equipollent force whose magnitude is the sum of the magnitudes of the given parallel forces (with due regard for sign), whose direction is parallel to the given forces, and whose point of application is determined. The point of application determined in this way depends only on the relative magnitudes of the forces and on the location of the points P_1, P_2, etc. It is sometimes termed the *center* of the set of parallel forces.

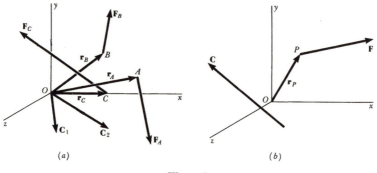

(a)   (b)

Fig. 2.6-3

Example 2.6-1

A body is subjected to forces $\mathbf{F}_A = 9\mathbf{i} - 2\mathbf{j} + 6\mathbf{k}$ *lb acting at* $\mathbf{r}_A = 5\mathbf{i} - 2\mathbf{k}$ *in.,* $\mathbf{F}_B = 3\mathbf{j} - 2\mathbf{k}$ *lb at* $\mathbf{r}_B = \mathbf{i} + \mathbf{j} - 2\mathbf{k}$ *in., and* $\mathbf{F}_C = -6\mathbf{i} + 4\mathbf{j} - \mathbf{k}$ *lb at* $\mathbf{r}_C = 4\mathbf{i}$ *in.* *Couples* $\mathbf{C}_1 = 4\mathbf{i} - 2\mathbf{j} + 5\mathbf{k}$ *lb-in and* $\mathbf{C}_2 = 3\mathbf{i} - \mathbf{j} + 2\mathbf{k}$ *lb-in are also applied.* *Replace the force set by an equipollent one consisting of a couple* **C** *and a force* **F** *acting at* $\mathbf{r}_P = \mathbf{i} + \mathbf{j} + \mathbf{k}$ *in.*

Solution: Figure 2.6-3a shows the original force set. The resultant force **F** is given by

$$\mathbf{F} = \mathbf{F}_A + \mathbf{F}_B + \mathbf{F}_C = 3\mathbf{i} + 5\mathbf{j} + 3\mathbf{k} \quad \text{lb};$$

we place **F** on a line of action through P (Fig. 2.6-3b). We must now compute **C**. Let us do this by computing the resultant moment of the two sets about P. For the resultant force set of Fig. 2.6-3b, we have

$$\mathbf{M}_P = \mathbf{C} + \mathbf{o} \times \mathbf{F} = \mathbf{C}.$$

For the original set, we have

$$\mathbf{M}_P = \mathbf{C}_1 + \mathbf{C}_2 + (\mathbf{r}_A - \mathbf{r}_P) \times \mathbf{F}_A + (\mathbf{r}_B - \mathbf{r}_P) \times \mathbf{F}_B + (\mathbf{r}_C - \mathbf{r}_P) \times \mathbf{F}_C$$

$$= (7\mathbf{i} - 3\mathbf{j} + 7\mathbf{k}) + (4\mathbf{i} - \mathbf{j} - 3\mathbf{k}) \times \mathbf{F}_A + (-3\mathbf{k}) \times \mathbf{F}_B + (3\mathbf{i} - \mathbf{j} - \mathbf{k}) \times \mathbf{F}_C$$

$$= (7\mathbf{i} - 3\mathbf{j} + 7\mathbf{k}) + \begin{vmatrix} \mathbf{i} & \mathbf{j} & \mathbf{k} \\ 4 & -1 & -3 \\ 9 & -2 & 6 \end{vmatrix} + \begin{vmatrix} \mathbf{i} & \mathbf{j} & \mathbf{k} \\ 0 & 0 & -3 \\ 0 & 3 & -2 \end{vmatrix} + \begin{vmatrix} \mathbf{i} & \mathbf{j} & \mathbf{k} \\ 3 & -1 & -1 \\ -6 & 4 & -1 \end{vmatrix}$$

$$= (7\mathbf{i} - 3\mathbf{j} + 7\mathbf{k}) + (-12\mathbf{i} - 51\mathbf{j} + \mathbf{k}) + (9\mathbf{i}) + (5\mathbf{i} + 9\mathbf{j} + 6\mathbf{k})$$

$$= 9\mathbf{i} - 45\mathbf{j} + 14\mathbf{k} \quad \text{lb-in.}$$

Setting the two expressions for $\mathbf{M}_P$ equal, we see that

$$\mathbf{C} = 9\mathbf{i} - 45\mathbf{j} + 14\mathbf{k} \quad \text{lb-in.}$$

Example 2.6-2

Replace the resultant force set of the last example by another with $\mathbf{F}$ *through the origin* O.

Solution: The force, $\mathbf{F}$, of the two sets is of course the same: $\mathbf{F} = 3\mathbf{i} + 5\mathbf{j} + 3\mathbf{k}$ lb. It remains to compute the couple $\mathbf{C}'$ that must be adjoined if $\mathbf{F}$ passes through O. We may compute $\mathbf{C}'$ by setting it equal to $\mathbf{M}_O = \mathbf{C}_1 + \mathbf{C}_2 + \mathbf{r}_A \times \mathbf{F}_A + \mathbf{r}_B \times \mathbf{F}_B + \mathbf{r}_C \times \mathbf{F}_C$ of the original set; or we may compute $\mathbf{C}'$ from the corresponding computation for the resultant set, $\mathbf{F}$ at $\mathbf{r}_P$ and $\mathbf{C}$, of the last example. For our new set, we have $\mathbf{M}_O = \mathbf{C}'$, of course; for the old set, we have (by Eq. 2.6-5)

$$\mathbf{M}_O = \mathbf{M}_P - \mathbf{r}_{PO} \times \mathbf{F}$$

$$= \mathbf{C} + \mathbf{r}_P \times \mathbf{F}$$

$$= 9\mathbf{i} - 45\mathbf{j} + 14\mathbf{k} + \begin{vmatrix} \mathbf{i} & \mathbf{j} & \mathbf{k} \\ 1 & 1 & 1 \\ 3 & 5 & 3 \end{vmatrix}$$

$$= 7\mathbf{i} - 45\mathbf{j} + 16\mathbf{k} \quad \text{lb-in.}$$

Therefore, $\mathbf{C}'$ should be chosen to be

$$\mathbf{C}' = 7\mathbf{i} - 45\mathbf{j} + 16\mathbf{k} \quad \text{lb-in.}$$

Example 2.6-3

Replace the set of forces and the couple shown by an equipollent set consisting of a single force and a couple about a parallel axis. Where does the line of action of the single force intersect the plane ABCD of the figure?

Solution: Choose cartesian axes as shown.   Then for the given force set (note that the couple vector is *not* added to the force vectors)

$$\mathbf{F} = -20\mathbf{j} + 30\left(\frac{12}{15}\mathbf{k} + \frac{9}{15}\mathbf{i}\right)$$

$$= 18\mathbf{i} - 20\mathbf{j} + 24\mathbf{k} \quad \text{lb,}$$

and

$$\mathbf{M}_B = 12\mathbf{k} \times (-20\mathbf{j}) + 156\left(\frac{-5}{13}\mathbf{j} + \frac{12}{13}\mathbf{k}\right)$$

$$= 240\mathbf{i} - 60\mathbf{j} + 144\mathbf{k} \quad \text{lb-ft.}$$

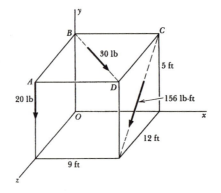

Fig. 2.6-4

The equipollent wrench will consist of the force $\mathbf{F}$ and the couple $\mathbf{C} = l\mathbf{F}$. According to Eq. 2.6-10, l is given by the expression

$$l = \frac{\mathbf{F} \cdot \mathbf{M}_B}{F^2} = \frac{(18)(240) + (-20)(-60) + (24)(144)}{(18)^2 + (-20)^2 + (24)^2} = 6.90 \quad \text{ft,}$$

so that

$$\mathbf{C} = 6.90(18\mathbf{i} - 20\mathbf{j} + 24\mathbf{k}) = 124\mathbf{i} - 138\mathbf{j} + 166\mathbf{k} \quad \text{lb-ft.}$$

To locate a point on the line of action of $\mathbf{F}$ we may use Eq. 2.6-11:

$$\mathbf{r} = \frac{\mathbf{F} \times \mathbf{M}_B}{F^2} = \frac{1}{1300} \begin{vmatrix} \mathbf{i} & \mathbf{j} & \mathbf{k} \\ 18 & -20 & 24 \\ 240 & -60 & 144 \end{vmatrix} = \frac{24}{1300} \begin{vmatrix} \mathbf{i} & \mathbf{j} & \mathbf{k} \\ 9 & -10 & 12 \\ 20 & -5 & 12 \end{vmatrix}$$

$$= \frac{24}{1300}(-60\mathbf{i} + 132\mathbf{j} + 155\mathbf{k}) = -1.11\mathbf{i} + 2.44\mathbf{j} + 2.86\mathbf{k} \quad \text{ft.}$$

Note that this **r** is measured from point B since moments were taken about B. Clearly this point of application does not lie in the plane $ABCD$, which is the plane $y=5$ ft from O and hence $y=0$ from B. But we can slide the force **F** along its line of action. That is, we can take for **r** the vector

$$\mathbf{r} = \frac{\mathbf{F} \times \mathbf{M}_B}{F^2} + c\mathbf{F}$$

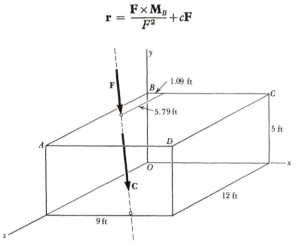

Fig. 2.6-5

where c is arbitrary. Let us give c such a value that the y-coordinate of **r** is 0 ft. Then, since

$$\mathbf{r} = -1.11\mathbf{i} + 2.44\mathbf{j} + 2.86\mathbf{k} + c(18\mathbf{i} - 20\mathbf{j} + 24\mathbf{k})$$

$$= (18c - 1.11)\mathbf{i} + (2.44 - 20c)\mathbf{j} + (2.86 + 24c)\mathbf{k},$$

we have

$$2.44 - 20c = 0, \qquad c = 0.122 \quad \text{ft/lb.}$$

With this value of c, $\mathbf{r} = 1.09\mathbf{i} + 5.79\mathbf{k}$ ft, measured from B. The resultant force set is a single force $\mathbf{F} = 18\mathbf{i} - 20\mathbf{j} + 24\mathbf{k}$ lb applied at a point $x = 1.09$ ft, $z = 5.79$ ft in the plane $ABCD$ together with a couple $\mathbf{C} = 124\mathbf{i} - 138\mathbf{j} + 166\mathbf{k}$ lb-ft. The force and couple vectors are parallel. This wrench is pictured in Fig. 2.6-5.

Example 2.6-4

The forces and couple shown in Fig. 2.6-6 act on a 10 in. by 20 in. angle. Find an equipollent single force and indicate where it should be applied along the x-axis.

Solution: The couple in the given force set has been indicated in the figure by a curved arrow. This is a conventional representation. It may be visualized as being produced, say, by an upward force slightly less than 8 in. from B and a downward force of equal magnitude slightly more than 8 in. from B. In any case, if we take x-, y-axes as shown, the axis of the couple will be along the (negative) z-axis and therefore at right

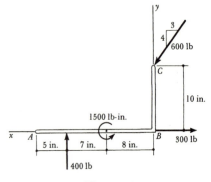

Fig. 2.6-6

angles to the plane of the forces. We may therefore expect to be able to find a resultant force set consisting of a single force. We have

$$\mathbf{F} = (-480\mathbf{j}+360\mathbf{i})+(400\mathbf{j})+(-300\mathbf{i}) = 100(0.6\mathbf{i}-0.8\mathbf{j}) \quad \text{lb},$$

$$M_B = -(360)(10)+(400)(15)-1500 = 900 \quad \text{lb-in}.$$

Of course $\mathbf{M}_B = M_B\mathbf{k}$; it is unnecessary to indicate the vectorial nature of the moment explicitly here since the forces can produce moment only

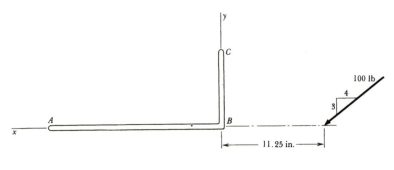

Fig. 2.6-7

about the z-axis. The positive direction of that axis must be into the paper (away from the reader) if (x, y, z) are to be right-handed. Therefore the positive sense of moment is clockwise. Now the force $\mathbf{F}$ is to be applied at a point on the x-axis so located as to produce a clockwise moment of

900 lb-in about *B*.   Let the location of this force be the point $(x, 0)$; then

$$-80x = 900 \qquad \text{or} \qquad x = -11.25 \quad \text{in.}$$

The equipollent force set is pictured in Fig. 2.6-7.

Example 2.6-5

Replace the three parallel forces shown in Fig. 2.6-8 by a single force, and indicate where it should be applied in order to be equipollent to them.

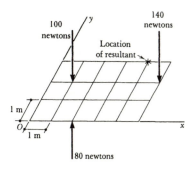

Fig. 2.6-8

Solution: Choosing axes as shown (z is positive upwards) we have

$$\mathbf{F} = (-100+80-140)\mathbf{k} = -160\mathbf{k} \quad \text{newtons},$$

$$\mathbf{M}_O = (-200\mathbf{i}+100\mathbf{j})+(-160\mathbf{j})+(-280\mathbf{i}+700\mathbf{j})$$

$$= -480\mathbf{i}+640\mathbf{j} \quad \text{newton-meters}.$$

If the equipollent force $\mathbf{F}$ is located at $\mathbf{r} = x\mathbf{i}+y\mathbf{j}$ we must have

$$(x\mathbf{i}+y\mathbf{j})\times(-160\mathbf{k}) = -480\mathbf{i}+640\mathbf{j},$$

$$160(x\mathbf{j}-y\mathbf{i}) = -480\mathbf{i}+640\mathbf{j}$$

so that $y=3$ meters and $x=4$ meters.   The location of the resultant, a downward force of magnitude 160 newtons, is shown in Fig. 2.6-8.

Example 2.6-6

Three equal forces $\mathbf{P}$ act at points A, B, C.   Show that they are equipollent to a single force $3\mathbf{P}$ acting at the point of intersection of the medians of the triangle formed by A, B, and C.

Solution: Since the forces are parallel we know they are equipollent to a single force. That force must be the sum of the individual forces, i.e., 3**P**. To find its point of application, first combine the forces at A and B. From Eq. 2.6-13 (or simply by taking moments about A or B) we see that the two forces at A and B may be replaced by a single force 2**P** located at the midpoint of AB. This force is now combined with the one at C. But the resultant force must lie on the line joining C to the midpoint of AB, i.e., on the median to the side AB of the triangle ABC. Similarly, if we first combine the forces at B and C, we find that their resultant force must act at the midpoint of the line BC and, when this is combined with the force at A, the final resultant must act at a point on the median to the side BC of the triangle ABC. The location of the equipollent force must therefore be at the common point of intersection of the three medians.

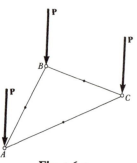

Fig. 2.6-9

2.7 Distributed Force Sets; Center of Mass and of Gravity

The forces considered thus far have all been idealized as concentrated forces with particular points of application and lines of action. This concept follows naturally from the equations of particle motion and equilibrium. However, when we come to consider the force exerted by a magnet on a large block of steel, or the effect of its own dead weight on the deformation of a beam, the representation of force at a point contact is no longer our intuitive interpretation of the physical situation. Instead we naturally suppose that the forces are distributed over a volume or an area or a length of the body in question and we speak of a *force density* (force per unit volume, or per unit area, or per unit length as the case may be). The resultant concentrated force and the resultant moment are then found by summing the individual elements of volume, area, or length, each multiplied by the force density at the location of the element. The summation entails the integration of a vector quantity.

Probably the commonest distributed force encountered in engineering is that exerted on a body as a result of the gravitational attraction of the earth. It is convenient in connection with this force (and also for the purposes of dynamics) to define the mass center of a mechanical system. We begin with a system consisting of n discrete particles of masses $m_1, m_2, \ldots m_n$ located at points $P_1, P_2, \ldots P_n$ whose position vectors, measured from some fixed point O, are $\mathbf{r}_1, \mathbf{r}_2, \ldots \mathbf{r}_n$. The mass center of this system is defined as the point P^* whose location is specified by the vector $\mathbf{r}^*$, where

$$\mathbf{r}^* = \frac{m_1\mathbf{r}_1 + m_2\mathbf{r}_2 + \ldots m_n\mathbf{r}_n}{m_1 + m_2 + \ldots m_n}; \qquad \textbf{2.7-1a}$$

or, in a somewhat more succinct notation,

$$\mathbf{r}^* = \frac{1}{m} \sum_{i=1}^{n} m_i\mathbf{r}_i, \qquad \textbf{2.7-1b}$$

where

$$m = \sum_{i=1}^{n} m_i = m_1 + m_2 + \ldots m_n \qquad \textbf{2.7-2}$$

denotes the total mass of the system. The definition is readily extended to a solid having a continuous distribution of mass. We have only to replace the summations in Eqs. 2.7-1 by integrations over the volume of the solid. Suppose that the density (mass per unit volume) of the body is denoted by the symbol ρ. Of course ρ is, in general, a function of spatial position; i.e., if cartesian coordinates are used, ρ is a function of x, y, z. We indicate this by writing $\rho = \rho(\mathbf{r}) = \rho(x, y, z)$. Then, if the element of volume is denoted dV, the mass of that element is $\rho\, dV$ and the total mass is

$$m = \int \rho\, dV. \qquad \textbf{2.7-3}$$

If cartesian coordinates are used, $dV = dx\, dy\, dz$ and Eq. 2.7-3 becomes the triple integral

$$m = \int \int \int \rho(x, y, z)\, dx\, dy\, dz. \qquad \textbf{2.7-4}$$

The position vector from O to the location of a typical volume element dV is $\mathbf{r}$ so that Eq. 2.7-1 becomes, for the continuous body,

$$\mathbf{r}^* = \frac{1}{m} \int \rho\mathbf{r}\, dV, \qquad \textbf{2.7-5}$$

where the integration is again extended over the entire solid. The definite integral of a vector quantity is another vector and may be defined by the condition that the component of the integral vector in any fixed direction in space is the integral of the component in that direction of the vector to be integrated.   Here, if fixed cartesian coordinates are used, $\mathbf{r} = x\mathbf{i} + y\mathbf{j} + z\mathbf{k}$ and

$$\mathbf{r}^* = \frac{1}{m} \int \int \int (x\mathbf{i} + y\mathbf{j} + z\mathbf{k})\rho \, dx \, dy \, dz. \qquad \textbf{2.7-6}$$

Since $\mathbf{i}$, $\mathbf{j}$, $\mathbf{k}$ are constant both in magnitude and in direction,

$$\mathbf{r}^* = \left(\frac{1}{m} \int \int \int x\rho \, dx \, dy \, dz\right)\mathbf{i} + \left(\frac{1}{m} \int \int \int y\rho \, dx \, dy \, dz\right)\mathbf{j}$$

$$+ \left(\frac{1}{m} \int \int \int z\rho \, dx \, dy \, dz\right)\mathbf{k}. \quad \textbf{2.7-7}$$

The mass center, as defined by Eq. 2.7-1 or Eq. 2.7-5, is in most mechanical problems taken to be the same as the *center of gravity* of a system.   The center of gravity of a system is that point where the resultant gravitational force on the system due to a second system— usually the earth—can be considered to act.   We wish to consider the circumstances under which we can identify the mass center and the center of gravity.

Let us consider, in fact, a set of forces of more general type than the gravitational forces acting on our system.   We shall treat once more the mechanical system consisting of n particles, with masses m_i located at points P_i with position vectors $\mathbf{r}_i$ relative to the fixed point O.   (Continuous systems may be treated by replacing the sums by appropriate integrals.)   Let the i-th particle be subject to a force $\mathbf{F}_i$ with line of action through O; i.e., $\mathbf{F}_i$ and $\mathbf{r}_i$ are parallel for each i, $i = 1, 2, \ldots n$.   Such a force is called a *central force*, and the point O is called the pole or *center of force*.   Since $\mathbf{F}_i$ and $\mathbf{r}_i$ are parallel, we must have

$$\mathbf{F}_i = k_i \mathbf{r}_i \qquad \textbf{2.7-8}$$

where the k_i are scalar proportionality factors.   Of particular interest are force functions in which the k_i are of the same functional form $k(m, r)$ for all particles and depend at most on the mass of the particle, its distance from O, and some characteristic "strength" factor for the center of force.   In particular, if $k = -\mu m/r^3$, $\mu > 0$, then the magnitude of $\mathbf{F} = k\mathbf{r}$ is $\mu m/r^2$ and the force law is of the "inverse-square" type, with O as a center of attraction of strength μ.   If a particle of mass M is placed at O with $\mu = GM$, then the force function is the newtonian gravitational force.

The resultant of the set of forces 2.7-8 is $\mathbf{F} = \sum k_i \mathbf{r}_i$; it must, of course, have line of action passing through the point O; since each force has zero moment about O. If we now ask for a second point r^{**} on the line of action of F such that F has the same form as each of the separate F_i, we are led by analogy with the definition 2.7-1 for the mass center to

$$F = k^{**}r^{**} = \sum k_i r_i, \qquad \textbf{2.7-9}$$

with one possible definition of k^{**} being

$$k^{**} = \sum k_i. \qquad \textbf{2.7-10}$$

Even assuming $\sum k_i \neq 0$, this definition for k^{**} is of utility only if we can interpret k^{**} in terms of simple physical quantities. For the mass center, we can interpret $\sum m_i$ as the total mass m of the system and think of an equivalent particle of mass m placed at r^*. For the inverse-square law, we have $k_i = -\mu m_i/r_i^3$; then we would want k^{**} to be of the form $-\mu m/(r^{**})^3$ and not of the form 2.7-10. Under these conditions, r^{**}, the center of gravity of the particle system relative to the attractive center O, is defined (from 2.7-9) by

$$\frac{r^{**}}{(r^{**})^3} = \frac{1}{m} \sum_{i=1}^{n} \frac{m_i r_i}{r_i^3}. \qquad \textbf{2.7-11}$$

The center of gravity does not appear, in general, to be even approximately the same as the center of mass. When does the r^{**} of 2.7-9 agree with the r^* of 2.7-1?

First of all, we can identify r^{**} with r^* (although r^{**} will not be given by Eq. 2.7-11 in this case) if all the particles are on the same line through O; certainly F and all the F_i pass through the mass center then. Second, if each k_i is strictly proportional to the mass m_i of the i-th particle, then 2.7-9 (with 2.7-10 for k^{**}) reduces to the mass center definition. Third, if we think of the points $P_1, P_2, \ldots P_n$ as fixed and the center of force O as remote from any of the P_i, the forces of the set may be made as nearly parallel as desired. The "center" r^{**} of a set of parallel forces was discussed in Section 2.6, and its location was shown to depend on the relative magnitude of the forces. If the magnitude of each is taken proportionate to the mass of the corresponding particle and if the distance dependence of the magnitude can be considered as constant because of the remoteness of the force center, then $r^{**} \cong r^*$. That is, if the particle-to-particle distances are all small compared to the mean distance to O, all the distances r_i can be replaced by a constant $r \cong r^{**}$. Under these conditions, Eq. 2.7-11, for example, reduces to

$$r^{**} \cong \frac{1}{m} \sum m_i r_i = r^*.$$

The gravitational attraction of the earth provides a force proportional to the mass of each particle attracted and, if all the interparticle distances or the dimensions of the volume occupied by a solid body are small compared to the radius of the earth, these forces may be regarded as parallel forces of the type just considered. It follows that for the purposes of technology the mass center may be regarded as the true center of gravity. We replace the effect of the earth's attraction by a concentrated force acting at the mass center in the direction opposite to the local vertical. The magnitude of this force, proportional to the mass of the body, is termed the *weight* of the body.

The foregoing observations are the basis for the usual practical method of measuring mass by means of an equal-arm balance. What is compared is actually the ratio of two weights, but since each is proportional to the corresponding mass, the mass of one of the bodies is found in terms of the mass of the other or "standard" body. Mass measured in this way, with the aid of a gravitational field, is sometimes termed *gravitational mass* in contradistinction to *inertial mass*, which would be determined by an experiment such as that described in Section 2.1. In point of fact, however, the most refined experiments have failed to detect any measurable difference between the two. In relativistic mechanics the identity of inertial and gravitational mass is a fundamental postulate; in newtonian mechanics it is simply a consequence of Newton's law of gravitation.

In technology the terms "center of mass" and "center of gravity" are used interchangeably. As has been pointed out, however, correspondence is based upon an approximation. The mass center of a body is a definite point in the body; it remains the same point in the body regardless of the body's position in space or its orientation. The center of gravity is also a definite point in the body; its location, however, depends, in general, upon the relative positions of the body and the attracting mass. Only in the case of a body with symmetry, like a sphere, does the position of the center of gravity remain unchanged when the focus of attraction is moved, and in this case the center of gravity coincides with the mass center exactly. Because its position depends upon the location of the focus of attraction, the center of gravity, when distinct from the mass center, is not a very useful concept; we generally prefer to replace the gravitational force, in cases where the distinction must be made, by a concentrated force at the mass center and a couple. The distinction is rarely of importance in engineering, but it is of some interest in celestial mechanics. Since the earth is not exactly a sphere, there is no one point in it through which the attraction of the sun (and moon) always acts. This attraction therefore exerts a variable torque about the earth's mass center which is responsible for a gradual change in the direction of the earth's axis of rotation. The effect is known in astronomy as the precession of the equinoxes.

We may also see from Eq. 2.7-1 that any part of the system may be replaced by a single particle of mass equal to the total mass of that part, located at the mass center of the part, without affecting the location of the mass center of the system as a whole. Suppose we replace the particles P_1, P_2,... P_k by a single particle of mass $m' = m_1 + m_2 + \ldots m_k$ located at P', the vector $\overline{OP'}$ being given by the expression

$$\mathbf{r}' = \frac{1}{m'} \sum_{i=1}^{k} m_i \mathbf{r}_i.$$

Then the mass center will be at the position

$$\mathbf{r}^* = \frac{1}{m} \left(m' \mathbf{r}' + \sum_{i=k+1}^{n} m_i \mathbf{r}_i \right) = \frac{1}{m} \sum_{i=1}^{n} m_i \mathbf{r}_i,$$

and this is the same as the mass center of the original system, given by Eq. 2.7-1. This result is of practical utility in the location of mass centers of irregular bodies.

If the origin of coordinates is taken at the mass center, we see from Eq. 2.7-7 that

$$\int \int \int x\rho \, dx \, dy \, dz = 0$$

and also that

$$\int \int \int y\rho \, dx \, dy \, dz = 0, \qquad \int \int \int z\rho \, dx \, dy \, dz = 0.$$

The integrals are sometimes termed the first moments of the mass with respect to the yz-, xz-, and xy-planes, respectively. The first moment of the mass with respect to any plane through the mass-center is zero. For other choice of origin, the coordinates of $\mathbf{r}^*$, which we may call x^*, y^*, z^*, are given by the expressions

$$x^* = \frac{\int \int \int x\rho \, dx \, dy \, dz}{\int \int \int \rho \, dx \, dy \, dz}, \quad y^* = \frac{\int \int \int y\rho \, dx \, dy \, dz}{\int \int \int \rho \, dx \, dy \, dz},$$

$$z^* = \frac{\int \int \int z\rho \, dx \, dy \, dz}{\int \int \int \rho \, dx \, dy \, dz}. \qquad \text{2.7-12}$$

If the density is uniform, ρ can be taken outside the integral sign. It then disappears from these expressions. Under these circumstances the mass center is simply a geometric property of the volume and is sometimes then called the *centroid* of the body. In the determination of the mass center it is not usually necessary to carry out all three integrations. Many of the objects of concern in technology have a uniform thickness, like a disk or a sheet (lamina) or a shell. In this case the volume density, ρ, may be replaced by a surface density (mass per unit area) and one of the integrations at least eliminated. If there is any plane of symmetry the centroid must be in that plane, and if there is any axis of symmetry the centroid must lie on that axis. Whenever the object can be regarded as built up of component parts each of which has a known mass center, the mass center of the assembly may be found by replacing each part by a single particle of mass equal to the total mass of the component and located at the mass center of the component. In many practical cases involving shapes which are not geometrically simple, it is ultimately necessary, however, to perform one or more of the integrations 2.7-12 approximately, either by graphical or numerical means.

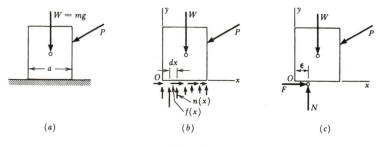

(a)   (b)   (c)

Fig. 2.7-1

Another form of distributed force commonly encountered in mechanics is that exerted at an interface, where solids come in contact either with other solids or with fluids. Consider, for example, a brick on a table. Suppose it to be subject to its own weight $W = mg$ and to an oblique force P, as shown in Fig. 2.7-1a. The distributed weight force may, as we have seen, be replaced by a concentrated force W (or mg) equal to the total weight of the brick and acting at the mass center of the brick. The reaction of the table on the brick may be regarded as distributed over the base. At a typical element of length, dx, as shown in Fig. 2.7-1b, there will be a force per unit length which may conveniently be separated into a normal component, $n(x)$, and a tangential or "frictional" component, $f(x)$. The resultant upward force exerted on the brick by the table is given by the expression

$$N = \int_0^a n(x)\, dx$$

and the total tangential force is given by

$$F = \int_0^a f(x)\, dx.$$

These forces are shown in Fig. 2.7-1c. They are the equipollent concentrated forces for the distributed force system shown in (b). Note that the point of application of the equipollent force has been taken at the point $x = \epsilon$. Taking moments about O, we see that for equipollence

$$\epsilon = \frac{1}{N} \int_0^a x n(x)\, dx = \frac{\int_0^a x n(x)\, dx}{\int_0^a n(x)\, dx}.$$

So long as we treat the brick as a rigid body we cannot hope to learn the actual distribution of the distributed force on the base of the brick. We therefore draw the free-body diagram as shown in Fig. 2.7-1c and use the equations of equilibrium to determine F, N, and ϵ. If the force P is increased gradually from zero it may eventually become large enough to cause the block to tip. When $P = 0$, clearly $F = 0$, $N = W$, and $\epsilon = a/2$. As P increases, ϵ decreases; and, just before the block begins to rotate about O, the reaction of the table is concentrated at the corner O. At this instant $\epsilon = 0$. Variations of the free-body diagram shown in Fig. 2.7-1c are, of course, quite allowable. For example, F and N may be replaced by a single force at an unknown angle to the axes. If the interface is a fluid-solid interface, the pressure on the solid will at all points be normal to the surface, at least so long as the fluid is in equilibrium. In this case there is no tangential component such as $f(x)$.

    The bodies on either side of the interface across which distributed forces are exerted need not be physically distinct. Consider

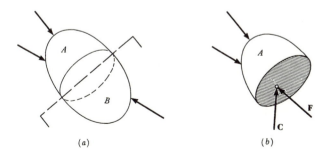

(a)   (b)

Fig. 2.7-2

a body of arbitrary shape loaded by arbitrary forces and suppose that a section is passed dividing the body into two parts, A and B, as shown in Fig. 2.7-2. The forces exerted by B on A are distributed over this section. They are equivalent to a concentrated force $\mathbf{F}$ and couple $\mathbf{C}$, as shown in Fig. 2.7-2. If the original set of forces in Fig. 2.7-2a is a null set, the whole body is in equilibrium and therefore part A, whose free-body diagram is shown in Fig. 2.7-2b, is in

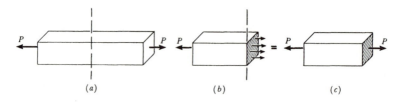

Fig. 2.7-3

equilibrium. Knowing the original set of external forces, we may therefore expect to be able to determine the "internal" forces $\mathbf{F}$ and $\mathbf{C}$ from the equations of equilibrium. Two cases deserve special mention. If the body in question is a two-force member (Section 2.4) the situation is as pictured in Fig. 2.7-3. The distributed force acting on any cross-section must be equipollent to a concentrated

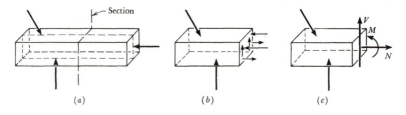

Fig. 2-7-4

force equal in magnitude to the load carried by the member. In other words, $\mathbf{F} = -\mathbf{P}$ and $\mathbf{C} = \mathbf{0}$. This fact is used continually in the design of framed structures. The second case of great interest to the structural designer is that in which the external forces all lie in one plane. This is pictured in Fig. 2.7-4. The distributed force on any cross-section will then be equipollent to a set consisting of a force at right angles to the section, a force tangent to the cross-section, and a couple whose vector is at right angles to the original set of external forces.   We may denote these vectors as $\mathbf{N}$, $\mathbf{V}$, and $\mathbf{M}$, respectively.   In the case $\mathbf{N} = \mathbf{V} = 0$ for all cross-sections, the structural element is said to be a beam in pure flexure or bending.   If $\mathbf{N}$ alone vanishes at all cross-sections, it is said to be a beam in flexure without tension, or simply a beam.

In engineering stress analysis, interest centers in the way in which the force distribution varies over any given cross-section. This is manifestly not revealed by a knowledge of the over-all force and moment resultants.   Any given resultant may be produced by a wide variety of distributions.   The determination of the way in which the traction at any point varies with the orientation of the cross-section and the location of the point requires additional physical hypotheses about the nature and behavior of the solid.   By supplementing the equations of equilibrium with other equations based on the load-deformation properties of the material and on the fact that these deformations, however small, must be such as will preserve geometric continuity, this point-to-point variation can be determined. That is in fact the primary task of continuum mechanics.   The interested student should refer to a text in deformable-body mechanics where this matter is treated in detail.*   In Chapter IV, however, we shall consider some cases of equilibrium under distributed load where separate equations for the deformations need not be considered.

A listing of mass centers of a variety of bodies of uniform mass distribution appears in the table following the Appendix.

Example 2.7-1

Locate the mass center of (a) a triangular plate, (b) a tetrahedron, and (c) a cone, all of uniform density.

Solution: (a) The mass center of a straight wire or rod of uniform mass per unit length lies at the midpoint of the rod.   If we think of the triangular plate ABC as built up of line elements parallel to one of the sides (say the side BC), the mass center of each element will be at its midpoint. It follows that the mass center of the triangular plate lies on the median from A, the line from A to the midpoint of the opposite side BC, as shown in Fig. 2.7-5.   If we decompose the plate into line elements parallel to the

---

* See S. H. Crandall and N. C. Dahl, eds., *An Introduction to the Mechanics of Solids* (New York: McGraw-Hill Book Company, Inc., 1959).   For a more advanced treatment, the reader may refer to S. Timoshenko and J. N. Goodier, *Theory of Elasticity*, 2nd ed. (New York: McGraw-Hill Book Company, Inc., 1951) or I. S. Sokolnikoff, *Mathematical Theory of Elasticity*, 2nd ed. (New York: McGraw-Hill Book Company, Inc., 1956).

side *AB* we find that the mass center lies on the median from *C*. In fact, the mass center of the triangular plate must lie at the intersection of the medians. (In Example 2.6-6 we found the same conclusion for three

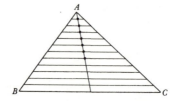

Fig. 2.7-5

equal mass particles at points *A*, *B*, *C*.) This intersection lies two-thirds of the way along the median from the vertex to the middle of the opposite side.

(b) Divide the tetrahedron into lamina parallel to one of the faces—say *ABC* in Fig. 2.7-6. The mass center of each triangular face lies at the intersection of the medians. It follows that the mass center of the tetrahedron as a whole lies along the line connecting the vertex *D* to the intersection of the medians of the opposite face. But it must then lie at the

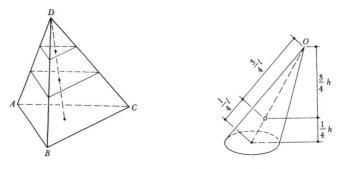

Fig. 2.7-6   **Fig. 2.7-7**

intersection of the lines from each of the four vertices to the intersections of the medians of the opposite faces. The mass center of the solid tetrahedron therefore coincides with the mass center of four equal particles at points *A*, *B*, *C*, *D*. It lies three-quarters of the way from any vertex to the intersection of the medians of the opposite face.

(c) Suppose that instead of a cone, we have a pyramid with a polygonal base. This polygon may be split up into triangles, each of which, together with the vertex *O*, determines a tetrahedron. The mass center of the pyramid therefore lies along the line connecting O to the centroid of the base, three-quarters of that distance from O, as shown in Fig. 2.7-7. Since a curved base may be approximated to any desired degree of approximation by a polygon, we conclude that the same result holds for a cone as for the pyramid, a result, that can also be derived, of course, by direct computation.

Example 2.7-2

Find (a) the mass center of a uniform rod bent to a circular arc of radius R subtending a central angle $2\theta_0$, and (b) the mass center of a uniform plate in the shape of a sector of a circle of radius R and central angle $2\theta_0$.

Solution: (a) Take origin at the center of the circular arc, with y as axis of symmetry. Then $x^* = 0$. Using polar coordinates, as shown in Fig. 2.7-8, a typical element of the rod will have length $R\,d\theta$ and mass

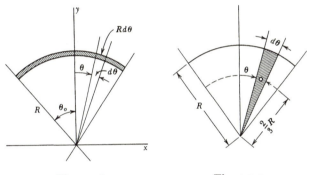

Fig. 2.7-8   **Fig. 2.7-9**

$\mu R\,d\theta$, where μ denotes the mass of the rod per unit length. The total mass of the rod is $2\mu R\theta_0$; consequently

$$y^* = \frac{1}{2\mu R\theta_0} \int_{-\theta_0}^{\theta_0} (R\cos\theta)(\mu R\,d\theta) = \frac{R\sin\theta_0}{\theta_0}.$$

The term $(R\cos\theta)$ in the integrand represents the y-coordinate of the mass center of the typical element. For a semicircular rod, $\theta_0 = \pi/2$ and $y^* = 2R/\pi$.

(b) Divide the sector into elementary "triangles" of which a typical one is shown in Fig. 2.7-9. The mass center of each element will, according to Example 2.7-1a, lie at a distance $2R/3$ from O. We see that the entire sector is equivalent to a circular arc of radius $2R/3$. From part (a) of this example, the y^*-coordinate of such an arc is

$$y^* = \left(\frac{2R}{3}\right) \frac{\sin\theta_0}{\theta_0}.$$

For a semicircular plate, $\theta_0 = \pi/2$ and $y^* = 4R/3\pi$.

Example **2.7-3**

A body of uniform density has the property that the area of any cross-section at right angles to some axis, y, has an area $A(y)$. If the solid is bounded by the planes $y = a, b$, show that

$$y^* = \frac{\int_a^b y A(y)\, dy}{\int_a^b A(y)\, dy}.$$

Solution: Take as typical element of the solid the part intercepted by two parallel planes distant dy, as shown in Fig. 2.7-10. The mass of a typical element is $\rho A\, dy$. Thus

$$m = \int_a^b \rho A(y)\, dy$$

and

$$y^* = \frac{1}{m} \int_a^b \rho y A(y)\, dy = \frac{\int_a^b y A(y)\, dy}{\int_a^b A(y)\, dy}$$

since ρ, being independent of y, may be taken out of the integral sign.

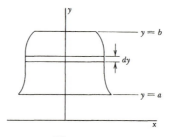

Fig. 2.7-10

If the solid is a solid of revolution with y as its axis of symmetry, $A(y) = \pi x^2$. Then if the bounding curve is $x = f(y)$, the formula above gives y^* immediately from the shape of the bounding curve. For a hemisphere of radius R with base on the plane $y = 0$ we have $a = 0$, $b = R$, $f(y) = (R^2 - y^2)^{1/2}$ and $A(y) = \pi(R^2 - y^2)$:

$$y^* = \frac{\pi \int_0^R y(R^2 - y^2)\, dy}{\pi \int_0^R (R^2 - y^2)\, dy} = \frac{\frac{1}{4} R^4}{\frac{2}{3} R^3} = \frac{3}{8}\, R.$$

Example **2.7-4**

Show that, if a uniform solid of revolution is generated by revolving an area about an axis in its plane, the volume of the solid is equal to the generating area multiplied by the length of the path of its centroid.

Solution: Take the y-axis as the axis of revolution and a typical area element in the cross-section $dx\,dy = dA$, as shown in Fig. 2.7-11. As the area is rotated about the y-axis this element sweeps out a total volume $(2\pi x)\,dA$ so that the volume of the solid is

$$V = 2\pi \int x\,dA,$$

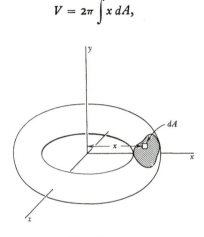

Fig. 2.7-11

the integration being extended over the area of the cross-section in the xy-plane. The x-coordinate of the centroid of this area is given by the expression

$$x^* = \frac{\rho \int x\,dA}{\rho \int dA} = \frac{V}{2\pi A}.$$

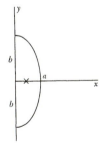

We have at once $(2\pi x^*)A = V$, which is the desired result. This theorem is one of two due originally to Pappus of Alexandria (A.D. 300) and discovered independently by Guldinus in the seventeenth century. The other of Pappus' theorems appears in the list of exercises.

If we know the volume of a solid of revolution and the area of its cross-section, this theorem may be used to find the location of the centroid of half the cross-section. For example, an ellipsoid of volume

Fig. 2.7-12

$4\pi a^2 b/3$ is generated by revolving a half ellipse about its major diameter, $2b$. The area of the semi-ellipse, shown in Fig. 2.7-12, is $(\pi ab)/2$. It follows that the centroid of the half ellipse is at

$$x^* = \frac{V}{2\pi A} = \frac{(4\pi a^2 b)/3}{(2\pi)(\pi ab)/2} = \frac{4a}{3\pi}.$$

It is interesting to note that this result, being independent of b, must agree with the result for a half-circle given in Example 2.7-2b.

Example 2.7-5

A beam is supported on rollers at one end and on a smooth pin joint at the other. It carries a distributed load and a concentrated load, as shown (Fig. 2.7-13). Find the resultant force and couple exerted by the material to the right of a section through the mid-span on the material to the left of the section.

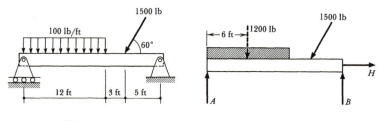

Fig. 2.7-13   Fig. 2.7-14

Solution: We note that the free-body diagram of the beam as a whole is as shown in Fig. 2.7-14. The left-hand reaction, A, is vertical because of the smooth rollers. The reactions A, B, and H are easily computed from the equilibrium equations; for this purpose the distributed load may

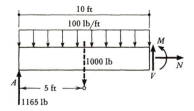

Fig. 2.7-15

be replaced by a concentrated force of 1200 lb (shown dotted in Fig. 2.7-14) at a point 6 ft from the left-hand end of the bearing. Then, taking moments about A, we have

$$20B - 7200 - (1300)(15) = 0, \qquad B = 1335 \text{ lb.}$$

Taking moments about B, we have

$$-20A + (1200)(14) + (1300)(5) = 0, \qquad A = 1165 \text{ lb.}$$

As a check we see that $A + B = 2500$ lb, the total downward load. Summing forces in the horizontal direction we have

$$H - 750 = 0, \qquad H = 750 \text{ lb.}$$

Next we proceed to the free-body diagram of that part of the beam to the left of a section through the midspan. This is shown in Fig. 2.7-15. The distributed forces exerted by the material to the right of the section have been replaced by their resultant, a force with components N and V and a couple M. It is these that we are asked to find. Again we replace the distributed 1000 lb/ft load by an equipollent concentrated force, this time of 1000 lb magnitude and located 5 ft from A. Now, summing forces vertically and horizontally, we have for equilibrium:

$$1165 - 1000 + V = 0, \qquad V = -165 \quad \text{lb;} \qquad N = 0.$$

Taking moments about point A, we find:

$$M + 10V - (1000)(5) = 0;$$
$$M + (10)(-165) - (1000)(5) = 0;$$
$$M = 6650 \text{ lb-ft.}$$

These are the wanted quantities.

Exercises

2.2-1: The resultant force on a particle at position O has zero moment about axes PQ, QR, and RP, where P, Q, R are three points not on the same straight line. If none of the three axes passes through O, can we correctly conclude that the particle is necessarily in equilibrium?
Ans.: No.

2.2-2: The four force vectors shown are drawn to a scale of 1 in. $\sim$ 5 lb. What is the resultant moment of the system about the origin? About the point $(2, 3, -1)$ in.?
Ans.: 100**k** lb-in.

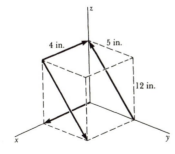

Exer. 2.2-2

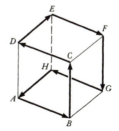

Exer. 2.2-3

2.2-3: Eight forces proportional in magnitude to the sides of a cube are given in direction as shown. What is the force sum? The moment sum about A? Can the force system be replaced by two forces along BC and GF? If it can be done, what are they?

2.2-4: A force $\mathbf{F} = 3\mathbf{i} - 4\mathbf{j} + 12\mathbf{k}$ newtons acts at point $\mathbf{r} = \mathbf{i} + \mathbf{j} - 3\mathbf{k}$ meters. A second force $(-\tfrac{1}{2}\mathbf{F})$ acts at $(-\tfrac{1}{2}\mathbf{r})$; a third force $(-\tfrac{1}{2}\mathbf{F})$ acts at $\mathbf{i} - \tfrac{1}{2}\mathbf{r}$. Find the moment vector of the resultant couple.
Ans.: $\mathbf{C} = -25.5\mathbf{j} - 8.5\mathbf{k}$ newton-meters.

2.2-5: A particle is in equilibrium in a plane under three forces:

$$\mathbf{F}_1 = -10\mathbf{j} \text{ lb},$$
$$\mathbf{F}_2 = U(\cos 120°\mathbf{i} + \sin 120°\mathbf{j}) \text{ lb},$$

and

$$\mathbf{F}_3 = T(\cos \theta \mathbf{i} + \sin \theta \mathbf{j}) \text{ lb}.$$

The angle θ lies in the range $0 \leq \theta \leq \pi/2$.
(a) Solve for the magnitudes T and U as functions of θ.
(b) For what θ will T be a minimum? What will be the values of T and U for this value of θ?
Ans.: (a) $T = (10 \sec \theta)/(\tan \theta + \sqrt{3})$ lb, $U = 20/(\tan \theta + \sqrt{3})$ lb;
(b) $\theta = 30°$.

2.2-6: A particle in equilibrium is subjected to four forces:

$$\mathbf{F}_1 = -10\mathbf{k} \text{ lb},$$
$$\mathbf{F}_2 = U[(4/13)\mathbf{i} - (12/13)\mathbf{j} + (3/13)\mathbf{k}] \text{ lb},$$
$$\mathbf{F}_3 = V[-(4/13)\mathbf{i} - (12/13)\mathbf{j} + (3/13)\mathbf{k}] \text{ lb},$$
$$\mathbf{F}_4 = W(\cos \theta \mathbf{i} + \sin \theta \mathbf{j}) \text{ lb},$$

where $0 \leq \theta \leq \pi$.
(a) Solve for U, V, and W as functions of θ.
(b) If U, V, W are all to be positive—say, that they are cord tensions—what is the allowable range of θ?
Ans.: (a) $U = 65(1 - 3 \cot \theta)/3$ lb; (b) $71.6° < \theta < 108.4°$.

2.3-1: A particle of mass m is in equilibrium in a vertical plane. It is suspended from two light inextensible cords as shown, one of length L, the other $3L$; the points of suspension A and B are on the same horizontal level at a distance $3L$ apart.

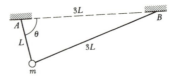

Exer. 2.3-1

(a) Draw a free-body diagram of the particle. Of the particle and the two cords together.

(b) Find the cord tensions for equilibrium.

Ans.: $T_A = 0.958\ mg$, $T_B = 0.169\ mg$.

2.3-2: Suppose, in the previous exercise, the cord to support B is replaced by a light ideal spring of natural length $2L$. What must the spring constant k be if the configuration of Exercise 2.3-1 is still an equilibrium configuration? Draw the free-body diagram of the particle, cord, and spring.

Ans.: $k = 0.169\ \dfrac{mg}{L}$.

2.3-3: An object of weight W at point O is suspended by three cables. The cable OC is horizontal. The other two, OA and OB, are attached to a wall which is at right angles to the horizontal cable, the points of attachment being 27 ft horizontally and 36 ft vertically distant from the point of suspension, as shown in the figure, and 28 and 108 ft from the vertical plane through the point of suspension and the horizontal cable. Find the tensions in the cables.

Ans.: $T_{OA} = 1.169W$, $T_{OB} = 0.669W$, $T_{OC} = 0.75W$.

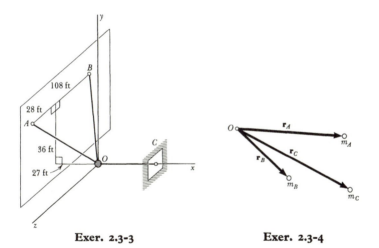

Exer. 2.3-3   **Exer. 2.3-4**

2.3-4: Three particles of masses m_A, m_B, m_C are located at $\mathbf{r}_A$, $\mathbf{r}_B$, $\mathbf{r}_C$; the only forces acting on each are due to their mutual gravitational attractions.

(a) Take all three masses as the system. Draw a free-body diagram showing external forces only; draw one showing external and internal forces.

(b) Do part (a) for a system consisting of A and C only; of B only.

2.3-5: The three particles of the previous problem are connected by light rigid rods and supported in a vertical plane by a cord and springs. A 10-lb load is applied to *A*. Draw free-body diagrams showing external forces only for systems consisting of (a) the three particles and rods; (b) particles *A* and *C* and their connecting rod; (c) particle *B* only.

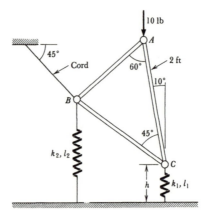

Exer. 2.3-5

Exercises 2.3-6 to 2.3-17: Draw free-body diagrams of the systems listed below.

2.3-6: System *ABCDEFG*. This is a pin-jointed plane truss of light rigid bars. The pin connection at *A* to the foundation is smooth, as is the two-sided roller and guide at *D*.

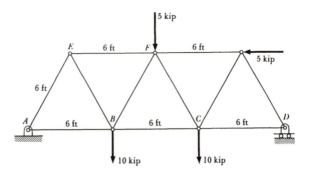

Exer. 2.3-6

2.3-7: The light rigid bar *BF* in the truss of Exercise 2.3-6.

2.3-8: The rigid triangle *BEF* (including the pins at the ends) of the truss of Exercise 2.3-6.

2.3-9: System *ABCDEFGHI.* This is a smoothly connected rigid space truss of light rigid bars, with the congruent triangular elements *ABC*, *DEF*, and *GHI* vertical. The truss is supported by a smooth ball-and-socket joint at *A*, light vertical cables at *C* and *I*, and a light horizontal rigid rod at *H*.

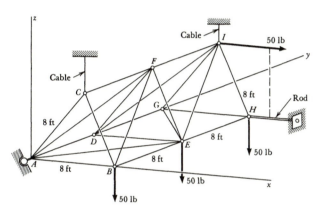

Exer. 2.3-9

2.3-10: The uniform rigid disk, of radius 2 ft and weight 10 lb. Suppose the contact with the horizontal plane to be (a) smooth, (b) rough.

2.3-11: The rigid top of weight *W*. Suppose the pivot at *O* to be (a) smooth, (b) rough.

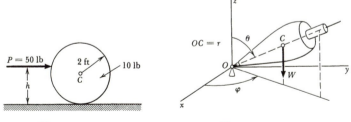

Exer. 2.3-10   **Exer. 2.3-11**

2.3-12: The A-frame shown. It consists of smoothly pinned, light rigid members *ABC*, *CDE*, and *BD*. The foundation pin at *A* and the roller at *E* are smooth. Load *P* is applied on the pin at *C*.

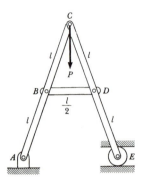

Exer. 2.3-12

2.3-13: Member *BD* only in Exercise 2.3-12.

2.3-14: Pin *C* only in Exercise 2.3-12.

2.3-15: Bar *ABC* of Exercise 2.3-12.

2.3-16: The system of Exercise 2.3-12 but with bar *BD* replaced by a light spring of constant *k* and unstretched length *l*.

2.3-17: All of the system of Exercise 2.3-16 except the spring.

Exercises 2.4-1 to 2.4-6: Write equilibrium equations for the following systems; *do not attempt to solve.*

2.4-1: The system of Exercise 2.3-6.

2.4-2: The system of Exercise 2.3-9.

2.4-3: The system of Exercise 2.3-10b.

2.4-4: The system of Exercise 2.3-12.

2.4-5: The system of Exercise 2.3-14.

2.4-6: The system of Exercise 2.3-15.

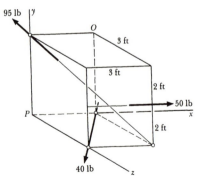

Exer. 2.6-1

2.6-1: Replace the force set shown by an equipollent resultant set consisting of a single force applied at O and a couple.
Ans.: $F = -27.2i+65.2j-20.6k$ lb; $C = -113i+3.37j-209k$ lb-ft.

2.6-2: Replace the force set of Exercise 2.6-1 by an equipollent set consisting of a single force applied at P and a couple.
Ans.: F is the same as the previous exercise; $C= -196i+65.1j+95.5k$ lb-ft.

2.6-3: Replace the force set of Exercise 2.6-1 by a wrench (force and couple about parallel axis). Where should the line of action of the force intersect the vertical plane through O and P?
Ans.: F is the same as in the previous two exercises; $C=1.40 F$ lb-ft, with F applied at $-1.27i+7.66j$ ft from P.

2.6-4: Which two of the following force sets are equipollent?

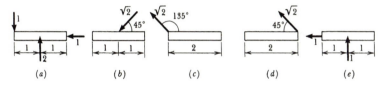

(a)　　(b)　　(c)　　(d)　　(e)

Exer. 2.6-4

2.6-5: Which of the following force and couple sets is not equipollent to any of the others?

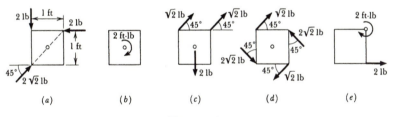

(a)　　(b)　　(c)　　(d)　　(e)

Exer. 2.6-5

2.6-6: Replace the set of four parallel forces shown by a single equipollent force, and indicate the point at which it should be applied. Each square is of 1 ft edge.
Ans.: $-170k$ lb at $(3/17, 9/17)$ ft.

2.6-7: What must be the ratio of the dimensions a and b of the rectangular parallelepiped shown if the force set is to be equivalent to a single force applied at P?
Ans.: $a/b = 5/8$.

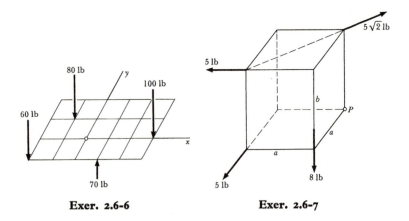

Exer. 2.6-6   **Exer. 2.6-7**

2.6-8: Forces of magnitude P act along the edges of a cube, parallel forces having the same direction. Replace these forces by a single force and a couple about a parallel axis. Show that this force passes through the center of the cube in the direction of a diagonal.

2.6-9: A set of forces and couples has the same resultant moment about each of three non-collinear points. What conclusions may be drawn about the nature of the force set? If the moment in question is zero, are the resultant force and couple both zero?
Ans.: $\mathbf{F} = \mathbf{0}$; yes.

2.6-10: Forces of magnitude P act along two of the non-intersecting edges of a regular tetrahedron of edge length $2a$. Replace these forces by an equipollent wrench.
Ans.: Force of magnitude $P\sqrt{2}$ acting at the center of the tetrahedron in a direction from the midpoint of one side to the center of the tetrahedron; parallel couple of magnitude Pa.

2.7-1: Particles of mass 30 kg, 50 kg, 80 kg are located at points having coordinates $(1, 2)$, $(5, 8)$, $(7, 0)$ meters, respectively, in a vertical plane. What are the coordinates of their mass center? If the particles lie in a horizontal plane, where is the mass center? If the particles are located near the surface of the earth, where is the center of gravity in each case?
Ans.: $x^* = 5.25$ m, $y^* = 2.875$ m in each case.

2.7-2: A set of n masses $m_1, m_2, \ldots m_n$ at points $P_1, P_2, \ldots P_n$ has mass center $\mathbf{r}^*$. Some of the masses, say $1, 2, \ldots s$, are shifted to new positions $P_1', P_2', \ldots P_s'$ where the vector $\overline{P_1 P_1'}$ is denoted by $\Delta \mathbf{r}_1$, the vector $\overline{P_2 P_2'}$ by $\Delta \mathbf{r}_2$, etc. The mass center shifts from P^* to a point whose directed distance from P^* is given by the vector $\Delta \mathbf{r}^*$. Show that

$$\Delta \mathbf{r}^* = \frac{1}{m} \sum_{i=1}^{s} m_i \, \Delta \mathbf{r}_i, \quad \text{where } m = \sum_{i=1}^{n} m_i.$$

2.7-3: Three particles of masses m_1, m_2, m_3 are distances a, a, $a/2$ apart in a vertical plane near the earth's surface, as shown. Where is the center of gravity located relative to m_2?

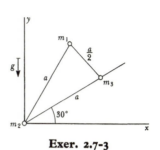

Exer. 2.7-3

2.7-4: The inner and outer radii of a uniform hemispherical shell are R_1 and R_0, respectively. What is the distance from the geometrical center of the shell to the mass center?
Ans.: $3(R_0+R_1)(R_0^2+R_1^2)/8(R_0^2+R_0R_1+R_1^2)$.

2.7-5: Show that if an arc of a plane curve is revolved about an axis that the arc does not intersect, the area of the resulting surface is equal to the length of the arc multiplied by the length of the path of the centroid of the arc. (This is the other theorem of Pappus to which reference was made in Example 2.7-4.)

2.7-6: A plate is cut from a piece of sheet metal in the form of a parabola, being bounded by the curves $y=h[1-(x^2/a^2)]$ and $y=0$. Find the mass center.
Ans.: $y^* = 2h/5$.

2.7-7: Find the mass center of a paraboloidal solid bounded by the plane $y=0$ and the surface formed by rotating the curve $y=h[1-(x^2/a^2)]$, $y \geq 0$, about the y-axis.
Ans.: $y^* = h/3$.

2.7-8: A container of mass m is filled with fluid of mass m'. Owing to a small hole in its bottom, fluid leaks out of the container. Show that the mass center of the system (fluid and container) is at a minimum height above the base of the container when the depth of fluid remaining in the container is the same as that height.

2.7-9: A child builds a staircase using blocks of length L, the overhang of each block being p. The staircase tumbles when a vertical line through the mass center of all but the bottom block falls outside the bottom block. Show that this occurs as soon as the total number of blocks used exceeds the number given by the expression L/p.

2.7-10: Show that if the child of the previous exercise has an unlimited number of blocks at his disposal he can, by shrewdly varying the amount of the overhang of each block, get the staircase to extend as far out from its base as he wishes.

2.7-11: Locate the mass center of the earth-fill dam whose cross-section is shown. The core has a density of 120 lb/ft³ while the outer fill has a density of 90 lb/ft³.

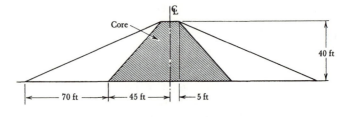

Exer. 2.7-11

2.7-12: Locate the mass center of the rotor whose cross-section is shown. The copper coils have a net density of 400 lb/ft³ while the steel frame has a density of 485 lb/ft³.

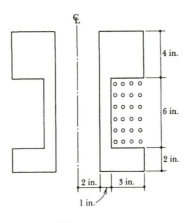

Exer. 2.7-12

2.7-13: Locate the mass center of the angle iron shown. Ignore the rounding of the corners.
Ans.: From corner of angle $x^* = 0.7$ in., $y^* = 2.2$ in.

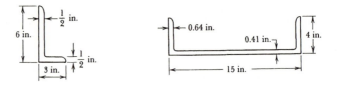

Exer. 2.7-13   **Exer. 2.7-14**

2.7-14: Locate the mass center of the channel section shown.
Ans.: 1.0 in. from base.

2.7-15: The gas in a spherical container is at an absolute pressure of five atmospheres. If the radius of the container is 10 ft, what is the magnitude of the single force equipollent to the net atmospheric and gas pressure acting on half the container?
Ans.: $2.66 (10)^6$ lb.

2.7-16: The simply-supported beam shown is subjected to a load distribution, as shown, of $p = p_0 \sin (\pi x/l)$ units of force per unit length. Replace the distribution by a single concentrated load; where does the load act?

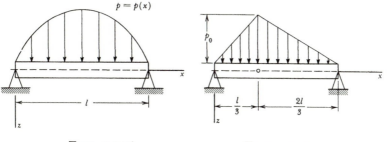

Exer. 2.7-16   Exer. 2.7-17

2.7-17: Solve Exercise 2.7-16 for the triangular load shown.

2.7-18: A circular rod of radius a is twisted by end loads τ lbs/in.^{2} in magnitude, directed perpendicularly to the radius line from the axis of the rod. Given that τ increases linearly with the radial distance r, replace the distributed force system by an equipollent resultant system.

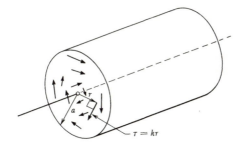

Exer. 2.7-18

CHAPTER **III**

Analysis of Statically Determinate States of Equilibrium

3.1 Statically Determinate and Statically Indeterminate Equilibrium

In the previous chapter the principles of mechanical equilibrium were developed. These lead to the conclusion that a mechanical system in equilibrium must satisfy certain mathematical relationships known as the equations of equilibrium. It is noteworthy that these equations are concerned only with the external forces acting on the system; they express the fact that these forces must have zero resultant in any direction and must produce zero moment about any axis. In the present chapter these equations are employed to investigate the equilibrium loads and equilibrium configurations of a number of systems commonly encountered in engineering practice. For the most part, discussion is restricted to loads applied at points (rather than to distributed loads, which are dealt with more fully in Chapter

IV) and to mechanical systems consisting of elements that deform so little under load that they may be regarded as rigid.

The first step in the analysis of the statical equilibrium of a mechanical system consists in isolating the system and drawing a free-body diagram that displays all the external forces and couples acting on the system—both those forces and couples whose magnitude and direction are known and those whose determination is the object of the analysis. The proper choice of a system, the decision as to what forces have to be considered in a meaningful analysis, and the representation of these forces as vectors is not always an easy matter. Good physical judgment is required at this point since we are really selecting an appropriate mathematical model of an engineering structure. Appropriate "model-making" is a skill developed by practice. Its possession is the hallmark of the well-trained engineer. Some of the more common representations of force systems were discussed in Chapter II and others are developed in the examples of this chapter.

Once a mechanical system has been isolated and the forces and couples acting on it have been identified, we may, supposing the system to be in equilibrium, write down the six independent scalar equations of equilibrium. Of course some of the equations may be what is technically known as *trivial*; that is, they may be of the form $0 = 0$. If, for example, the set of external forces and couples consists of forces all lying in one plane (say the xy-plane) they can have no moment about the x- and y-axes and no components in the z-direction. The equations that express these facts will then be trivial. We need not even bother to write them down. In this example, then, there could be no more than three non-trivial equations of equilibrium. The unknowns in the equations of equilibrium will be either forces of constraint which are not known initially and whose determination is the object of the analysis or else geometrical quantities, such as lengths or angles, which define the configuration of the system. These unknowns are constants so that the equations of equilibrium take the form of simultaneous algebraic (or trigonometric) equations. In dynamics, on the other hand, the governing equations of motion are differential equations. If the unknowns are concentrated forces or couples, the equations of equilibrium will be linear. When the equations of equilibrium can be solved so as to determine all the external forces and couples exerted by the constraints in order to hold the system in equilibrium, the system is said to be *statically determinate*. When they do not provide all the information needed for this purpose, the system is said to be *statically indeterminate* or *hyperstatic*.

Before we proceed to investigate the implications of statical determinacy and indeterminacy, the qualifying phrase "supposing the system to be in equilibrium" appearing at the outset of the preceding paragraph deserves brief reconsideration. To see what is implied by this reservation, consider a very simple illustration. A uniform bar 2 ft long, weighing 10 lb, is free to rotate about a smooth hinge or pin at one end, as shown in Fig. 3.1-1a. At the other end the bar

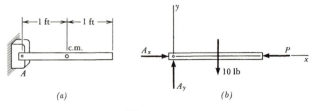

Fig. 3.1-1

carries an unknown horizontal force, P. Suppose we ask to know the magnitude of P required to maintain the bar in equilibrium in a horizontal position. The free-body diagram of the bar is shown in Fig. 3.1-1b. The sum of the moments of the external forces about the z-axis at A is -10 ft-lb and no possible values of P, A_x, and A_y

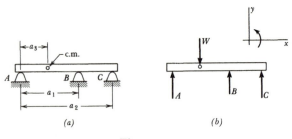

Fig. 3.1-2

will make the resultant moment vanish. The equations of equilibrium, therefore, will have no solution; i.e., there are no values of P, A_x, A_y which satisfy all of them. The physical interpretation of this situation is obvious; the horizontal position of the bar is not a possible equilibrium position under the applied loads and constraints. The system must move under the action of the forces applied, and the equations of dynamics must be employed.   While it often may be easy to see by inspection of the free-body diagram whether a given configuration is or is not a possible equilibrium configuration, complex situations do arise in which only a careful analysis will reveal the existence or lack of existence of complete solutions of the equations of equilibrium.

The distinction between statically determinate and statically indeterminate states of equilibrium is a fundamental one in engineering.   It too may be illustrated by means of a simple example.   Consider a horizontal beam resting on three smooth supports at the same elevation, as shown in Fig. 3.1-2a.   Its free-body diagram is shown in Fig. 3.1-2b.   Since there are no horizontal forces, there are only two non-trivial equations of equilibrium:

$$\sum F_y = A + B + C - W = 0,$$

$$\sum M_A = Ba_1 + Ca_2 - Wa_3 = 0. \qquad \textbf{3.1-1}$$

The three unknown reactions, A, B, and C, cannot be found from the two equations of equilibrium.   Nor will it help matters to attempt to find a third equation by taking a new set of axes or a new origin for the moment equation.   If, for example, we take moments about an axis through B, we find

$$\sum M_B = -Aa_1 + C(a_2 - a_1) + W(a_1 - a_3) = 0. \qquad \textbf{3.1-2}$$

But this is the same thing as the first of Eqs. 3.1-1 multiplied by $-a_1$ and added to the second of Eqs. 3.1-1.   It therefore provides no information not already inherent in Eqs. 3.1-1.   Equations 3.1-1 and 3.1-2 are said to be *linearly dependent*.   We cannot hope, therefore, to find all of the unknown reactions from the equations of equilibrium alone.   Statical indeterminacy is the price we pay in this instance for idealizing our system and considering it to be rigid.   In order to determine a solution, the equations of equilibrium must be supplemented by further information drawn from the mechanics of deformable bodies.   If we can determine by computation or experiment that removal of the support at B would cause the beam to sag by a distance of 0.1 inch at B and that an upward force of 1 lb at B would produce an upward motion of 0.01 inches at B, we can, assuming displacement to be proportional to load, conclude that the actual reaction at the central support must be $B = 10$ lb.   This additional information, together with the equations of equilibrium, readily determines the three reactions.

Another example of a statically indeterminate structure is a light bar connected by smooth pins to a large rigid body, as pictured in Fig. 3.1-3a. The three non-trivial equations of equilibrium ($\sum F_x = 0$, $\sum F_y = 0$, $\sum M_z = 0$) will not determine the four unknown components of the reactions at A and B. All they will tell us is that these end forces must be equal in magnitude and directed in opposite senses along the line joining the pins, i.e., that the bar is a two-force member. Since there are more unknowns than equations of equilibrium, we conclude that the bar is statically indeterminate. This example may serve to illustrate one of the important properties of a statically indeterminate structure: it may be in a state of self-strain even when no external load is applied. Suppose that the pins are at a distance l apart and that the holes in the bars are drilled at a distance

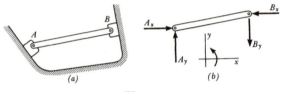

Fig. 3.1-3

$l - \epsilon$ apart. Then the bar will be in an initial state of tension whose magnitude can only be found by a consideration of the elastic properties of the bar. Similarly, a temperature change in the bar will produce a large stress due to the suppression by the attached body of the small change in length that would otherwise occur in free thermal expansion or contraction. In a statically determinate structure, on the other hand, small changes in the dimensions have only a minor influence on the loads carried by the members.

The engineer designing a structure or machine must frequently decide at the outset whether to employ a statically determinate or a statically indeterminate design. The former choice is to be preferred, for example, in the design of precision instruments whose readings are not supposed to reflect the ambient temperature, or for long bridges over three or more piers when the piers may be subject to slight settlement. On the other hand, a designer may deliberately select a statically indeterminate design in order to take advantage of the additional rigidity inherent in the extra constraint. The products of a technological civilization represent the designer's response to the sometimes conflicting requirements of function, economy, and reliability. It is always instructive to observe the ways in which these requirements have been answered.

For a system consisting of a single rigid body, such as the bar of the two previous paragraphs, statical determinacy or indeterminacy is relatively easy to recognize. When systems consist of many bodies connected together, even when the over-all structure is itself rigid, the determinacy of the problem is not as readily apparent. The

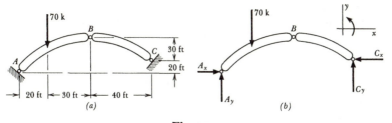

Fig. 3.1-4

equations for the entire structure may appear to lead to indeterminacy; but if we split the system into its parts and examine the equilibrium of all the separate parts, it may transpire that all the unknown external forces can be found. A simple illustration of this situation is provided by the so-called three-hinged arch pictured in Fig. 3.1-4a. We wish to find the reactions due to the presence of a 70,000 lb load located as shown. (Note the abbreviation: 70,000 lb = 70 kilopounds, written 70 k here; sometimes "kip" is used.) The free-body diagram of the arch, shown in Fig. 3.1-4b, contains four unknowns, just as did the free-body diagram of Fig. 3.1-3b. And, indeed, the equilibrium equations for the structure are

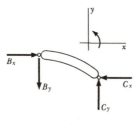

$$\sum F_x = A_x - C_x = 0,$$
$$\sum F_y = A_y + C_y - 70 = 0,$$
$$\sum M_B = 50A_x - 50A_y - 30C_x$$
$$+ 40C_y + 2100 = 0.$$

3.1-3

Fig. 3.1-5

If we now consider the segment *BC* alone, its free-body diagram (Fig. 3.1-5) introduces two new unknowns (the components of the force at *B* exerted by *AB* on *BC*) and three equations of equilibrium:

$$\sum F_x = B_x - C_x = 0,$$
$$\sum F_y = C_y - B_y = 0,$$
$$\sum M_B = 40C_y - 30C_x = 0.$$

3.1-4

These six equations suffice to determine the six unknowns. We find

$$A_x = 16 \text{ k}, \qquad B_x = 16 \text{ k}, \qquad C_x = 16 \text{ k},$$

$$A_y = 58 \text{ k}, \qquad B_y = 12 \text{ k}, \qquad C_y = 12 \text{ k}. \qquad \textbf{3.1-5}$$

The presence of the third hinge at B changes the arch from a statically indeterminate to a statically determinate structure.

The program of statical analysis in engineering entails five distinct steps: (a) the isolation of a system that is in equilibrium; (b) the identification of all the external forces and couples acting on that system by means of a free-body diagram; (c) the writing of the equations of equilibrium; (d) the solution of the equations of equilibrium; and (e) the interpretation of the solution. Physical judgment and engineering judgment enter at the last step to at least as great an extent as at the first two. At the lowest level of judgment one must always ask whether the numerical solution obtained is a reasonable one. A reaction of 100,000 lb, when all the other forces involved are of about 100 lb magnitude, should probably be viewed with suspicion. Directions of vectors must be examined: a supposed cable tension should not turn out to be a compression, for example. More complex judgments arise when the results of the analysis are applied to considerations of design. A bar may buckle under the compressive load it must carry, or a cable may snap if the tension in it is too great. Deformation of the structural elements may not be negligible if the loads are too great. The more complex questions of this character cannot be dealt with in a first course to any appreciable extent; our primary concern must be the development of analytical method and technique. The student should be conscious, however, that his solution to a problem of mechanical analysis must be scrutinized for its meaningfulness. The solution itself is often only the first step in a complex engineering process.

3.2 Analysis of the Equilibrium of Systems of Rigid Bodies

In Section 3.1 the general program of statical analysis was described and a distinction was made between those cases in which the equations of equilibrium are sufficient by themselves to determine all the unknown forces of constraint and those cases in which they are not. In this section we consider efficient methods for the formulation and solution of the equations for statically determinate systems.

After a system has been isolated and the external forces acting on it have been identified, how shall the axes be chosen? Certainly it is never incorrect to write $\sum \mathbf{F} = \mathbf{0}$ and $\sum \mathbf{M}_O = \mathbf{0}$, where O is the origin of coordinates. There may, however, be better choices than the origin of coordinates as the base point for the moment equation. It may even be better to write a number of moment equations for different points rather than work with the force equation directly. By choosing our points or axes for the moment equation judiciously, the number of unknown quantities that appear in each equation can be held to a minimum. Often, axes can be chosen so that only one unknown appears in each equation. Considering the frailty of humans and the ease with which algebraic mistakes can be made in solving simultaneous equations, this is the proper choice of axes where possible. The labor of solving the equations of equilibrium is therefore greatly reduced if we pause, before writing any equations at all, to identify axes that are parallel or perpendicular to a number of unknown force components, or axes which such force components intersect.

The formalism of vector algebra is quite helpful in three-dimensional problems. It relieves much of the geometrical computation of skew moment arms and oblique components that would otherwise be necessary. In two-dimensional problems, where the forces all lie in a plane, vector notation is only a minor convenience and we can usually dispense with it. The most important function of vector notation is not that of a problem-solving aid. In any particular problem there is likely to be some special geometrical aspect that makes the use of a particular coordinate system advantageous. The vectorial notation is of value primarily because it is the natural language of statics and dynamics—and of electrodynamics and fluid mechanics as well—in which the principles of these subjects find their most succinct form of expression.

It is important to realize, as was pointed out in Section 3.1, that there are not more than six independent scalar equilibrium equations for a given mechanical system (three for a strictly planar model). Once one has the appropriate number of independent equations for a given system, no amount of additional equation writing for that system will lead to information not already contained in (if difficult at times to extract from) the original set. If, however, the system consists of n bodies interconnected in some way, we may profitably write the equations of equilibrium for any $n - 1$ subdivisions of the original system in addition to the equilibrium equations for the system as a whole. In this procedure the third law of motion plays a fundamental role. When the original system is split up, new unknown forces must be introduced to represent the mechanical interaction between the parts of the system. We need not, however, introduce a complete set of new unknowns for each subsystem; the third law requires us to use, except for a reversal of direction, the same force for each pair of subsystems at their mutual point of contact. Therefore, when we increase the number of equilibrium equations we do not increase the number of new unknowns to the same extent.

Complex systems may be classified according to the amount of restraint placed on them. As we have seen, a rigid body can be completely constrained against motion in space if three of its points are fixed by means of six independent constraints against rotation and translation. Similarly, a body whose motion is restricted to a plane can be completely constrained if two points are fixed. More complex systems can also be constrained if each of the parts is constrained, directly or through connections to the other points. If a system is underconstrained, i.e., still has one or more possible modes of motion, it is called a *mechanism*; if it is completely constrained against motion, it is called a *frame*. Since mechanisms are open to the possibility of motion under the applied forces, equilibrium problems for mechanisms usually involve position or angle coordinate unknowns as well as force unknowns. Only certain configurations of the mechanism can be equilibrium configurations. These configurations will, in general, be different for different sets of applied loads. Frames, on the other hand, are fixed in space, or in a plane, by the constraint conditions. The equilibrium configuration is therefore known in advance and the only unknowns appearing in the analysis are the forces carried by the different parts of the frame. It is important to examine constraint conditions carefully, not only in order to represent the forces of constraint properly, but also in order to decide whether complete constraint is present or not.

We now turn to specific examples illustrating these points.

Example 3.2-1

The 100 lb boom supports a 500 lb load at its end and is in turn supported in a vertical plane by a smooth pin connection and a light inextensible cable (Fig. 3.2-1a). The boom may be treated as a uniform bar. Dimensions are shown. Find the cable tension and the pin reaction.

Solution: In Figs. 3.2-1b, c, and d, three different free-body diagrams are shown. In the first, we have split the system into two parts, the boom and the load (treated as a particle) separately. In the second and third, the load is part of the system; and in the third, the cable is part of the system also. The pin reaction is represented by its horizontal and vertical components (P_x, P_y), assumed positive in the directions shown. The cable tension is of unknown magnitude T in a known direction. The weight force of the boom is placed at its midpoint, since we treat the boom as uniform.

In Fig. 3.2-1b, the third law of motion has been used to write Q as the magnitude of the two forces, one on the boom and the opposite one on the load, exerted by the supporting cord on each member. In truth, one should also show the cord, with opposing forces of magnitude Q on it. It is clear, however, that $Q = 500$ lb for equilibrium of the load, and we

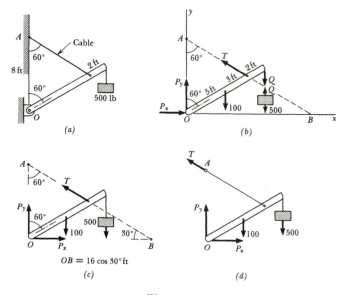

Fig. 3.2-1

would ordinarily go immediately to one of the last two diagrams or replace Q on the first diagram by 500 lb, without the formal intermediate step of solving for it.

We may now write the equilibrium equations for the system of Fig. 3.2-1c. The force equations of equilibrium are

$$\sum F_x = P_x - T \cos 30° = 0,$$

$$\sum F_y = P_y + T \sin 30° - 600 = 0. \qquad \textbf{3.2-1}$$

The moment equation should be taken about the pin, since that eliminates the unknowns (P_x, P_y) from consideration. The moment is about the z-axis at O, but, since this is a plane problem, we suppress designation of the axis and speak simply of the "moment about O." The moment of **T** is computed by the $M_z = xT_y - yT_x$ formulation. The tension force **T** is split into components $T_x = - T \cos 30°$, $T_y = T \sin 30°$, with $x = 8 \cos 30°$, $y = 8 \sin 30°$ being the coordinates of its point of application:

$$\sum M_O = (8 \cos 30°)(T \sin 30°) - (8 \sin 30°)(- T \cos 30°)$$

$$- (5 \cos 30°)(100) - (10 \cos 30°)(500) = 0,$$

or

$$\sum M_O = 16T \sin 30° \cos 30° - 5500 \cos 30° = 0. \qquad \textbf{3.2-2}$$

If we first solve the moment equation for T, and then solve the force equations, we find

$$T = 687.5 \text{ lb}, \qquad P_x = 343.75 \sqrt{3} \text{ lb}, \qquad P_y = 256.25 \text{ lb},$$

or, to the three-figure accuracy that we shall ordinarily use,

$$T = 688 \text{ lb}, \qquad P_x = 595 \text{ lb}, \qquad P_y = 256 \text{ lb}. \qquad \textbf{3.2-3}$$

Since all answers are positive, the directions of the forces are as shown on the free-body diagram.

The computation of the moment of the tension force can be done much more simply if we use the fact that "transmissibility" holds; i.e., we may compute the moment by considering the force to act at any point along its line of action. By considering **T** to act at A in Fig. 3.2-1c—or using the system of Fig. 3.2-1d—we find its moment from its x-component only:

$$-(8)(- T \cos 30°) = 4\sqrt{3}T.$$

Also, note that P_x may be determined independently of the other two unknowns if moments are written about A; and P_y, if moments are written about B (Fig. 3.2-1c):

$$\sum M_A = 8P_x - (5 \cos 30°)(100) - (10 \cos 30°)(500) = 0,$$

$$\sum M_B = - (16 \cos 30°)P_y + (11 \cos 30°)(100) + (6 \cos 30°)(500) = 0. \quad \textbf{3.2-4}$$

The only problem here is in the determination of the location of point B where the lines of action of T and P_x intersect. The three equations of equilibrium $\sum M_O = 0$, $\sum M_A = 0$, $\sum M_B = 0$ (Eqs. 3.2-2 and 3.2-4) are completely equivalent to the original set (Eqs. 3.2-1 and 3.2-2) $\sum F_x = 0$, $\sum F_y = 0$, $\sum M_O = 0$. If the three moment equations are used, the force equations should be automatically satisfied, and hence may be used as a check on the accuracy of the computations.

Example 3.2-2

Consider the boom and load of the last example again, but now supported by two cables as shown in Fig. 3.2-2a; instead of a pin at O, the support is a ball-and-socket joint. Write governing equilibrium equations and solve.

Solution: The boom and load are still in the vertical *xy*-plane. The support reaction now has a *z*-component, P_z. The two cable tensions are denoted by T_1 and T_2. The free-body diagram is given in Fig. 3.2-2b. Note that there are five scalar unknowns: P_x, P_y, P_z, T_1, and T_2. There are also five non-trivial independent equilibrium equations. Since all forces, known and unknown, have lines of action intersecting the boom, the moment about the line *OE* must automatically vanish. Thus some linear combination of any six equilibrium equations must reduce to the trivial o = o form; only five are independent.

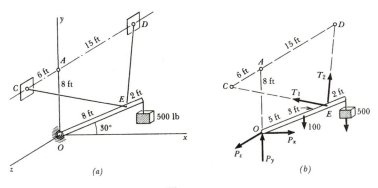

Fig. 3.2-2

Here vectorial methods are of aid in the computation. We may write the two weight forces as $-100\mathbf{j}$, $-500\mathbf{j}$ and the reaction force at *O* as $\mathbf{P} = P_x\mathbf{i} + P_y\mathbf{j} + P_z\mathbf{k}$ immediately. The vectorial form of the tension forces requires the computation of the unit vectors from *E* to *C* and *D*:

$$\mathbf{r}_{EC} = \mathbf{r}_C - \mathbf{r}_E = (8\mathbf{j} + 6\mathbf{k}) - (4\sqrt{3}\mathbf{i} + 4\mathbf{j}) = -4\sqrt{3}\mathbf{i} + 4\mathbf{j} + 6\mathbf{k} \text{ ft};$$

$$\mathbf{r}_{ED} = \mathbf{r}_D - \mathbf{r}_E = (8\mathbf{j} - 15\mathbf{k}) - (4\sqrt{3}\mathbf{i} + 4\mathbf{j}) = -4\sqrt{3}\mathbf{i} + 4\mathbf{j} - 15\mathbf{k} \text{ ft};$$

$$\mathbf{e}_{EC} = \mathbf{r}_{EC}/|\mathbf{r}_{EC}| = -\frac{2\sqrt{3}}{5}\mathbf{i} + \frac{2}{5}\mathbf{j} + \frac{3}{5}\mathbf{k};$$

$$\mathbf{e}_{ED} = \mathbf{r}_{ED}/|\mathbf{r}_{ED}| = -\frac{4\sqrt{3}}{17}\mathbf{i} + \frac{4}{17}\mathbf{j} - \frac{15}{17}\mathbf{k}.$$

Therefore, we may write

$$\mathbf{T}_1 = T_1\left(-\frac{2\sqrt{3}}{5}\mathbf{i} + \frac{2}{5}\mathbf{j} + \frac{3}{5}\mathbf{k}\right), \qquad \mathbf{T}_2 = T_2\left(-\frac{4\sqrt{3}}{17}\mathbf{i} + \frac{4}{17}\mathbf{j} - \frac{15}{17}\mathbf{k}\right).$$

We now proceed to the writing of the equations of equilibrium. Writing the moment equation about *O* is an obvious first step, since the unknown **P** is eliminated from consideration. Whether one proceeds directly to the scalar equations by writing the moments about the x-, y-, and z-axes in turn or by writing the vector equation first is a matter of taste; we choose the latter procedure. We have

$$\sum \mathbf{M}_O = \mathbf{0} = 5\left(\frac{\sqrt{3}}{2}\mathbf{i}+\frac{1}{2}\mathbf{j}\right) \times (-100\mathbf{j})+10\left(\frac{\sqrt{3}}{2}\mathbf{i}+\frac{1}{2}\mathbf{j}\right) \times (-500\mathbf{j})$$

$$+\mathbf{r}_E \times \mathbf{T}_1 + \mathbf{r}_E \times \mathbf{T}_2$$

$$= -2750\sqrt{3}\mathbf{k}+\frac{4T_1}{5} \begin{vmatrix} \mathbf{i} & \mathbf{j} & \mathbf{k} \\ \sqrt{3} & 1 & 0 \\ -2\sqrt{3} & 2 & 3 \end{vmatrix}$$

$$+\frac{4T_2}{17} \begin{vmatrix} \mathbf{i} & \mathbf{j} & \mathbf{k} \\ \sqrt{3} & 1 & 0 \\ -4\sqrt{3} & 4 & -15 \end{vmatrix}$$

$$= -2750\sqrt{3}\mathbf{k}+\frac{4T_1}{5}(3\mathbf{i}-3\sqrt{3}\mathbf{j}+4\sqrt{3}\mathbf{k})$$

$$+\frac{4T_2}{17}(-15\mathbf{i}+15\sqrt{3}\mathbf{j}+8\sqrt{3}\mathbf{k}).$$

Collecting terms and setting each scalar component of the resultant vector equal to zero, we have

$$\sum M_x^O = \frac{12}{5}T_1-\frac{60}{17}T_2 = 0,$$

$$\sum M_y^O = -\frac{12}{5}\sqrt{3}T_1+\frac{60}{17}\sqrt{3}T_2 = 0, \qquad \textbf{3.2-5}$$

$$\sum M_z^O = \frac{16}{5}\sqrt{3}T_1+\frac{32}{17}\sqrt{3}T_2-2750\sqrt{3} = 0.$$

From the first, we find $T_1=(25/17)T_2$; from the third, then,

$$T_2 = (17)(2750)/(112) = 417 \text{ lb}, \qquad T_1 = (25)(2750)/(112) = 614 \text{ lb}.$$

Notice that the $\sum M_y$ equation is automatically satisfied when the $\sum M_x$ equation is, and conversely. The two are not independent. This is a reflection of the fact that the moment about axis OE must vanish, and that the axis OE lies in the xy-plane through O.

Now that the tensions have been determined, we can write the force equations to determine **P**. In this case, these equations are relatively straightforward:

$$\sum \mathbf{F} = \mathbf{P}+\mathbf{T}_1+\mathbf{T}_2-600\mathbf{j} = \mathbf{0},$$

or

$$\sum F_x = P_x - \frac{2}{5}\sqrt{3}T_1 - \frac{4}{17}\sqrt{3}\,T_2 = 0,$$

$$\sum F_y = P_y + \frac{2}{5}\,T_1 + \frac{4}{17}\,T_2 - 600 = 0, \qquad \textbf{3.2-6}$$

$$\sum F_z = P_z + \frac{3}{5}\,T_1 - \frac{15}{17}\,T_2 = 0.$$

However, note that P_x can be determined easily independently of the others by writing an additional moment equation—an equation about the z-azis through A, i.e., the axis DAC. Since $\mathbf{T}_1$, $\mathbf{T}_2$, and $P_y\mathbf{j}$ have lines of action that intersect the axis DAC, they have no moment about it; since $P_z\mathbf{k}$ is parallel to the axis, it has zero moment about the axis. Only the known weight forces and $P_x\mathbf{i}$ have moment:

$$\sum M_z^A = \sum \mathbf{M}_A \cdot \mathbf{k} = 8P_x - (5\cos 30°)(100) - (10\cos 30°)(500) = 0. \quad \textbf{3.2-7}$$

This is, of course, the same equation as the first of Eqs. 3.2-4 in the last example, and P_x has the same value, 595 lb. If Eq. 3.2-7 is taken as a basic equilibrium equation, then the first of 3.2-6 can no longer be so considered but may be used as a check. Similarly, since axis AE extended intersects both x- and y-axes, none of the forces except $P_z\mathbf{k}$ can have moment about it; therefore, $P_z = 0$. [This result may also be found from the last of Eqs. 3.2-6, once we know that $T_1 = (25/17)T_2$.]

Finally, P_y may be determined from the second of Eqs. 3.2-6:

$$P_y = 600 - \frac{2}{5}\,T_1 - \frac{4}{17}\,T_2 = 600 - \frac{14}{17}\,T_2 = 600 - 344 = 256\text{ lb},$$

the same as the P_y of Example 3.2-1.

To summarize, we see that our preliminary analysis showed the problem to be statically determinate, with five unknowns and five independent equations. The five independent equations can be chosen in a number of ways. In particular, the four moment equations $\sum M_x^O = \sum M_z^O = \sum M_z^A = \sum M_{AE} = 0$ and the force equation $\sum F_y = 0$ form one such set; $\sum M_x^O = \sum M_z^O = 0$ and $\sum F_x = \sum F_y = \sum F_z = 0$ form another.

Example 3.2-3

A rigid plane frame (Fig. 3.2-3a) is constructed of two light rigid bars, AC and BC, pinned smoothly together at C and to the foundation at A and B. The frame is subjected to the 50 lb load at C as shown. Discuss the equilibrium of the frame.

Solution: A free-body diagram of the frame is shown in Fig. 3.2-3b, together with the coordinate system to be used. Before analyzing the equilibrium state, let us assure ourselves that this is a rigid frame. Each bar has at most three degrees of freedom in the plane, so the system could have six. Fixing A and B by pins fixes four of the six coordinates needed; fixing point C then indeed fixes the angles between the bars and the x-axis.

There are four unknowns, A_x, A_y, B_x, and B_y, as shown—or, equivalently, the magnitudes $A=[A_x^2+A_y^2]^{1/2}$, $B=[B_x^2+B_y^2]^{1/2}$ and the directions $\theta_A=\arctan (A_y/A_x)$, $\theta_B=\arctan (B_y/B_x)$ from the x-axis to the lines of the forces—and but three independent equations:

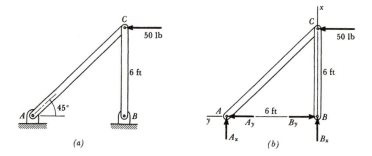

(a)   (b)

Fig. 3.2-3

$$\sum F_x = A_x+B_x = 0,$$
$$\sum F_y = A_y-B_y+50 = 0,\qquad\qquad \textbf{3.2-8}$$
$$\sum M_B = (6)(50)-(6)(A_x) = 0.$$

We see that $A_x= -B_x=50$ lb; that is, we have guessed wrong on the direction of B_x. However, *we do not change the diagram*; we leave it as it is and work with $B_x= -50$ lb. The second of Eqs. 3.2-8 provides a relation

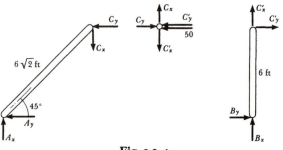

Fig. 3.2-4

between A_y and B_y, but that is all; no more information can be gained by writing any more equations. If ABC were a single rigid body, we would be finished. But let us split the system into its component parts and see what happens.

Now consider three subsystems: the bar AC, the bar BC, and the pin at C. (We are purposely making the analysis more complex than is necessary in order to show features that are taken for granted later on.) In Fig. 3.2-4, free-body diagrams are shown with the 50 lb load applied to the pin. Note that the reaction forces at A and B are shown in exactly the same way

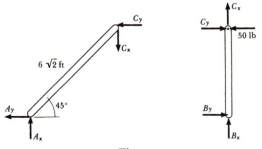

Fig. 3.2-5

here as they were in Fig. 3.2-3b. The force exerted by the pin on AC is denoted by components C_x and C_y with assumed directions as shown; the oppositely directed force is exerted on the pin by the bar. Similarly, C'_x and C'_y denote the force components for the reactions between the pin and bar BC.

The analysis may now be completed in a number of ways. The three equations for AC, coupled with $A_x = 50$ lb from our previous solution,

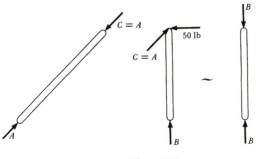

Fig. 3.2-6

determine C_x, C_y, and A_y. The two pin equations then can be used for C'_x, C'_y; and the second of Eqs. 3.2-8, for B_y. The equations for BC may then be used as checking equations.

Let us examine alternative ways of analysis, and particularly a simple method based on the "two-force member" concept. First, it should be noted that the solution for the forces at *A* and *B* would be the same wherever the 50 lb force is considered to act: on the pin or on either one of the bars. The values of the C_x, C_y, C'_x, C'_y forces will change; but not the values of A_x, A_y, B_x, and B_y. Second, we need not separate the pin as a separate element; we can consider it as part of one of the bars. If the pin is part of *BC*, then the free-body diagrams of the subsystems appear as in Fig. 3.2-5. The analysis is correspondingly simpler, with two unknowns fewer appearing.

Finally, all of this has been much too complicated from the beginning. Examining Fig. 3.2-5, we see that both bars are loaded only at their ends. Replacing the forces at each point by the resultant at that point, we see that each member is a two-force member, and that we may therefore determine the lines of action of the unknown forces, if not their magnitudes, for equilibrium (Fig. 3.2-6). Indeed, recognizing that each bar is a two-force member, we need not split our system at all, but can put in the reactions at *A* and *B* in the proper directions to begin with (Fig. 3.2-7). Moment

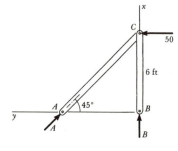

Fig. 3.2-7

equilibrium about pin *C* is now automatically satisfied. Only two equations remain for the two unknowns *A* and *B*:

$$\sum F_x = B + A(\sqrt{2}/2) = 0,$$

$$\sum F_y = 50 - A(\sqrt{2}/2) = 0,\qquad\text{3.2-9}$$

from which we find

$$A = 50\sqrt{2}\text{ lb},\qquad B = -50\text{ lb}.\qquad\text{3.2-10}$$

The sign of *B*, of course, signifies that we have chosen the wrong direction for *B* originally; bar *BC* is really in tension, while *AC* is in compression.

A frame such as this, consisting of two-force members, is called a *truss*; it consists of members joined at their ends and loaded only at the joints, with the weights of the bars being ignored or replaced by equipollent loads at the joints. Analysis of trusses is the subject of the next section.

Example **3.2-4**

The crank-connecting rod-piston mechanism is observed to be at rest in the position shown (Fig. 3.2-8). The uniform 5 ft crank *OA* weighs 20 lb; the 10 ft connecting rod *AB* is light; and the piston *P*, which slides in the smooth but tightly fitting vertical cylinder, weighs 10 lb. All pin connections are smooth. What must be the force of compression on the gas in the cylinder?

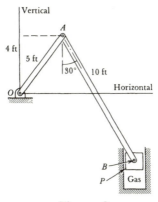

Fig. 3.2-8

Solution: This is a single-degree-of-freedom system; knowing the angle between, say, the horizontal and the crank *OA* determines the position of the whole system, but that angle is free to change since the system is not completely constrained. We are here concerned with finding the forces necessary to hold the mechanism in equilibrium in a certain position, and, although probable, it is not at all clear that a compressive force on the gas, and hence on the piston, will in fact accomplish this.

There is another aspect to this problem that deserves consideration before the exact formulation is made. We are asking here for only one of the forces, not all, needed for equilibrium, and we shall therefore attempt to set up an analysis that will determine that force as quickly as possible without finding too many of the other unknown constraint forces.

In Fig. 3.2-9, we draw a free-body diagram of the whole system. *F* denotes the force magnitude we wish to determine; *N*, the resultant normal reaction from the walls of the cylinder on the piston; and Q_x, Q_y, the components of the pin reaction at *O*. A set of axes is shown, as well as all dimensions of interest. Here we have four unknowns, and only the three independent equations for a plane system. Since there is no equation governing *F* alone immediately apparent, let us write no equations at all but proceed directly to consideration of subsystems. Here there are three: the piston *P*, the crank *OA*, and the connecting rod *AB*. The pins need

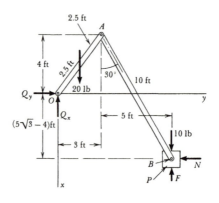

Fig. 3.2-9

not be considered separately, but can both be considered as part, say, of
AB. Before drawing the free-body diagrams of the parts, however, we
make a preliminary analysis of what to expect. The piston is to be treated
as a particle; each of the other bodies, as rigid. Furthermore, *AB*—since
we are considering it as "light," i.e., neglecting its weight—may be treated
as a two-force member. The reactions at *A* and at *B* on *AB* must therefore
be negatives of one another for equilibrium; only the piston and crank need
be considered (Fig. 3.2-10a, b).

We see that, to determine *F*, only two equations need be written. The
moment equation $\sum M_O = 0$ on *OA* determines *R*, the magnitude of the
thrust carried by *AB*. The vertical force equation of equilibrium $\sum F_x = 0$
on *P* then determines *F*. We have, on *OA*,

$$\sum M_O = (3)(R \cos 30°) + (4)(R \sin 30°) - (1.5)(20) = 0,$$

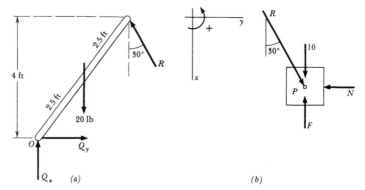

Fig. 3.2-10

$$(2 + 1.5\sqrt{3})R = 30; \qquad\qquad \textbf{3.2-11}$$

and on P:

$$\sum F_z = 10 + R\cos 30° - F = 0,$$

$$F = 10 + R\cos 30° = 10 + \frac{\sqrt{3}}{2}\left(\frac{30}{2 + 1.5\sqrt{3}}\right),$$

$$F = 15.6 \text{ lb}. \qquad\qquad \textbf{3.2-12}$$

Example 3.2-5

A bent shaft ABC is carried in smooth bearings at D and E, as shown. At C a 96 lb-in. torque is applied about the axis BC (*clockwise as viewed from C*

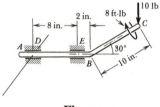

Fig. 3.2-11

to B) and a force of 10 lb acts at right angles to the plane of ABC. What torque Q must be applied to the bar at A, about the axis AB, in order to maintain equilibrium? What forces will be exerted on the bearings?

Solution: The free-body diagram of the shaft is shown in Fig. 3.2-12. Note that vectors representing both forces and couples appear in the free-

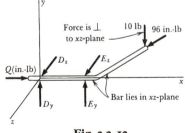

Fig. 3.2-12

body diagram. These must not be confused and the mistake of adding couple vectors to force vectors must not be made. (Some people denote the couple vector by a double-headed or double-shafted arrow in order to lessen the danger of making this error.)   There is a real advantage in vector representation here because the couples are completely specified by their vectors and these enter the moment equations very simply.   The equations of equilibrium are

$$\sum \mathbf{F} = \mathbf{D} + \mathbf{E} - 10\mathbf{j} = 0,$$

$$\sum \mathbf{M}_D = Q\mathbf{i} + 8\mathbf{i} \times \mathbf{E} + (18.7\mathbf{i} - 5\mathbf{k}) \times (-10\mathbf{j}) + (-83\mathbf{i} + 48\mathbf{k}) = 0. \qquad \textbf{3.2-13}$$

We can now write $\mathbf{D} = D_y\mathbf{j} + D_z\mathbf{k}$ and $\mathbf{E} = E_y\mathbf{j} + E_z\mathbf{k}$ and separate these vector equations into scalar ones by equating the $\mathbf{i}$, $\mathbf{j}$, $\mathbf{k}$ components separately to zero.   As a matter of fact the moment equation gives the wanted torque, Q, at once.   The term $\mathbf{i} \times \mathbf{E}$ cannot have an $\mathbf{i}$-component. Therefore,

$$Q - 50 - 83 = 0 \qquad \text{and} \qquad Q = 133 \text{ lb-in.} \qquad \textbf{3.2-14}$$

Note that Q must be a clockwise torque, as viewed from A to B; this follows from the positive sign of Q in our analysis, the sense in which the $\mathbf{Q}$ vector was taken in the free-body diagram, and from the right-hand rule.   Note also that if left-handed coordinate axes had mistakenly been employed, we would have got a numerically incorrect answer for Q.

Returning now to the bearing reactions, $\mathbf{E}$ is readily found from the moment equation and then $\mathbf{D}$ from the force equation.   From the moment equation we have

$$\mathbf{i} \times \mathbf{E} = 17.3\mathbf{k}.$$

This makes it obvious that $\mathbf{E}$ is a vector $17.3\mathbf{j}$, but if the right-hand side of the above expression contained a $\mathbf{j}$-component as well as a $\mathbf{k}$-component the determination of $\mathbf{E}$ would be less transparent.   We can, of course, always write $\mathbf{E} = E_y\mathbf{j} + E_z\mathbf{k}$ and use the scalar forms of the vector expression. But it is neater to utilize vector algebra:

$$(\mathbf{i} \times \mathbf{E}) \cdot \mathbf{j} = 17.3\mathbf{k} \cdot \mathbf{j} = 0,$$

$$(\mathbf{j} \times \mathbf{i}) \cdot \mathbf{E} = -\mathbf{k} \cdot \mathbf{E} = -E_z = 0.$$

This determines one of the components of $\mathbf{E}$.   The other is found similarly:

$$(\mathbf{i} \times \mathbf{E}) \cdot \mathbf{k} = 17.3\mathbf{k} \cdot \mathbf{k},$$

$$(\mathbf{k} \times \mathbf{i}) \cdot \mathbf{E} = 17.3,$$

$$\mathbf{j} \cdot \mathbf{E} = E_y = 17.3.$$

The force $\mathbf{E}$ is, therefore,

$$\mathbf{E} = 17.3\mathbf{j} \text{ lb.} \qquad \textbf{3.2-15}$$

The force equation now yields $\mathbf{D}$:

$$\mathbf{D} = 10\mathbf{j} - \mathbf{E} = -7.3\mathbf{j} \text{ lb.} \qquad \textbf{3.2-16}$$

A final word: we are asked to find the forces exerted *on the bearings*. In the above equations the symbols **D** and **E** represent forces exerted by the bearings on the shaft. We conclude that the force exerted by the shaft on the bearing at D will be 7.3 lb in the direction of the positive y-axis (i.e., upward) and that the force exerted by the shaft on the bearing at E will be 17.3 lb in the direction of the negative y-axis (i.e., downward).

Example 3.2-6

An open-ended cylindrical tube of radius R and weight W', containing two smooth spheres of radius r and weight W, is placed on a smooth table with its axis vertical. As shown in Fig. 3.2.13a, $\frac{1}{2}R < r < R$. Show that this arrangement will tip over if r/R is less than $1 - (W'/2W)$.

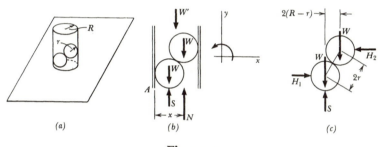

(a)   (b)   (c)

Fig. 3.2-13

Solution: We take as our system the tube and the two spheres. The free-body diagram of this system is shown in Fig. 3.2-13b. Note that this view is a cross-section containing the axis of the cylinder and the centers of the two spheres. The system is subject to the weight forces W' and W; since the mass center of the tube is on its axis, W' is shown acting along the vertical axis of the tube. The tube also comes in contact with the table that supports it. This contact is made all around the base. Its resultant is an upward force, N, acting at an unknown distance, x, from the left-hand edge, A, of the cylinder. Since the table is supposed to be smooth, there is no horizontal component. The system also comes in contact with the table at the point where the lower sphere touches the table. Here there is a normal contact force denoted S. The equations of equilibrium for the system are

$$\sum F_y = S + N - W' - 2W = 0,$$

$$\sum M_A = Nx + Sr - W'R - Wr - W(2R - r) = 0. \qquad \textbf{3.2-17}$$

These are the only non-trivial equations of equilibrium because there are no horizontal forces. Since there are three unknowns, S, N, and x, we cannot hope to find them from these two equations. We therefore pass on to a consideration of one of the possible subsystems. Here we are faced with a choice. Shall we isolate the upper sphere, the lower sphere, or the two of them? In the long run it does not matter what choice we make because a systematic investigation of the subsystems will always lead to the same conclusion. But human analysis is fallible, and it therefore pays to take the most direct path. Our choice is guided by the following considerations. Since the spheres are smooth, the forces acting on each will all intersect at the center of that sphere; therefore we will get only two non-trivial equations of equilibrium for each sphere. But, with each sphere, there are two new unknown forces, the force exerted by the other sphere and the force exerted by the cylinder; therefore neither sphere, taken separately, will immediately yield a new relation between S, N, and x. Since we must eventually consider both spheres, it will be best to do so at once. The free-body diagram for the pair of spheres is shown in Fig. 3.2-13c. We see at a glance that $S = 2W$. Returning to the force equation for the original system, we find $N = W'$. Substituting these values of S and N into the moment equation, we find

$$x = R + \frac{2W}{W'}(R - r). \qquad\qquad \textbf{3.2-18}$$

The analysis now requires interpretation. Will the arrangement be stable or will it tip? In order for the arrangement to be in equilibrium, the reaction N must lie somewhere inside the cross-section of the cylinder; that is, x must be greater than zero and less than $2R$. Now the largest value of r produces the smallest value of x. This largest value is $r = R$ (if r is any bigger the spheres won't fit in the cylinder); therefore the minimum value of x is $x = R$. This is certainly greater than zero and less than $2R$, so there is no danger of the cylinder tipping over to the left. On the other hand, x will be greater than $2R$ if

$$\frac{2W}{W'}(R - r) > R. \qquad\qquad \textbf{3.2-19}$$

Since both sides of this inequality are positive, we may multiply them by any positive number without altering the inequality. Multiplying by $W'/2WR$ we have

$$1 - \frac{r}{R} > \frac{W'}{2W}.$$

Again, both sides of the inequality are positive numbers. If we multiply them each by the same negative number, we reverse the sign of the inequality. Multiplying by -1, we have

$$\frac{r}{R} - 1 < -\frac{W'}{2W}.$$

And now adding $+1$ to each side we see that x will be greater than $2R$ if

$$\frac{r}{R} < 1 - \frac{W'}{2W}. \qquad\qquad \textbf{3.2-20}$$

The cylinder will tip over to the right if this condition is fulfilled. In particular, we see that if the tube is light enough it will always tip. That is, for a given r, R, and W, we can always pick W' small enough so that $1 - \dfrac{W'}{2W}$ is as close to one in value as we wish and hence greater than r/R.

A different insight into the conditions for tipping is gained by studying the forces exerted by the spheres on the cylinder wall. This is left to the exercises for the reader.

3.3 Analysis of Trusses

In this section and the next, methods previously developed for the analysis of statically determinate structures are applied in areas of technological importance, i.e., the determination of the loads carried by truss structures and frames. This analysis is an essential concomitant of design: once the loads to be carried are known, the

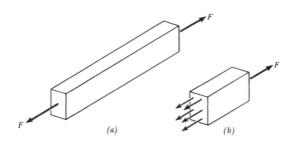

Fig. 3.3-1

member can be proportioned to bear them safely. We begin with the simplest and most commonly encountered of all structures constructed of subunits, the *truss*. A truss is a structure, either planar or three-dimensional, made of members each of which is a two-force body. That is, each member of the truss is attached to other members, or to the foundation, at two and only two points, and each is loaded only at those points. Furthermore, the term is restricted to structures that may be considered to be rigid when all the members are treated as rigid bodies. A truss is therefore a special kind of frame. It is not a mechanism.

In practice the members or subunits of a truss, known as *bars*, are long and straight, being made up, in the case of metal trusses, of lengths of the common rolled sections—channels, angles, and wide-flange sections, with or without cover plates. Forces (and not couples) are applied to the bars at their ends, which are called *joints*.

Since the bar is a two-force member, these forces must be equal in magnitude and opposite in direction. There are two possible situations that may arise, depending on whether the forces applied to the bar are directed toward or away from the middle of the bar. The latter case is pictured in Fig. 3.3-1a, the forces being denoted by the symbol F. If we pass a section through the bar and construct the free-body diagram of either of the two parts thus created, the action of the part removed on the part being considered consists in a stress distributed over the cross-section, as shown in Fig. 3.3-1b. This distributed force is equipollent to a concentrated force equal in magnitude and opposite in direction to the force F applied to the uncut end. We say that the bar is carrying a tensile load of magnitude F. In the other possible case the forces F are directed toward the middle of the bar and we say that the bar carries a compressive load of magnitude F. Some engineers make a practice of assuming all initially unknown forces to correspond to tensile loads; then, if analysis shows them to have negative magnitude, the minus sign serves to indicate compression.

The simple nature of the distributed force acting on any cross-section of a truss bar serves to explain the structural effectiveness of this type of construction as well as to facilitate the determination of the loads carried by each member. Because the distributed forces on any cross-section are, to a close approximation, uniform in their distribution over the cross-section (at least away from the ends of the bar), the bar is simply stretched or compressed. If the bar were not straight, or if a load were applied at a third point of the bar in a direction transverse to the axis of the bar, the bar would bend as well as stretch. The distributed forces acting on any cross-section would be equipollent to a force and a couple instead of simply to an axial force. This situation has been described in Section 2.7, and will be discussed further in Section 4.3, where the forces in beams are discussed. For the present we note that the designer of a truss is at some pains to ensure that loads shall be applied to the bars only at their ends and that these loads shall not consist of couples of any appreciable magnitude. To apply loads only at the joints, the designer arranges, where necessary, for a subsidiary system of floor beams and stringers to carry load to the joints, rather than allow this load to be applied directly to the bars. Of course the dead weight of the truss members themselves cannot be carried in this way. However, since the truss is intended to carry a load other than its own weight, the dead load of any member is usually small enough to permit it to be taken into account, where necessary, by concentrated forces equal to half the weight of the member applied at each end of the member. The joints themselves are sometimes known as *pins*, a nomenclature that survives from days when the members were solid bars of rectangular cross-section with enlarged ends (eye-bars) in which holes were drilled for steel rods. The rods, or pins, passed through such holes in all the members meeting at a common joint. If the pins had been smooth and the holes a loose fit, they would indeed have prevented the application of any couple to the ends of the members. The contact pressures used, however, were so high that these joints never were the smooth, freely turning joints that the theory presupposes. Today most joints are made by riveting, welding, or bolting the adjacent members to a common gusset plate. We can still treat the joint as a smooth pin, however, because the member is so long,

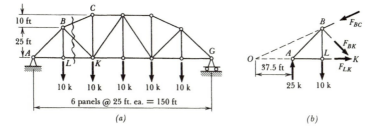

Fig. 3·3-2

compared with its lateral dimensions, that the connection is able to exert little restraint against rotation. Loads determined on these assumptions are known as *primary stresses*. They form the usual basis of truss design. Where unusually short, stubby members are encountered, it is sometimes necessary to add to the primary stresses the *secondary stresses* due to rotational restraint at a joint. This task, however, would lead us beyond the scope of the present text.

We now have a picture of a load-carrying framework consisting of a number of straight two-force members connected together at their ends by joints that may be regarded as smooth pins. How shall we find the loads carried by the members? The basic method for finding the load carried by any particular member is to pass a curve (a surface for space trusses), called a *section*, through the truss. The section cuts that member (and as few others as possible) and divides the truss into two separate parts. The load carried by the member under investigation then appears as an external force in the free-body diagram of either part of the truss, and is found from the equations of equilibrium.   As a typical instance of this procedure we may consider the determination of the load carried by the diagonal bar BK of the planar bridge truss shown in Fig. 3.3-2a.   (This type of truss is known as a Warren truss with a curved upper chord.)   The truss carries a load of 10,000 lb = 10 k at each joint or *panel point* of the lower chord.   The end reactions are found by considering the equilibrium of the truss as a whole.   In the present case they consist of 25 k upward reactions at A and G.   The section passing through BK and separating the truss into two parts is shown in Fig. 3.3-2a by means of a wavy line.   We may take as our system either the part of the truss to the left or the part to the right of this section.   Clearly it will be simpler in this instance to isolate the part to the left, since fewer forces will then enter the analysis.   The free-body diagram of this section is shown in Fig. 3.3-2b.   Note that because of our assumptions concerning the two-force nature of the members, the action of the part of the truss to the right of the section on the part under consideration is represented completely by three forces of unknown magnitude whose directions are those of the cut members.   We find any one of these forces most efficiently by writing the moment equation about the point of intersection of the other two.   In the case of F_{BK} we take moments about point O:

$$\sum M_O = -\left(\frac{\sqrt{2}}{2} F_{BK}\right)(87.5) + (25)(37.5) - (10)(62.5) = 0,$$

$$F_{BK} = 5.05 \text{ k.} \qquad\qquad\qquad\qquad \textbf{3.3-1}$$

The diagonal member BK therefore carries a load of 5050 lb tension. It is easy to see that the load is tension and not compression: a glance at the free-body diagram shows that F_{BK} is tending to stretch the member, not to compress it.   To secure such ease of interpretation it is important that the section be drawn through members and not through the joint.   Each of the three unknown forces appearing in the free-body diagram can be obtained by means of a single equation. For example, the load carried by the lower chord member LK is found by writing the equilibrium equation of moments about point B:

$$\sum M_B = -(25)(25) + (25)(F_{LK}) = 0;$$

$$F_{LK} = 25 \text{ k tension.} \qquad\qquad\qquad \textbf{3.3-2}$$

There are two types of section that may be used in analyzing a truss: those that completely surround a single joint and those that, like the section of the previous paragraph, do not.   To find the load carried by member *LK* of Fig. 3.3-2a, we could start, for example, by taking a section enclosing joint *A*.   Such a section is shown by wavy lines in Fig. 3.3-3a, and the free-body diagram of the part of the truss enclosed is shown in Fig. 3.3-3b.

The equations of equilibrium are

$$\sum F_x = F_{AL} - 0.707 F_{AB} = 0$$

and   **3.3-3**

$$\sum F_y = 25 - 0.707 F_{AB} = 0,$$

whence we conclude that

$$F_{AB} = 35.3 \text{ k compression}, \qquad F_{AL} = 25 \text{ k tension} \qquad \textbf{3.3-4}$$

Next we take a section completely enclosing the joint *L*.   It is shown by a wavy line in Fig. 3.3-3a, and the free-body diagram of the part of the truss enclosed is shown in Fig. 3.3-3c.   Since we now know that

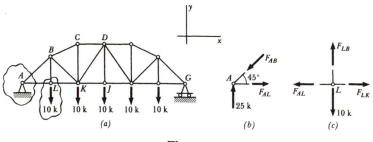

Fig. 3.3-3

$F_{AL} = 25$ k tension, the summation of forces in the *x*-direction tells us at once that $F_{LK} = 25$ k tension.   This, of course, is the same conclusion reached previously in Eq. 3.3-2.   The process of analysis in which one proceeds, joint by joint, through the truss, each time taking a section that completely encloses a joint, is sometimes known as the "method of joints."   The alternative procedure, using sections such as that shown in Fig. 3.3-2 which do not completely enclose a single joint, is then called the "method of sections."   This distinction seems unnecessary, since both "methods" are variations of the same basic procedure.   For the beginner it may be appropriate to remark that the "method of joints" has certain drawbacks.   If we want to find the load carried by a member of the truss remote from a suitable starting joint such as *A* in Fig. 3.3-3, we have to proceed, joint by joint, through the truss until we come to the member of interest. During this lengthy process a good deal of extraneous geometry and algebra is apt to be encountered. An error along the way perpetuates itself. Even ordinary round-off and slide rule error tends to be cumulative. For these reasons it is usually better to pass a section through the member of interest directly. On the other hand, the "method of joints" will sometimes provide a quick answer. A section surrounding joint J, for example, tells us at once that member DJ of Fig. 3.3-3 must carry a tensile load of 10,000 lb. In practice, a judicious combination of the two "methods" is most effective.

We now have a procedure for finding the loads carried by members of a truss. Will this method of analysis always serve? The question is closely related to the question of statical determinacy and is therefore of no little interest. In attempting to answer it we look first at the over-all equilibrium of a rigid planar truss. The complete truss is attached to a foundation and, when the truss as a whole is taken as an isolated system, the forces exerted by the foundation must be represented by certain unknown vectors. Let the number of unknown independent components—say the rectangular cartesian components—of these forces be denoted by the symbol C. For example, in Fig. 3.3-3 there are two unknown foundation force components at A and one at G, so that $C = 3$. Now suppose that there are B bars in the truss. The complete statical analysis of the truss will require the determination of the load carried by each bar. There will be $B + C$ unknowns in all. If we imagine the truss to be analyzed by taking sections that surround each of the J joints of the truss, we will get a set of J free-body diagrams in each of which the forces are planar and concurrent. There will therefore be $2J$ independent equations of equilibrium available. (The equilibrium equations obtained by considering the truss as a whole are simply appropriate linear combinations of the corresponding ones for the J joints and therefore do not furnish any further independent equations.) Now consider the number N where

$$N = B + C - 2J. \qquad \textbf{3.3-5}$$

If N is positive, there are more unknowns than equations and the truss is statically indeterminate. It may have more bars than are necessary for rigidity, in which case it is said to have one or more *redundant* members or to be *internally redundant*, or it may have too many foundation constraints, in which case it is said to be *externally redundant*. In any event, when $N > 0$, while it may be possible to find some of the foundation forces and some of the bar loads, they cannot all be determined by means of the equations of equilibrium alone.

If N is negative there are more equilibrium conditions than there are quantities at our disposal to satisfy them.   The structure is then a mechanism.   It may move as a whole (if $C < 3$), or parts of it may move relative to other parts, but it will not be in equilibrium, except possibly for special configurations.   Finally, if N is zero the number of unknowns is exactly equal to the number of equations of equilibrium. If a planar truss is to be statically determinate—what is sometimes known as a *just-rigid* truss—it must have $N = 0$.   On the other hand, the fact that $N = 0$ does not ensure that the truss is statically determinate.   There exist what are known as *critical forms* in which the equations of equilibrium are inconsistent with any finite set of loads.

These matters deserve some amplification.   The basic planar truss element is a triangle and the simplest possible truss is that shown in Fig. 3.3-4a.   In this case $N = 0$.   The structure is statically determinate; both the reactions and the bar loads can be found by

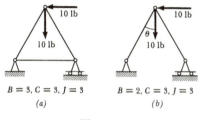

$B = 3, C = 3, J = 3$ $B = 2, C = 3, J = 3$

(a)   (b)

Fig. 3.3-4

means of the equations of equilibrium.   If the horizontal bar is removed, as in Fig. 3.3-4b, $N = -1$.   The structure is now a mechanism; it can only be in equilibrium for particular values of the angle θ. If the right-hand roller support of (a) is replaced by a pin, like the left-hand support, C becomes 4 and the truss is statically indeterminate.   Now if we start with the original statically determinate truss of Fig. 3.3-4a and add to it another triangular element obtained by connecting some point in the plane to two points of the original triangle, the new structure has two more bars and one more joint, so that N remains zero.   Any truss that can be regarded as built from an original triangular element by successive additions of two bars with a common joint will be statically determinate once its foundation reactions are known.   Such trusses are called *simple trusses*.   The Warren truss of Figs. 3.3-2 and 3.3-3 is an example of a truss of this kind.   On the other hand, if we connect two triangular elements by means of three parallel members, as shown in Fig. 3.3-5, we get what was referred to in the last sentence of the previous paragraph as a *critical form.* Although $N = 0$ the truss is statically indeterminate. We can recognize the possibility of self-straining fairly easily in this case because the center horizontal bar could be made to carry a tensile load (say by being fabricated too short) and the upper and lower horizontal bars could carry compressive loads half as great without disturbing the equilibrium of the truss as a whole. Such a state of load could be superposed on any other set that satisfied the equations of equilibrium. If a member of a statically determinate truss such as the triangular truss of Fig. 3.3-4a is fabricated too long or too

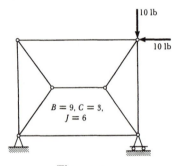

Fig. 3.3-5

short, the geometry of the truss changes slightly but no initial state of load is induced. For this reason, and because, as will be shown, bar loads in trusses of near-critical form are extremely high (theoretically infinite), engineers invariably avoid such forms. The simple truss built up of all-triangular elements is free of this possible design defect.

The analysis of a three-dimensional truss follows the same lines as that of the planar truss. In investigating statical determinacy, as there are now $3J$ independent equations of equilibrium, we take

$$N' = B + C - 3J. \qquad \textbf{3.3-6}$$

Again, B is the number of bars, J the number of joints, and C the number of external constraints; the critical number for the latter is six, since a rigid body can have six degrees of freedom in three-dimensional motion. If N' is less than zero, the truss is a mechanism; if N' is greater than zero, the truss is statically indeterminate. If the truss is statically determinate, $N' = 0$. Just as the triangle is the basic unit for a planar truss, so the tetrahedron or six-bar truss is the basic statically determinate spatial truss. Spatial trusses that can be regarded as built up from an original tetrahedral unit by successive additions of three non-coplanar bars with a common joint will be statically determinate.

Returning now to the planar truss, we see that in order for it to be statically determinate, the number of bars, B, joints, J, and

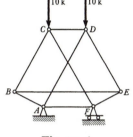

Fig. 3.3-6

foundation constraint unknowns, C, must be related by the equation $B + C = 2J$. When this condition is fulfilled, there may still be difficulty in the analysis should we encounter a situation where there is no two-member joint and in which no section separating the truss into two parts and cutting three non-parallel bars can be found.

Consider, for example, the non-simple truss shown in Fig. 3.3-6. In this figure the joints are, as is conventional, indicated by small circles; where these are not present the bars overlap without being connected. Though admittedly a textbook example, this truss does present awkward features in its analysis. Since $B = 9$, $C = 3$, and $J = 6$, the condition $B + C = 2J$ is fulfilled; $N = 0$. The foundation reactions at A and R are easily found by our considering the equilibrium of the truss as a whole. But there is no joint at which only two bars are attached and no section through the truss cutting only

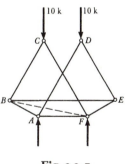

Fig. 3.3-7

two members. Neither the "method of sections" nor the "method of joints" can be applied directly. The simplest way to treat this situation is to remove one of the bars—say CD—and keep the truss just rigid by introducing a new bar, say BF. The situation is then as shown in Fig. 3.3-7, with the new bar shown by a broken line. The truss of Fig. 3.3-7 is called a *replacement truss* for that of Fig. 3.3-6. Now the analysis is easily carried out—say, by taking sections surrounding joints C, D, E, F, and A in that order. The resulting loads may be denoted T'_{DE}, $T'_{BE}, \ldots$, including T'_{BF}, using the convention that tension is a positive load. Next, consider the replacement truss of Fig. 3.3-7 with oppositely directed unit loads, applied at the joints where a bar has been removed, in the direction of the bar, and tending to bring the two loaded points together. This loading is shown in Fig. 3.3-8. Again we easily find the corresponding bar loads T''_{DE}, $T''_{BE}, \ldots$, including T''_{BF}. Finally, consider the replacement truss simultaneously carrying the external loads shown in Fig. 3.3-7 and those of Fig. 3.3-8, the latter being increased by a factor, x. Since the equations of equilibrium are linear, the loads carried by the members will be $T'_{DE} + xT''_{DE}$, $T'_{BE} + xT''_{BE}, \ldots$, including $T'_{BF} + xT''_{BF}$.

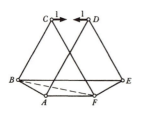

Fig. 3.3-8

That is, the solution for the combined loading is a linear superposition of the solutions for the separate loadings. If x is now given the value $-(T'_{BF}/T''_{BF})$, the member BF will carry no load. It may therefore be removed from the truss. We have evidently restored the original truss of Fig. 3.3-6 with the member CD replaced by forces of magnitude x. We conclude that the actual loads carried by the members of the truss of Fig. 3.3-6 are

$$T_{DE} = T'_{DE} - (T'_{BF}/T''_{BF})T''_{DE},$$

$$T_{BE} = T'_{BE} - (T'_{BF}/T''_{BF})T''_{BE}, \text{ etc.,} \qquad \textbf{3.3-7}$$

and finally

$$T_{CD} = -T'_{BF}/T''_{BF}.$$

This procedure will serve to complete the analysis in those cases where the ordinary methods break down. For the reasons previously discussed, such cases are not often encountered in practice. The technique is of value because it at once reveals the presence of a critical form. Such a form must be associated with a case in which the "method of sections" and the "method of joints" are inapplicable and for which the method of the present paragraph is appropriate. If T''_{BF}, the load carried by the replacement bar when the removed bar is replaced by unit forces, should be very small, the stresses given by Eq. 3.3-7 will be extremely large. When $T''_{BF} = 0$, we have a critical form. When a truss of critical form is actually fabricated, using normal construction materials, it distorts under a small load until the shape is no longer quite critical. The loads carried by the members are then large but finite. For obvious reasons, critical truss forms are avoided in good engineering design.

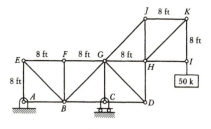

Fig. 3.3-9

Example 3.3-1

Find the force in members DG and CD of the pin-jointed simple truss of Fig. 3.3-9, carrying a 50,000 lb load at I.

Solution: A combination of the two basic methods is useful here. Indeed, we need not solve for the reactions at *A* and *C* at all, and even if the truss were overconstrained, we could still solve for the information desired.

A cut to the right of *GC* divides the truss into two parts; a free-body diagram of section *DHJKI* is given in Fig. 3.3-10. There are four unknowns, and we cannot expect to solve the problem. However, F_{CD} can be determined immediately since the lines of action of the other three unknowns pass through *G*. Therefore,

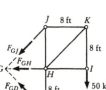

$$\sum M_G = -8F_{CD} - (16)(50) = 0,$$

$$F_{CD} = -100 \text{ k} = 100 \text{ k } (C). \qquad \textbf{3.3-8}$$

The other two equations for the section do not permit a complete determination of the other three unknowns. Vertical equilibrium does give us F_{GD} in terms of F_{GJ}:

Fig. 3.3-10

$$F_{GD}\frac{\sqrt{2}}{2} - F_{GJ}\frac{\sqrt{2}}{2} - 50 = 0,$$

$$F_{GD} = F_{GJ} + 50\sqrt{2}.$$

To find F_{GJ}, use the method of joints starting at I, and then proceeding to K and J. The free-body diagrams are shown in Fig. 3.3-11. For I, we have

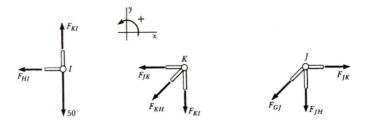

Fig. 3.3-11

$$\sum R_y = F_{KI} - 50 = 0, \qquad F_{KI} = 50\,\text{k }(T);$$

for K, we have

$$\sum R_y = -F_{KI} - F_{KH}\frac{\sqrt{2}}{2} = 0,$$

$$\sum R_x = -F_{JK} - F_{KH}\frac{\sqrt{2}}{2} = 0,$$

and

$$F_{JK} = -F_{KH}\frac{\sqrt{2}}{2} = F_{KI} = 50\,\text{k }(T);$$

for J, we find

$$\sum R_x = F_{JK} - F_{GJ}\frac{\sqrt{2}}{2} = 0,$$

$$F_{GJ} = F_{JK}\sqrt{2} = 50\sqrt{2}\,\text{k }(T).$$

Therefore, bar DG carries a load

$$F_{GD} = F_{GJ} + 50\sqrt{2} = 100\sqrt{2}\,\text{k }(T). \qquad \textbf{3·3-9}$$

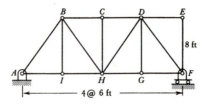

Fig. 3.3-12

Example **3.3-2**

The plane truss of Fig. 3.3-12 is subjected to two different load systems: (a) downward loads of 100 lb at joints I, C, and G; (b) downward loads of 100 lb at joints B, H, and D. What are the constraint loads at A and F? What are the loads carried by the members BI, CH, DG, and GH in the two cases?

Solution: The free-body diagrams of the whole truss for the two loadings are shown in Fig. 3.3-13. The external constraints are the same for both cases, the over-all equations of equilibrium being

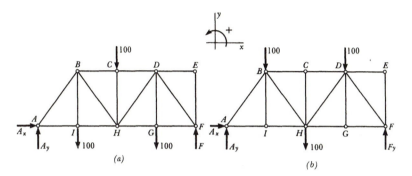

Fig. 3.3-13

$$\sum R_x = A_x = 0,$$

$$\sum M_A = 24F_y - 600 - 1200 - 1800 = 0, \qquad \textbf{3.3-10}$$

$$\sum M_F = -24A_y + 1800 + 1200 + 600 = 0,$$

with solutions

$$A_x = 0, \qquad A_y = F_y = 150 \text{ lb}. \qquad \textbf{3.3-11}$$

The forces carried by the individual struts may be different, however. Examine joints I, C, and G for the two cases. For vertical equilibrium, we must have bars BI and DG carrying 100 lb tensile loads and bar CH carrying a 100 lb compressive load in case (a); all three bars must carry zero load in case (b). Bar GH carries the same load in both cases. Isolate section DEFG and take moments about D (see Fig. 3.3-14); in both cases,

$$\sum M_D = 6F_y - 8F_{GH} = 0,$$

and

$$F_{GH} = \frac{3}{4} F_y = 112.5 \text{ lb } (T).$$

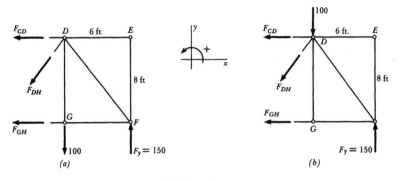

Fig. 3.3-14

Recognizing that certain bars carry zero load often simplifies the analysis. For instance, bars *DE* and *EF* of the truss for both the loadings given here carry zero load, as we can see from equilibrium of joint *E*. The basic theorems for plane trusses are:

    (1) If two bars that are not collinear meet at an unloaded joint, then neither bar can carry load.

    (2) If three bars, two of which are collinear, meet at an unloaded joint, then the third bar oblique to the line of the other two must carry no load.

Similar three- and four-bar, unloaded-joint theorems can be stated for space trusses.

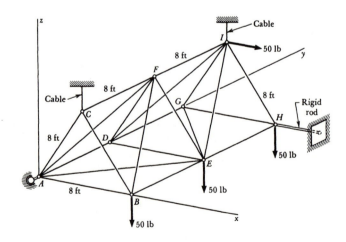

Fig. 3.3-15

It should not be thought that such zero-load bars are not needed in the truss; they generally are, to keep the structure rigid. Removal of BI in the truss of Fig. 3.3-13b will turn the structure into a linkage mechanism.

Example **3.3-3**

Find the forces of constraint on the space truss of Fig. 3.3-15 *and the load carried by bar BF.*

Solution: A free-body diagram of the whole truss is given in Fig. 3.3-16. To determine the forces of constraint, we first take moments about A:

$$\sum \mathbf{M}_A = (4\mathbf{i} + 4\sqrt{3}\mathbf{k}) \times T_1\mathbf{k} + (4\mathbf{i} + 16\mathbf{j} + 4\sqrt{3}\mathbf{k}) \times (50\mathbf{i} + T_2\mathbf{k})$$
$$+ 8\mathbf{i} \times (-50\mathbf{k}) + (8\mathbf{i} + 8\mathbf{j}) \times (-50\mathbf{k}) + (8\mathbf{i} + 16\mathbf{j}) \times (-H\mathbf{i} - 50\mathbf{k})$$
$$= (16T_2 - 1200)\mathbf{i} + (1200 + 200\sqrt{3} - 4T_1 - 4T_2)\mathbf{j} + (16H - 800)\mathbf{k}$$
$$= 0. \qquad\qquad \textbf{3.3-12}$$

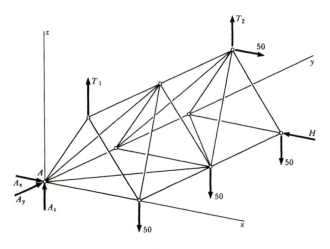

Fig. 3.3-16

Therefore,

$$\sum M_x^A = 16T_2 - 1200 = 0, \qquad T_2 = 75 \text{ lb};$$
$$\sum M_y^A = 1200 + 200\sqrt{3} - 4T_1 - 4T_2 = 0, \quad T_1 = 225 + 50\sqrt{3} \cong 312 \text{ lb};$$
$$\sum M_z^A = 16H - 800 = 0, \qquad H = 50 \text{ lb}. \qquad \textbf{3.3-13}$$

To determine the forces at A, we write the resultant force equilibrium equation:

$$\sum \mathbf{R} = (A_x\mathbf{i} + A_y\mathbf{j} + A_z\mathbf{k}) + T_1\mathbf{k} + T_2\mathbf{k} - 150\mathbf{k} + 50\mathbf{i} - H\mathbf{i}$$
$$= (A_x + 50 - H)\mathbf{i} + A_y\mathbf{j} + (A_z + T_1 + T_2 - 150)\mathbf{k} = 0. \quad \textbf{3.3-14}$$

Therefore,

$$\sum R_x = A_x + 50 - H = 0, \qquad A_x = H - 50 = 0;$$

$$\sum R_y = A_y = 0;$$ 3.3-15

$$\sum R_z = A_z + T_1 + T_2 - 150 = 0,$$

$$A_z = 150 - (T_1 + T_2) = -150 - 50\sqrt{3} \cong -237 \text{ lb.}$$

For determining the force carried by strut BF, the "method of joints" does not suggest itself as a likely procedure. We would like to start at a joint with only three unknowns; in this case, joint H would be the only place to start. Then joint G could be considered, followed by joints

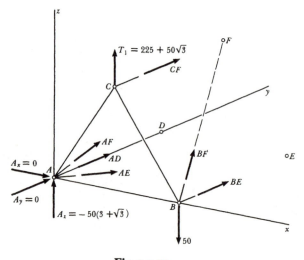

Fig. 3.3-17

I, D, E, and finally F. The "method of sections" can be used by cutting to the positive y-side of joints A, B, and C. The free-body diagram is shown in Fig. 3.3-17; there are six unknowns, not all of which intersect at the same point or intersect the same line, so that we should be able to determine BF. In fact, only one equation need be written, that for moments about the y-axis through A. Choosing A as base point eliminates three of the unknowns, AD, AE, and AF; and taking only the y-component of $\sum \mathbf{M}_A$ eliminates CF and BE, which are parallel to that axis. Since the position of F relative to B is $-4\mathbf{i} + 8\mathbf{j} + 4\sqrt{3}\mathbf{k}$ feet, the vector for the strut force is

$$(BF)\left(-\frac{\sqrt{2}}{4}\mathbf{i} + \frac{\sqrt{2}}{2}\mathbf{j} + \frac{\sqrt{6}}{4}\mathbf{k}\right). \qquad 3.3-16$$

Thus

$$\sum M_y^A = -4T_1 + (8)(50) - (8)\left(\frac{\sqrt{6}}{4}\right)(BF)$$

$$= -(900 + 200\sqrt{3}) + 400 - 2\sqrt{6}(BF) = 0,$$

and

$$BF = -\frac{500 + 200\sqrt{3}}{2\sqrt{6}} \cong 173 \text{ lb } (C). \qquad \textbf{3.3-17}$$

The student may wish to compute the rest of the bar forces on this section to gain practice in the use of three-dimensional equilibrium equations.

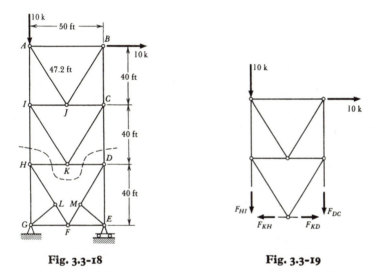

Fig. 3.3-18   **Fig. 3.3-19**

Example **3.3-4**

Find the load carried by members CD and KI of the truss shown in Fig. 3.3-18.

Solution: Teaching experience shows that students have difficulty finding where to start the analysis of *K*-type bracing such as that shown. Yet the structure is statically determinate ($B = 23$, $C = 3$, $J = 13$) and the method of analysis is conventional. Perhaps the mental block arises from always tending to think in terms of a straight section when, in fact, no such restriction is implied by the theory. If we take the section indicated by the dotted line, and draw the free-body diagram (Fig. 3.3-19) of the material above it, we have, on taking moments about point *H*,

$$\sum M_H = -50F_{DC} - 800 = 0, \qquad F_{DC} = -16 \text{k}, \qquad \textbf{3.3-18}$$

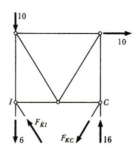

Fig. 3.3-20

so the load carried by DC is 16,000 lb compression. If moments are taken about D,

$$\sum M_D = 50F_{HI} + 500 - 800 = 0, \qquad F_{HI} = 6 \text{ k}. \qquad \textbf{3.3-19}$$

The load carried by HI is 6000 lb tension.   It is now easy to find the load carried by the diagonal KI.   A horizontal section through the middle panel yields the free-body diagram shown in Fig. 3.3-20.   Taking moments about point C, we find

$$\sum M_C = -F_{KI}\left(\frac{40}{47.2}\right)(50) + 800 - 400 = 0,$$

$$F_{KI} = 9.43 \text{ k}. \qquad \textbf{3.3-20}$$

Member KI carries a compressive load of 9430 lb.

Example 3.3-5

Find the foundation reactions for the Wichert truss shown in Fig. 3.3-21.

Solution: The Wichert truss is a designer's attempt to obtain the advantages of a continuous structure without the disadvantages of statical indeterminacy.   It obviously does have the extra stiffness that comes from

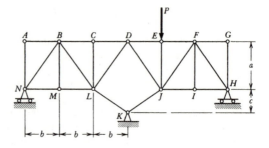

Fig. 3.3-21

having the left-hand span assist in carrying load applied to the right-hand span (and vice-versa). In this respect it is superior to two trusses, one spanning the gap from N to K and the other that from K to H. If the foundation reactions can be found, the truss will be statically determinate

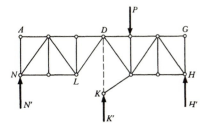

Fig. 3.3-22

and the bar loads easily found. We find that $B = 24$, $C = 4$, and $J = 14$, so $B + C = 2J$, and unless the truss is a critical form it will indeed be statically determinate.

The analysis can be performed in several ways. If we use the replacement-truss procedure described in the text immediately before Example 3.3-1, we will remove one bar, say LK, and add one, say KD, in such a way as to leave the structure just rigid and easily analyzed. When this has been done the truss has the appearance shown in Fig. 3.3-22; the new bar

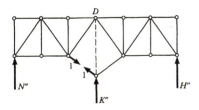

Fig. 3.3-23

is represented by a broken line. The reactions are easily found by considering two bodies, $ANLD$ and $DKHG$:

$$N' = 0, \qquad K' = \frac{2}{3}P, \qquad H' = \frac{1}{3}P. \qquad\qquad \textbf{3.3-21}$$

Also, the load in the replacement bar is

$$T'_{DK} = -\frac{2}{3}P. \qquad\qquad \textbf{3.3-22}$$

Next we remove the load P and insert unit forces at K and L in the direction of the missing bar, as shown in Fig. 3.3-23. The directions of the unit forces are always taken so as to pull the points to which they are applied together. Now if we consider the equilibrium of the left-hand half of the truss and take moments about D, we find

$$N'' = \frac{a+c}{3\sqrt{b^2+c^2}}.$$ 3.3-23

Similarly, by considering the equilibrium of the right-hand side of the truss and taking moments about D, we find

$$H'' = \frac{a+c}{3\sqrt{b^2+c^2}}.$$ 3.3-24

Since $N'' + H'' + K'' = 0$, we at once conclude that

$$K'' = \frac{-2(a+c)}{3\sqrt{b^2+c^2}},$$ 3.3-25

and, on taking a section surrounding joint K, we find that

$$T''_{DK} = \frac{2(a-2c)}{3\sqrt{b^2+c^2}}.$$ 3.3-26

We are now in a position to test for the presence of a critical form. This will occur if and only if $T''_{DK}=0$, that is, when $a=2c$. A truss with these particular proportions will buckle. In any other case, however, the actual state of the truss is a linear combination of the single- and double-primed states. That is,

$$N = N'+xN'', \quad K = K'+xK'', \quad H = H'+xH'',$$ 3.3-27

and

$$T_{DK} = T'_{DK}+xT''_{DK}.$$

But T_{DK} must be made zero because this bar is not present in the original truss, so

$$x = -\frac{T'_{DK}}{T''_{DK}} = P\frac{\sqrt{b^2+c^2}}{a-2c}.$$ 3.3-28

Therefore,

$$N = 0 + \left(P\frac{\sqrt{b^2+c^2}}{a-2c}\right)\left(\frac{a+c}{3\sqrt{b^2+c^2}}\right) = \frac{1}{3}P\frac{a+c}{a-2c},$$

$$K = \frac{2}{3}P - \left(P\frac{\sqrt{b^2+c^2}}{a-2c}\right)\left(\frac{2(a+c)}{3\sqrt{b^2+c^2}}\right) = \frac{2}{3}P\left(1-\frac{a+c}{a-2c}\right),$$ 3.3-29

$$H = \frac{1}{3}P + \left(P\frac{\sqrt{b^2+c^2}}{a-2c}\right)\left(\frac{a+c}{3\sqrt{b^2+c^2}}\right) = \frac{1}{3}P\left(1+\frac{a+c}{a-2c}\right).$$

These are the foundation reactions. We note, incidentally, that

$$T_{LK} = T'_{LK} + xT''_{LK} = 0 + (x)(1) = P\frac{\sqrt{b^2+c^2}}{(a-2c)}.$$

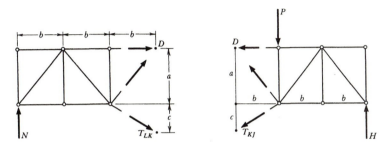

Fig. 3·3-24   Fig. 3·3-25

It is instructive to look at an alternative approach. If we consider the equilibrium of the part of the truss (Fig. 3.3-24) to the left of a vertical section cutting the members *CD, LD,* and *LK,* we find:

$$\sum M_D = -3bN + \frac{b}{\sqrt{b^2+c^2}} T_{LK}(a+c) = 0,$$

$$N = T_{LK}\frac{a+c}{3\sqrt{b^2+c^2}}. \qquad \textbf{3·3-30}$$

Now if we consider the part of the truss (Fig. 3.3-25) to the right of a vertical section cutting members *DE, DJ,* and *KJ,* we find

$$\sum M_D = 3bH - Pb - \frac{b}{\sqrt{b^2+c^2}} T_{KJ}(a+c) = 0,$$

$$H = T_{KJ}\frac{a+c}{3\sqrt{b^2+c^2}} + \frac{1}{3}P. \qquad \textbf{3·3-31}$$

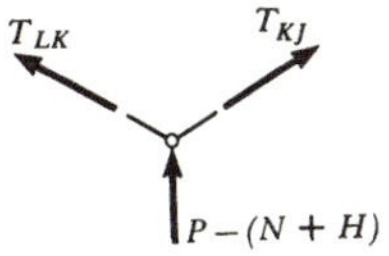

Turning now to a section surrounding the center reaction joint *K,* we have a free-body diagram as shown in Fig. 3.3-26. Note that the upward reaction at *K* has been written as $P-(N+H)$, as is clear from the equilibrium of the truss as a whole. It is clear also from the diagram that

Fig. 3·3-26

summing forces in the x-direction will show that $T_{LK} = T_{KJ}$. Summing forces in the y-direction, we have

$$P - (N+H) + T_{LK} \frac{c}{\sqrt{b^2+c^2}} + T_{KJ} \frac{c}{\sqrt{b^2+c^2}} = 0. \qquad \textbf{3.3-32}$$

If we substitute the expressions for H and N from Eqs. 3.3-30 and 3.3-31 into Eq. 3.3-32 and set $T_{LK} = T_{KJ}$, we find

$$P - T_{KJ} \frac{2(a+c)}{3\sqrt{b^2+c^2}} - \frac{1}{3} P + T_{KJ} \frac{2c}{\sqrt{b^2+c^2}} = 0,$$

$$T_{KJ} = P \frac{\sqrt{b^2+c^2}}{a-2c}. \qquad \textbf{3.3-33}$$

Now it is simple to find H and N:

$$H = \frac{1}{3} P\left(1 + \frac{a+c}{a-2c}\right), \qquad N = \frac{1}{3} P \frac{a+c}{a-2c}. \qquad \textbf{3.3-34}$$

Of course the reaction at K is simply $P - (N+H)$. The results agree with the previous analysis. They hint at the presence of a critical form when $a = 2c$. In this method, essentially, we write all the equations of equilibrium and solve them as simultaneous equations, the solution being aided in the present case by symmetry.

3.4 Analysis of Frames and Mechanisms

Unlike a truss, a general frame or mechanism may include members that are not two-force members. Such a member, even if it is a straight bar, will be subject to flexure and shear as well as to axial tension or compression. The distinction is so fundamental that

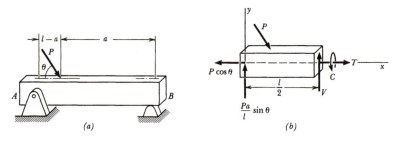

Fig. 3.4-1

it deserves a specific illustration. Consider a bar such as that chosen in Fig. 3.4-1a. The reactions at the supports A and B are readily found from the equations of equilibrium for the bar as a whole. At A they consist of a force $P \cos \theta$ directed to the left and an upward force $Pa \sin \theta / l$. If a vertical section is passed through the middle of the bar, cutting it in two, the free-body diagram of the left-hand side will be as shown in Fig. 3.4-1b. The equations of force equilibrium applied to this left-hand section of the bar are

$$\sum F_x = T + P \cos \theta - P \cos \theta = 0,$$

$$\sum F_y = -P \sin \theta + \frac{Pa}{l} \sin \theta + V = 0. \qquad \textbf{3.4-1}$$

The equation of moment equilibrium can be written (ignoring the bar thickness)

$$\sum M_A = C - P(l-a) \sin \theta + Vl/2 = 0. \qquad \textbf{3.4-2}$$

Here (T, V, C) are the resultant set corresponding to the distributed forces on the cross-section (see Section 2.7). When we solve these equations, we find that

$$T = 0, \qquad V = P \frac{l-a}{l} \sin \theta, \qquad C = \frac{P(l-a)}{2} \sin \theta. \qquad \textbf{3.4-3}$$

The lesson we learn from this illustration is that, when a three-force member is cut by a transverse section, the forces distributed over the cut face are equipollent to a force and a couple, neither of which is, in general, zero. The presence of the couple is, of course, associated with the fact that the bar flexes or bends. In the present example, the resultant force at the particular section is a transverse shear force only, and no axial component exists at the section. In a truss, on the other hand, the members are straight two-force members and only the axial component, T, is present. As a result of this distinction, the simplicity inherent in truss analysis largely disappears when we come to consider frames. There is no longer any point in passing a section through a member so as to isolate some part of the system; such a step would introduce too many new unknown quantities. Where, as is usually the case, it is necessary to consider the equilibrium of a part of the frame, the members of which the part is composed must be isolated intact.

For the foregoing reasons, the analysis of frames follows closely the methods of Section 3.2. A complex system is resolved into a number of constituent parts; the third law of motion is used to reduce the number of unknown forces acting on the parts; finally, force and moment equilibrium equations are solved to yield the desired information. There is no universal guide to the best method of choosing subsystems, though it may be remarked that the beginner usually errs by choosing subsystems that consist of too few elements and thus becomes entangled in a computation in which a large number of subsystems are involved. These are matters best learned by experience gained by working through examples such as those that follow, which deal with some of the standard elements encountered in engineering.

Example **3.4-1**

The four-bar linkage mechanism (Fig. 3.4-2) consists of three identical light, rigid bars pinned smoothly together and to the foundation (which is the "fourth bar"). For what angles θ can the linkage be in equilibrium? What are the forces acting on each member?

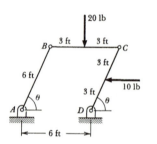

Fig. 3.4-2

Solution: Figs. 3.4-3a, b, c, and d are free-body diagrams of the whole truss and of bars *AB*, *BC*, and *CD*, respectively. Note that *AB* is "light" and therefore is a two-force member, which is assumed to carry a compressive load. Note that in drawing the free-body diagram for bar *AB* we make use of the fact that *AB* is a two-force member to determine the direction of force *P*. Then, when drawing the free-body diagram of bar *BC*, we utilize the third law of motion and make the force at *B* exerted by *AB* on *BC* equal in magnitude and opposite in direction to the force at *B* exerted by *BC* on *AB*. A similar consideration governs the forces at *C* exerted on bar *CD* by bar *BC*. Later on it may turn out that C_x, say, is negative. This would reverse the direction of both forces labeled C_x; but the third law would still be satisfied. We see that in this instance we

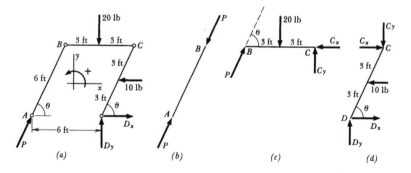

Fig. 3.4-3

shall have six unknowns (P, C_x, C_y, D_x, D_y, and θ) and six non-trivial equations of equilibrium—three from each of the free-body diagrams of Fig. 3.4-3c and d. We have already satisfied the equations of equilibrium for bar AB by our choice of the direction of the force P, and we do not get any further independent equations of equilibrium by considering the free-body diagram for the mechanism as a whole, since these equations are simply linear combinations of those for the three bars. Nevertheless, it is expeditious to begin with the equilibrium equations for the mechanism as a whole (Fig. 3.4-3a):

$$\sum M_A = 6D_y - (3 \sin \theta)(-10) + (6 \cos \theta + 3)(-20) = 0,$$

$$\sum M_D = (-6)(P \sin \theta) + (6 \cos \theta - 3)(-20) - (3 \sin \theta)(-10) = 0, \quad \textbf{3.4-4}$$

$$\sum R_x = D_x + P \cos \theta - 10 = 0.$$

The solutions of these for D_x, D_y, and P as functions of θ are

$$D_y = 10 + 20 \cos \theta - 5 \sin \theta,$$

$$P = 10 \csc \theta - 20 \cot \theta + 5, \quad \textbf{3.4-5}$$

$$D_x = 10 - 10 \cot \theta + 20 \cot \theta \cos \theta - 5 \cos \theta.$$

As a check, note that $\sum R_y = D_y + P \sin \theta - 20 = 0$, as it should.

Turning now to the horizontal member BC (Fig. 3.4-3c), we find

$$\sum M_C = -6P \sin \theta + 60 = 0, \qquad P = 10 \csc \theta. \quad \textbf{3.4-6}$$

Combining this with the second of Eqs. 3.4-5, we determine θ:

$$P = 10 \csc \theta = 10 \csc \theta - 20 \cot \theta + 5,$$

$$\theta = \text{arccot} \, (0.25) \cong 76° \text{ or } 256°. \quad \textbf{3.4-7}$$

There are two possible equilibrium configurations. We find that D_x, D_y, and P are given by

$$\theta = 76°: D_x = 7.5 \text{ lb}, \qquad D_y = 10 \text{ lb}, \qquad P = \frac{10\sqrt{17}}{4} \cong 10.3 \text{ lb};$$

$$\theta = 256°: D_x = 7.5 \text{ lb}, \qquad D_y = 10 \text{ lb}, \qquad P = -10.3 \text{ lb}. \quad \textbf{3.4-8}$$

The last answer means, of course, that AB must be in tension when the linkage is in its second equilibrium position.

We must now determine the components of the force at C. Completing the equilibrium equations for bar BC, we find

$$\sum M_B = 6C_y - 60 = 0, \qquad C_y = 10 \text{ lb};$$

$$\sum R_x = P \cos \theta - C_x = 0, \qquad C_x = 10 \cot \theta = 2.5 \text{ lb}. \quad \textbf{3.4-9}$$

Since these are both positive, whichever equilibrium position has been chosen, we know that the directions assumed on the free-body diagrams are correct.   As a check on our computations, the three equations for bar *CD* may be written.

The answer desired may be given in the following way.   The linkage is in equilibrium under the given loads when $\theta=\text{arccot}$ (0.25), i.e., when $\theta=76°$ or $256°$.   In both of these positions, bar *CD* is subjected to forces 7.5 lb to the right and 10 lb upward at *D*, and to forces 2.5 lb to the right and 10 lb downward at *C*—besides the applied 10 lb load.   The bar *AB* supports a 10.3 lb compressive load at each end for $\theta=76°$, and 10.3 lb tensile loads for $\theta=256°$.   Bar *BC* is subjected to the 20 lb central load; to a 2.5 lb load to the left and a 10 lb upward load at *C*; and to a 10.3 lb load at 76° counterclockwise from *BC* at pin *B* in both cases.

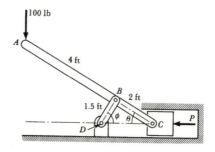

Fig. 3·4-4

Example **3·4-2**

The device shown in Fig. 3.4-4 is to be used as a hand-operated crushing machine.   Ignoring friction, how much crushing force, P, will the device exert when the angle θ is 30 degrees ?

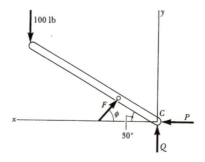

Fig. 3·4-5

Solution: The bar DB is a two-force member. Therefore the free-body diagram for AC has the form shown in Fig. 3.4-5. Note that the force component Q exerted on the bar AC by the crushing block is equal to that exerted on the crushing block by the foundation if the weight of the block is ignored. The angle ϕ is found with the aid of the sine law of trigonometry:

$$\sin \phi = \frac{2}{1.5} \sin 30° = \frac{2}{3}$$

$$\phi = 41.8°, \qquad \cos \phi = \frac{\sqrt{5}}{3}.$$

We find F by taking moments about the origin and then find P from the force equilibrium equations:

$$\sum M_C = -(100)(6)\left(\frac{\sqrt{3}}{2}\right) + \left(\frac{2}{3}F\right)(\sqrt{3}) + \left(\frac{\sqrt{5}}{3}F\right)(1) = 0, \quad F = 274 \text{ lb};$$

$$\sum F_z = P - \frac{\sqrt{5}}{3}F = 0, \qquad P = 204 \text{ lb}. \qquad \textbf{3.4-10}$$

In practice the dimensions and angles would probably be read from a blueprint, rather than determined by trigonometry, and the designer would want P as a function of θ.

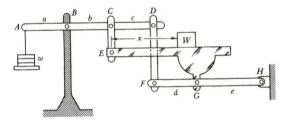

Fig. 3.4-6

Example **3.4-3**

The platform scale of Fig. 3.4-6 is to be constructed so that the balance weight, w, is independent of the position x of the load W on the platform. (Why is this desirable?) What should be the relationship between the dimensions b, c, d, and e in order to achieve this? With this design what is the relationship between w, W, a, and b? All members are light and rigid, and the pins at B through H are smooth.

Solution: In this mechanism, the basic element is the lever, such as AD, with fixed fulcrum point, such as B; moment equations of equilibrium are the basic ones used in the analysis. Here we are given an equilibrium position, with bars AD and FH and the platform floor horizontal, and bars CE and DF vertical. If all dimensions are given, the free-body diagram of the system involves four unknowns, the two components of pin reaction at B and the two at H.

Isolate as subsystems bars AD, FH, and the platform. Since CE and DF are light and loaded only at their ends (there being no interference between DF and the platform), equilibrium requires that they be two-force members. In Fig. 3.4-7, the free-body diagrams are shown. For the system of Fig. 3.4-7c, we write moments about H:

$$\sum M_H = -(d+e)T_2 + eG_y = 0. \qquad \textbf{3.4-11}$$

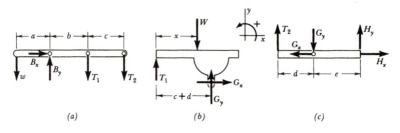

(a)   (b)   (c)

Fig. 3.4-7

For the platform (Fig. 3.4-7b), we find

$$\sum M_G = -(c+d)T_1 + [(c+d)-x]W = 0, \qquad \textbf{3.4-12}$$

$$\sum R_y = T_1 + G_y - W = 0; \qquad \textbf{3.4-13}$$

and, for AD (Fig. 3.4-7a),

$$\sum M_B = aw - bT_1 - (b+c)T_2 = 0. \qquad \textbf{3.4-14}$$

From these four equations we must determine what we wish to find; any other equations will involve other unknowns. The horizontal force equilibrium equations serve to show that $B_x = G_x = H_x = 0$; the other two vertical force equilibrium equations determine B_y and H_y, once the other forces are known. In the four equations, we have four unknowns: T_1, T_2, G_y, and w, if W, x, and all bar dimensions are given. What we wish to do is to find dimensions such that x disappears from the problem.

Eliminate G_y from Eqs. 3.4-11 and 13:

$$eW - eT_1 - (d+e)T_2 = 0;$$

now remove T_2 from consideration:

$$T_2 = \frac{e(W-T_1)}{d+e} = \frac{aw - bT_1}{b+c}. \qquad \textbf{3.4-15}$$

Solving for T_1, we find

$$\frac{eW}{d+e} - \frac{aw}{b+c} = \left(\frac{e}{d+e} - \frac{b}{b+c}\right)T_1,$$

$$T_1 = \frac{e(b+c)W - a(d+e)w}{e(b+c) - b(d+e)}. \qquad\qquad 3.4\text{-}16$$

From Eq. 3.4-12, then,

$$\frac{e(b+c)W - a(d+e)w}{e(b+c) - b(d+e)} = \left(1 - \frac{x}{c+d}\right)W$$

and

$$w = \frac{e}{a}\left(\frac{b+c}{d+e}\right)W - \frac{e(b+c) - b(d+e)}{a(d+e)}\left[1 - \frac{x}{c+d}\right]W$$

$$= \frac{b}{a}W + \frac{x}{c+d}\left[\frac{e(b+c)}{a(d+e)} - \frac{b}{a}\right]W.$$

Therefore,

$$\frac{w}{W} = \frac{b}{a} + \frac{x}{a(c+d)}\left[\frac{ce - bd}{d+e}\right]. \qquad\qquad 3.4\text{-}17$$

Therefore, if x is to disappear from the problem, the dimensions of the scale must be such that

$$ce = bd \quad\text{or}\quad \frac{b}{c} = \frac{e}{d}. \qquad\qquad 3.4\text{-}18$$

Using this condition, we find that $w/W = a/b$ and

$$T_1 = W, \qquad B_y = w + W = W\left(1 + \frac{b}{a}\right), \qquad T_2 = G_y = H_y = 0.$$

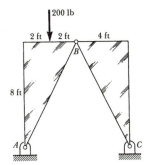

Fig. 3.4-8

Example **3.4-4**

An arch is constructed of two identical 90 lb uniform rigid plates. It supports a 200 lb load as shown in Fig. 3.4-8. Find the forces on the plates at the smooth pins A, B, and C.

Solution: In Fig. 3.4-9, we show free-body diagrams of the whole structure and each part. The c.m. of each plate has been located from the fact that each plate is uniform in its mass distribution, so that the c.m. coincides with the centroid of area, and from the fact that the medians of

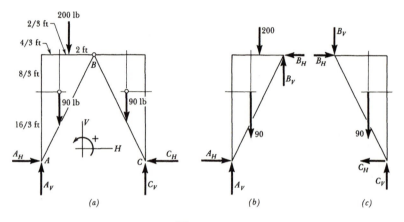

Fig. 3.4-9

a triangle intersect at the centroid at distances one-third of the altitude from the base on which the altitude is dropped. Equilibrium of the whole arch is governed by the equations

$$\sum M_A = 8C_V - \left(\frac{4}{3}\right)(90) - (2)(200) - \left(\frac{20}{3}\right)(90) = 0,$$

$$\sum M_C = -8A_V + \left(\frac{4}{3}\right)(90) + (6)(200) + \left(\frac{20}{3}\right)(90) = 0,$$

$$\sum R_H = A_H - C_H = 0,$$

from which we find

$$C_V = 140 \text{ lb}, \qquad A_V = 240 \text{ lb}, \qquad A_H = C_H. \qquad \textbf{3.4-19}$$

For plate BC (Fig. 3.4-9c), the equations are

$$\sum R_H = B_H - C_H = 0$$

$$\sum M_B = 4C_V - \left(\frac{8}{3}\right)(90) - 8C_H = 0,$$

$$\sum R_V = C_V - 90 - B_V = 0,$$

from which

$$B_V = 50 \text{ lb}, \qquad C_H = B_H = A_H = 40 \text{ lb.} \qquad \textbf{3.4-20}$$

All of the forces are thus determined. Since all answers are positive, the directions shown in Fig. 3.4-9 are all correct.

Example 3.4-5

A windlass used to hoist a 100 lb weight consists of a 20 lb cylindrical drum of radius 2 ft and a crank AOBDE that is light; the drum is rigidly keyed to the shaft that passes through the geometric center O of the drum.

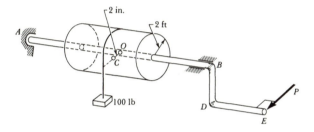

Fig. 3.4-10

The shaft is supported horizontally in a step bearing at A and a sleeve bearing at B. In the position shown in Fig. 3.4-10, the handle BDE is in the vertical plane through AB and the eccentric c.m., C, of the drum is 2 inches off the center line horizontally to the left as one looks from B to A. The cord and weight lie in the vertical plane through OC. What horizontal force, P, must be applied perpendicular to the handle in the given position to lift the weight slowly? What are the bearing reactions at A and B? Dimensions are AO = 4 ft, OB = 3 ft, BD = DE = 1 ft.

Solution: Before we formulate this three-dimensional mechanism problem, we note three things. First, the smooth sleeve bearing (which we have encountered before in Chapter II) is represented by a force normal to its axis; the step bearing also can sustain thrust along the line *AB*. Second, this is a mechanism; the hoist itself is a rigid body that can rotate about the horizontal axis *AB*. Third, we are introducing a new area of application of the equilibrium equations here, i.e., to a machine that is actually moving, but "slowly." The effect of "slowly" is not only to require a very small, i.e., negligible, angular acceleration—the phrase "slowly varying angular velocity" would do that—but also to require that the rotational angular velocity itself be small. The full dynamical theory of such fixed-axis rotation problems shows that couples proportional to the square of the angular speed (which tend to rock the rotating body on its supports) can be present even if the angular acceleration vanishes.

We take as our system the hoist; Fig. 3.4-11 is a free-body diagram complete with axis system with origin A. Taking moments about A, we have (since equilibrium of the weight requires $T = 100$ lb)

$$\sum \mathbf{M}_A = \left(4\mathbf{k}+\frac{1}{6}\mathbf{i}\right) \times 20\mathbf{j}+(4\mathbf{k}+2\mathbf{i}) \times 100\mathbf{j}$$
$$+7\mathbf{k} \times (B_z\mathbf{i}-B_y\mathbf{j})+(8\mathbf{k}+\mathbf{j}) \times P\mathbf{i}$$
$$= (7B_y-480)\mathbf{i}+(7B_z+8P)\mathbf{j}+\left(\frac{610}{3}-P\right)\mathbf{k} = \mathbf{0},$$

from which we find

$$P = 203.3 \text{ lb}, \qquad B_z = -232.4 \text{ lb}, \qquad B_y = 68.6 \text{ lb}. \qquad \textbf{3.4-21}$$

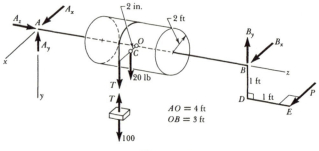

Fig. 3.4-11

We may now write the equations of moment equilibrium about B to determine A_z and A_y independently of B_z and B_y; in this simple problem, force equilibrium serves equally well:

$$\sum \mathbf{R} = P\mathbf{i}+(B_z\mathbf{i}-B_y\mathbf{j})+20\mathbf{j}+100\mathbf{j}+(A_z\mathbf{i}-A_y\mathbf{j}+A_z\mathbf{k})$$
$$= (P+B_z+A_z)\mathbf{i}+(120-B_y-A_y)\mathbf{j}+A_z\mathbf{k} = \mathbf{0}.$$

We find $A_z=0$, as we expect, since there is no applied thrust in the z-direction. Further, we have that

$$A_y = B_y-120 = -51.4 \text{ lb},$$
$$A_z = -(P+B_z) = 29.1 \text{ lb}. \qquad \textbf{3.4-22}$$

The signs of the answers tell us that we have shown the right directions for P, A_z, and B_y and the wrong directions for A_y and B_z in Fig. 3.4-11. The answers should not be given by tampering with the free-body diagram at this stage but, probably best, in vector form relative to the chosen axis system:

$$\mathbf{A} = A_z\mathbf{i}-A_y\mathbf{j}+A_z\mathbf{k} = 29.1\mathbf{i}+51.4\mathbf{j} \text{ lb},$$
$$\mathbf{B} = B_z\mathbf{i}-B_y\mathbf{j} = -232.4\mathbf{i}-68.6\mathbf{j} \text{ lb}, \qquad \textbf{3.4-23}$$
$$\mathbf{P} = 203.3\mathbf{i} \text{ lb}.$$

Example **3.4-6**

A wall bracket frame (Fig. 3.4-12) of light rigid bars supports a 5 lb pulley of 5 in. radius at its center of gravity C and a 50 lb weight by a light inextensible cable, which is fastened to pin A and which passes over the pulley. All pins are smooth. Find the forces on the pulley and on bars CF and AC.

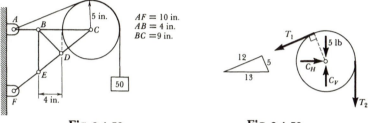

Fig. 3.4-12   **Fig. 3.4-13**

Solution: Isolate the pulley as system, or rather the pulley and the portion of cable resting against the pulley (Fig. 3.4-13). The forces acting are the weight of the pulley, 5 lb down at C; the two cable tensions, T_1 and T_2, tangent to the pulley; and the pin force at C on the pulley, represented by its horizontal and vertical components, C_H and C_V. The basic effect of

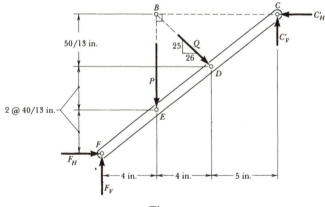

Fig. 3.4-14

the pulley, of course, is to cause a change in the direction of the cable, and hence of the tension force, around the pulley (and, if friction is present, sometimes a change in tension magnitude). Moment equilibrium about C requires that $T_1 = T_2$. Equilibrium of the 50 lb weight requires that

$T_2 = 50$ lb; therefore, $T_1 = T_2 = 50$ lb. The pin reaction on the pulley is then determined:

$$\sum R_H = C_H - \frac{12}{13} T_1 = 0, \qquad C_H = \frac{600}{13} \cong 46.2 \text{ lb;}$$

$$\sum R_V = C_V - 5 - T_2 - \frac{5}{13} T_1 = 0, \qquad C_V = \frac{965}{13} \cong 74.2 \text{ lb.} \quad \textbf{3.4-24}$$

The forces for the equilibrium of the pulley are now completely determined.

Turning now to bar CF, we first recognize that the light bars BE and BD are loaded only at their ends, so that they are two-force members.

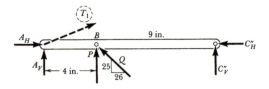

Fig. 3.4-15

Suppose BE carries a compressive load P; BD, a compressive load Q. The pin at F exerts a force with components F_H and F_V on the bar. The pin at C is our next concern. Suppose the forces on CF are C_H' and C_V'; then the free-body diagram of CF is Fig. 3.4-14. We have six scalar

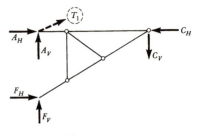

Fig. 3.4-16

unknowns and only three equations. Before writing any equations it is best to go on to some other system to see whether the problem is statically determinate. We have used the two-force members completely. We have only three other choices: AC, the whole system, or the frame without the pulley. Since the pulley has been considered completely, let us look at bar AC (Fig. 3.4-15) and at the frame without the pulley (Fig. 3.4-16). On bar AC, the forces P and Q due to the cross-members appear at B.

So does a reaction (C_H'', C_V'') due to the pin at C.   At A, there is also a pin reaction force (A_H, A_V).   If the cable is attached to the pin, no force T_1 should be shown.   If the cable is attached to the bar, T_1 should be shown. We will not consider it in our equations, but will put it in as dotted in Figs. 3.4-15 and 3.4-16 to show how it would appear.   If it is considered to be there, the only effect is to change the values of A_H and A_V; the total resultant force at point A will be the same either way.   Four more un-knowns have been introduced, and three equations: a total of ten unknowns, and only six equations.

On the whole frame, we have the forces at A and F and the reverse of the pulley reaction at the pin.   It would seem wise to determine A_H, F_H, and a relation between A_V and F_V from this system before con-sidering either of the separate bars.

Finally, we note that all of the forces due to the pin at C are not independent.   We have three members—two bars and the pulley—coming together at C.   For equilibrium of the pin (Fig. 3.4-17), we must have $C_H'' = C_H - C_H'$, $C_V'' = -(C_V + C_V')$.   The important thing to note here, and to watch in one's own work, is the proper use of the action-reaction law at

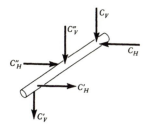

Fig. 3.4-17

a point where three or more members come together.   It is worthwhile to examine the pin at B to see how the two-force members simplify the analysis.

At this point our unknowns are A_H, A_V, F_H, F_V, P, Q, and one pair of forces at C, say C_H', C_V': eight in number.   We have used up all the information that equilibrium of the pulley, the pin at C, and the two bars BD and BE can give us.   The three equations on the whole frame (Fig. 3.4-16), with C_H and C_V known, will determine F_H, A_H, and a relation between F_V and A_V.   Expressing F_V in terms of A_V, we have left five unknowns.   The three equations for AC (Fig. 3.4-15) can be written, with C_H'' and C_V'' expressed in terms of C_H' and C_V'.   Now we must be cautious. Three more equations can be written for CF, but we cannot expect all three to be independent.   Indeed, if they were, we would have more equations than unknowns.   Moreover, we have put the whole system in equilibrium as well as each part except CF.   Since the system must "fit together" we cannot expect any fresh information from these equations and should not be surprised if the system has two degrees of redundancy—that it is statically indeterminate.

Let us write the equations. For the free-body diagram of Fig. 3.4-16, we have

$$\sum M_A = 10F_H - 13C_V = 0, \qquad F_H = \frac{13}{10} C_V = 96.5 \text{ lb};$$

$$\sum R_H = A_H + F_H - C_H = 0, \qquad A_H = -50.3 \text{ lb}; \qquad \textbf{3.4-25}$$

$$\sum R_V = A_V + F_V - C_V = 0, \qquad F_V = C_V - A_V = 74.2 - A_V \text{ lb}.$$

Turning to bar AC (Fig. 3.4-15), we may write

$$\sum M_C = -13A_V - 9\left(P + \frac{25}{\sqrt{25^2 + 26^2}} Q\right) = 0,$$

$$\sum M_B = -4A_V + 9C_V'' = -4A_V - 9(C_V + C_V') = 0, \qquad \textbf{3.4-26}$$

$$\sum R_H = A_H - \frac{26}{\sqrt{25^2 + 26^2}} Q - C_H''$$

$$= A_H - \frac{26}{\sqrt{25^2 + 26^2}} Q - (C_H - C_H') = 0,$$

wherein the equilibrium equations for pin C have been utilized. From these, we find

$$C_V' = -C_V - \frac{4}{9} A_V = -74.2 - \frac{4}{9} A_V,$$

$$P + \frac{25}{\sqrt{1301}} Q = -\frac{13}{9} A_V, \qquad \textbf{3.4-27}$$

$$C_H' - \frac{26}{\sqrt{1301}} Q = C_H - A_H = 96.5.$$

For bar CF (Fig. 3.4-14), we have

$$\sum M_B = -4F_V + 10F_H + 9C_V' = 0,$$

$$\sum R_H = F_H + \frac{26}{\sqrt{1301}} Q - C_H' = 0, \qquad \textbf{3.4-28}$$

$$\sum R_V = F_V + C_V' - P - \frac{25}{\sqrt{1301}} Q = 0.$$

From these, we obtain

$$\sum M_B = -4(C_V - A_V) + 13C_V - 9C_V - 4A_V \equiv 0, \qquad \textbf{3.4-29}$$

that is, the first equation is satisfied identically;

$$\sum R_H = 96.5 + \frac{26}{\sqrt{1301}} Q - C_H' = 0, \qquad \textbf{3.4-30}$$

which is the same as the third of Eqs. 3.4-27; and

$$\sum R_V = (C_V - A_V) + \left(-C_V - \frac{4}{9} A_V\right) - P - \frac{25}{\sqrt{1301}} Q = 0, \qquad \textbf{3.4-31}$$

which is the same as the second of Eqs. 3.4-27. There are two redundancies, all other unknown forces being expressible in terms of, say, A_V and Q:

$$F_V = 74.2 - A_V,$$

$$C_V' = -74.2 - \frac{4}{9} A_V, \qquad \textbf{3.4-32}$$

$$P = -\frac{13}{9} A_V - \frac{25}{\sqrt{1301}} Q.$$

The structure may be made statically determinate if the pin at A or F is replaced by a roller, in which case A_V or F_V will be zero; and if brace BE is removed, in which case P will be zero, or brace BD is removed, so that Q does not appear in the problem.

We have used this statically indeterminate problem to show two things: that the solution methods of rigid-body statics are still applicable as far as they will take us, and that the partial solution of a statically indeterminate problem may indicate what must be done to make a structure just determinate.

Exercises

3.2-1: A uniform L-section of weight 75 lb is pinned smoothly at the elbow P so that the body may rotate in a vertical plane. Its motion is further restrained by a vertical cable attached to end A of the 6 ft arm PA and to a ceiling so that arm PA makes a 35° angle with the horizontal. Find the cable tension and the pin force on the body for equilibrium.

Ans.: $T = 15.5$ lb, $P_x = 0$, $P_y = 59.5$ lb.

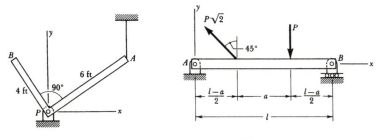

Exer. 3.2-1   Exer. 3.2-2

3.2-2: Find the reactions at the pin A and the roller support B on the light beam loaded as shown.

Ans.: $A_x = P$, $A_y = -Pa/l$, $B_y = Pa/l$.

3.2-3: What horizontal force, P, is required to hold the 10 lb, 4 in. by 8 in. uniform block shown on the smooth $30°$ incline? What is the normal reaction, N, at the slope? What is the distance along the slope from the lowest corner to the line of action of N?

Ans.: $P = 5.77$ lb, $N = 11.5$ lb, $x = 1.5$ in.

3.2-4: A flat triangular plate of plan area A is to be cut from a sheet of thickness t and mass density ρ. The plate is to be supported on spring mounts at its vertices so that a horizontal position is the position of equilibrium. We should like to use three identical spring mounts of known properties. Does it make any difference in what way the triangular area is cut? That is, must we choose an equilateral triangle or some other special shape if identical mounts are used?

3.2-5: A rocker arm can be treated as a uniform rigid bar of mass m and length l pinned smoothly at its c.m. Its motion is restrained by an ideal torsional spring of constant K (moment/radian) which is undeformed when the bar is horizontal. (That is, the spring provides a restoring torque of K moment units per radian of angular deformation.)

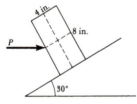

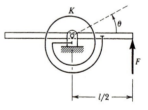

Exer. 3.2-3   Exer. 3.2-5

(a) Suppose a force $F=F_1$ is applied to the end of the arm and remains perpendicular to the arm as it rotates; find the angle $\theta=\theta_1$ with the horizontal that the bar will assume when it is in equilibrium.

Ans.: $\theta_1 = F_1 l/2k$.

(b) Suppose a force $F=F_2$ is applied to the end of the arm and remains vertically upward in direction as the arm rotates; what equation governs the angle $\theta=\theta_2$ for equilibrium?

Ans.: $\theta_2 \sec \theta_2 = F_2 l/2k$.

(c) If $F_1=F_2$, which is larger, θ_1 or θ_2? If $\theta_1=\theta_2$, which is larger, F_1 or F_2?

3.2-6: A 500 lb, 4 ft by 4 ft horizontal uniform slab is being slowly lifted in a triangular sling as shown. Find the tensions in the cables AP, BP, and CP.

Ans.: $T_B = 2T_A = 224$ lb;
$T_C = 229$ lb.

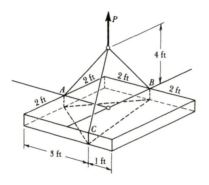

Exer. 3.2-6

3.2-7: A uniform semicircular arch of weight W and mean radius R is supported by a smooth pin at one end and by a smooth roller at the other. It supports two normal loads $90°$ apart, each equal to the weight of the arch in magnitude. Find the support reactions as functions of the angle θ shown, $0° < \theta < 90°$.

Ans.: $B = A_y = W(1+\sin \theta+\cos \theta)/2$, $A_x = W(\cos \theta-\sin \theta)$.

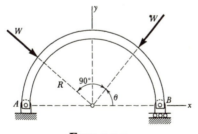

Exer. 3.2-7

3.2-8: A smooth ring weighing 1 lb is threaded on a light cord, the ends of which are attached to the ceiling of a room at points A and B. A horizontal force acts on the ring. Find the magnitude of this force if the plane of the string makes an angle of $45°$ with the vertical and the bisector of the angle subtended by the string at the ring makes an angle of $60°$ with AB.

Ans.: 1.29 lb.

3.2-9: The figure shows a plan view of a bent shaft held in smooth bearings at A and B. A couple of magnitude 277 lb-in. is applied to BC in a clockwise direction looking from B to C. A force of 50 lb acts at C at right angles to the plane of ABC and is directed so as to produce a clockwise moment about AB looking from B to A. What torque about AB and what bearing reactions are required to maintain equilibrium?

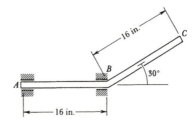

Exer. 3.2-9

Ans.: 160 lb-in., clockwise viewing from A to B; reaction at $A = 52$ lb in same direction as 50 lb force and reaction at $B = 102$ lb, opposite to direction of 50 lb force.

3.2-10: A particle weighing 10 lb is suspended from a point A on the ceiling by means of a cord 6 in. long. What is the greatest distance it can be pushed from the vertical through A by a force having a magnitude of 5 lb?

Ans.: 3 in.

3.2-11: The cord attached to the sphere has a length equal to the radius of the sphere. If the pull on the other cord is n times the weight of the sphere, what is the angle between the vertical and the cord attached to the sphere?

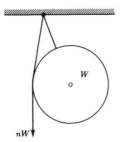

Exer. 3.2-11

Ans.: $\arcsin\left(\dfrac{n}{2+2n}\right)$.

3.2-12: A pipe ABC has a 90° bend at B; $AB = 3$ ft, $BC = 4$ ft. Two machinists, one at each end, thread the pipe, each exerting a 24 lb-ft clockwise torque on the 9 in. long handles of a die holder. At the moment of interest, the pipe and die handles are horizontal. A support at B provides a vertical constraint force there. From above, the direction BC is 90° counterclockwise from AB. What forces must the man at A exert on the die handles?

Ans.: Left hand: 20 lb up; right hand: 12 lb down.

3.2-13: A uniform light rod ABC is freely pivoted at B (0, 0, 0) and held by cables AD and CE attached to A (-3, 0, 0) and C (4, 0, 0) and the fixed points D (0, -2, -6) and E (0, 4, -2). A load of 10 lb acts at C in the direction of the negative *y*-axis. Find the tensions in the strings and the magnitude of the reaction at *B*.

Ans.: $T_{AD} = 20/3$ lb; $T_{CE} = 90/7$ lb; $|\mathbf{B}| = 11.99$ lb.

3.2-14: The tail-gate of a truck may be regarded as a uniform rectangular plate *ABCD* weighing 50 lb. It is held in an inclined position by hinges at *C* and *D* and by a chain connecting *B* to a point 5 ft above *D*. The chain is 6.3 ft long and *AB*=4 ft, *BC*=3 ft. What is the load carried by the chain?

Ans.: 31.5 lb.

3.2-15: A rectangular picture of width 2*w* and depth *d* is hung from a smooth peg by means of a cord attached to the two upper corners of the picture frame. The length of the cord is 2*l*. Show that if the depth of the picture is less than $2w^2/(l^2-w^2)^{\frac{1}{2}}$, the picture can hang at an oblique angle to the horizontal.

3.3-1: The diagram shows a Howe truss, a common type for timber roof construction. Find the reactions and the loads carried by members *a*, *b*, *c*, and *d*. Are these loads tensile or compressive? Which members obviously carry no stress for the loading shown?

Ans.: $G_y = 2230$ lb; $C_y = 4910$ lb; $C_x = -3570$ lb; *a*: 4000 lb (*C*); *b*: 2230 lb (*T*); *c*: 0; *d*: 4460 lb (*T*); members *e* and *k*: 0.

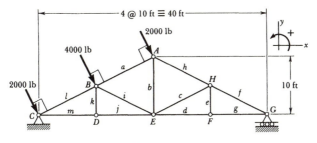

Exer. 3.3-1

3.3-2: The diagram shows a Warren truss. Find the load carried by members *a*, *b*, and *c*. Are they tensile or compressive?

Ans.: *a*: 3.33 k (*C*); *b*: 1.08 k (*T*); *c*: 2.91 k (*T*).

3.3-3: The diagram shows a Warren truss with verticals. Find the load carried by members *a*, *b*, *c*, *d*, *e*. All upper chord forces are 600 lb and all lower chord forces are 2100 lb.

Ans.: *a*: 600 lb (*C*); *b*: 4360 lb (*C*); *c*: 8100 lb (*T*); *d*: 600 lb (*C*); *e*: 8640 lb (*C*).

3.3-4: Because of a rise in temperature, the arch shown expands and develops a reaction at *A* whose horizontal component is 60 k acting to the

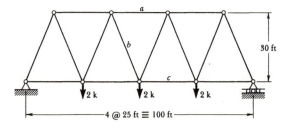

Exer. 3.3-2

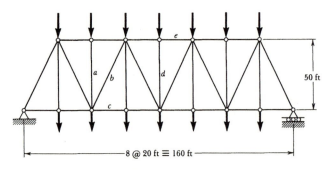

Exer. 3.3-3

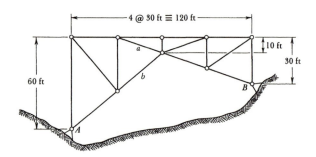

Exer. 3.3-4

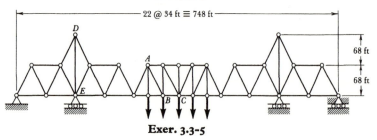

Exer. 3.3-5

right. Calculate the other reactions and the tensile or compressive loads carried by members *a* and *b*.

Ans.: H_B = 60 k left; V_B = 15 k down; V_A = 15 k up; *a*: 190 k (*C*); *b*: 117 k (*C*).

3.3-5: The cantilever bridge shown is symmetrical about its centerline. The suspended span is known as a Pratt truss. Its dead weight is 120,000 lb and may be considered divided into 5 equal loads as shown. Find the loads carried by members *AB* and *BC*. Tension or compression?
Ans.: *AB*: 40,200 lb (*T*); *BC*: 48,000 lb (*T*).

3.3-6: In Exercise 3.3-5, what is the load carried by the member *DE*? Is it tension or compression? Do you think that the design of this member would present any difficulty?
Ans.: *DE*: 240,000 lb (*C*).

3.3-7: Each lower chord panel point of the *K*-truss shown carries a load of 5000 lb and each upper panel point carries 2000 lb. Find the tensile or compressive loads carried by members *a*, *b*, *c*, and *d*.
Ans.: *a* = 12,450 lb (*C*); *b* = 2350 lb (*C*); *c* = 2350 lb (*T*); *d* = 12,450 lb (*T*).

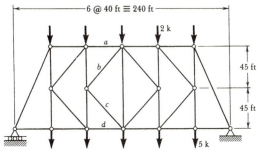

Exer. 3.3-7

3.3-8: Find the loads, tensile or compressive, carried by members *a*, *b*, *c*, *d* of the truss shown.
Ans.: *a* = 7.5 k (*T*); *b* = 42.8 k (*T*); *c* = 7.5 k (*C*); *d* = 15 k (*T*).

3.3-9: Find the loads carried by members *a*, *b*, *c*, and *d* of the truss shown.
Ans.: *a* = 225 k (*C*); *b* = 75 k (*C*); *c* = 125 k (*C*); *d* = 0.

3.3-10: Design your own truss to carry a downward load of 30,000 lb at the center of a 250 ft span. For ease of handling, no member is to exceed 40 ft in length. Find the greatest load carried by any member.

3.3-11: Show that a statically determinate planar truss with the usual three foundation constraint quantities (*C*=3) must have at least one joint at which only two bars are connected or else one joint where only three bars are connected.

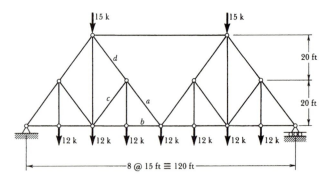

Exer. 3.3-8

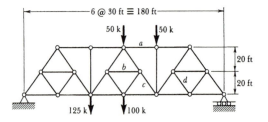

Exer. 3.3-9

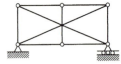

Exer. 3.3-12

3.3-12: Show that the truss pictured is a critical form. The bars are not connected at the center where they overlap.

3.3-13: The hexagonal truss shown is made of bars 6 ft long. Find the loads carried by members AB and CD. Would the truss be statically determinate if there were a bar connecting F and G?

Ans.: $F_{AB} = 25$ lb compression; $F_{CD} = 32.7$ lb tension.

3.3-14: Show that the hexagonal truss pictured is a critical form. There is no joint at the center; the bars merely overlap.

3.3-15: A regular tetrahedral truss is subject to the parallel loads shown. Show that the truss as a whole is in equilibrium and find the loads carried by all six members.

Ans.: $AB = AC = AD = P\sqrt{6}/6$ compression; $BC = CD = DB = P\sqrt{6}/18$ tension.

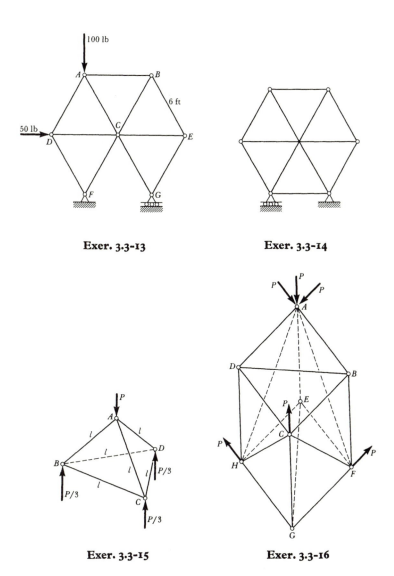

Exer. 3.3-13

Exer. 3.3-14

Exer. 3.3-15

Exer. 3.3-16

3.3-16: A spatial truss $ABCDEFGH$ has the form of a cube. It is made of 12 bars of length l and six of length $l\sqrt{2}$. ("Hidden" bars are shown by broken lines.) It is subjected to 6 forces of magnitude P directed along the edges of the cube as shown.   Show that the truss as a whole is in equilibrium and that it satisfies the necessary conditions for statical determinacy. Find the loads carried by the six diagonal members.

Ans.: $AF = AH = EG = P/\sqrt{2}$ compression; $BD = CF = CH = P/\sqrt{2}$ tension.

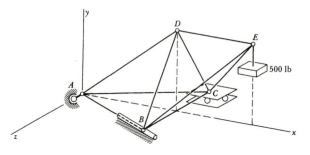

Exer. 3.3-17

3.3-17: The truss shown in the accompanying figure supports a vertical load of 500 lb at joint E.   Joint A is supported in a smooth pivot; joint B, in a smooth sleeve bearing with axis along the line AB; and joint C, on smooth rollers between fixed horizontal guide planes parallel to plane ABC.   Find the external reactions and the bar forces.   Bars AB, BC, AC,

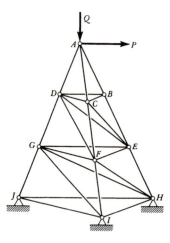

Exer. 3.3-18

BD, CD, and *DE* are each 6 ft long; joints *A, D,* and *E* are in the same vertical plane, with *D* above the midpoint of *BC*.

Ans.: A $= -577\mathbf{j}$ lb; **B** $= \mathbf{C} = 539\mathbf{j}$ lb; $BE = 408$ lb (C); $BD = 333$ lb (C); $AD = 816$ lb (T).

3.3-18: (a) State precisely the conditions under which one can conclude that one or more bars meeting at an unloaded joint in a space truss carry no load.
(b) Apply one or both of these basic theorems step by step to the loaded truss of the accompanying figure to show that only the slant legs, and not the horizontal or diagonal cross-bracing, of the tower are carrying load.

3.4-1: Three masses are connected by light inextensible cords, which pass over small smooth pulleys distant $2a$ apart on the same horizontal level. Find the sag b for equilibrium of the central mass. For what ratios M/m of the mass of the center body to the mass of either of the other bodies is equilibrium possible?

Ans.: $\dfrac{b}{a} = \dfrac{M/2m}{[1-(M/2m)^2]^{1/2}}; \dfrac{M}{m} < 2.$

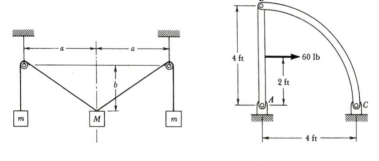

| Exer. 3.4-1 | Exer. 3.4-2 |

3.4-2: The smoothly pinned arch shown consists of a 30 lb vertical bar *AB* and a light quarter-circle *BC*. Find the forces acting on member *AB* when the 60 lb horizontal load is applied at its center.

Ans.: at *A*: 30 lb to left; at *B*: $30\sqrt{2}$ lb up and to the left at $45°$.

3.4-3: The rhomboidal frame *ABCD* consists of light rigid bars pinned smoothly together. The foundation pin at *A* is smooth, as are the rollers in the two-sided guide at *D*. A torque of 100 lb-ft is applied to member *AB*. Find the support reactions and the forces on each member.

Ans.: Support reactions: 25 lb up at *A*, 25 lb down at *D*; bars *BC* and *CD* carry $50\sqrt{3}/3$ lb tensile load; bar *AD* carries a $25\sqrt{3}/3$ lb tensile load; bar

AC carries a 50 lb compressive load; bar *AB* is subject to a 50√3̄/3 lb force to the right at *B* and a resultant 50√3̄/3 lb force to the left at *A*, besides the 100 lb-ft torque.

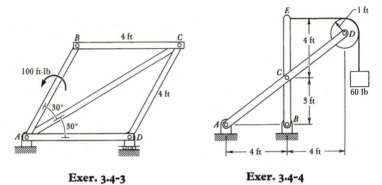

Exer. 3.4-3   Exer. 3.4-4

3.4-4: The vertical plane frame shown supports the 60 lb weight.   The uniform pulley weighs 50 lb, while the bars *BE* and *AD* weigh 10 lb/ft. All pins and the pulley are smooth, and the cord is light and inextensible. Find the forces on bar *ACD* at the pins *A*, *C*, and *D*.

Ans.: *A*: 80 lb left, 125 lb down; *C*: 140 lb right, 335 lb up; *D*: 60 lb left, 110 lb down.

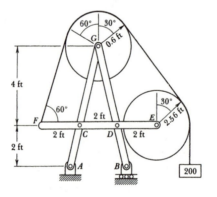

Exer. 3.4-5

3.4-5: The symmetrical *A*-frame, pulleys, and cord shown support a 200 lb load.   Dimensions are as shown, all contacts are smooth, and all weights may be ignored except the 200 lb load.   Find the forces on the horizontal member *FCDE*.

Ans.:

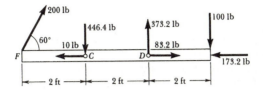

3.4-6: The weight of the study lamp shown is held in equilibrium by the spring. The tension of the spring is k multiplied by the length of the spring. Show that, if the designer makes the proper choice of k, the lamp will be balanced for all angles θ, and find what this value of k should be in terms of W, a, b, and h.

Ans.: $k = Wb/ah$.

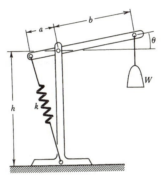

Exer. 3.4-6

3.4-7: The crankshaft of a single-cylinder engine is shown in outline.
(a) What horizontal force P must be applied to the handle in order to produce a compression force C of 100 lb in the cylinder?
(b) Determine the bearing reaction at B when this force is applied.

Ans.: (a) $P = 40.7$ lb; (b) $B_y = 45.3$ lb; $B_z = -82.1$ lb.

3.4-8: A circular table weighing 50 lb stands on a horizontal floor, supported by four equal legs placed symmetrically around its edge. What is the largest weight that can safely be placed on the table?

Ans.: 120 lb.

3.4-9: (a) In a so-called four-ball testing rig, three smooth, light ball bearings are placed, just touching, inside a cylinder and a fourth ball is placed symmetrically on top of them, touching all three as shown. A vertical load is then applied to the upper ball. What is the force exerted by any one of the lower ball bearings on the cylinder wall?

Ans.: $P\sqrt{18}/18$.

(b) This is the starting pressure in the test. As the test proceeds, the cylinder is spun about a vertical axis. What effect do you think this has on the force in question? Do you see any disadvantage to using four or five balls in the cylinder and thus testing a larger number at once?

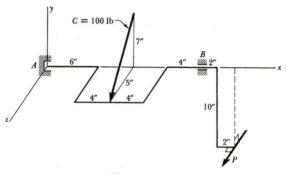

Exer. 3.4-7

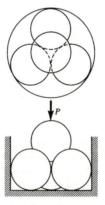

Exer. 3.4-9

3.4-10: Here we show two smooth spheres of weight w and radius r placed inside a cylinder of radius R and weight W which is open at both ends and rests on a horizontal plane (see text Example 3.2-6). Of course $2r > R$.
(a) What is the pressure exerted by either one of the spheres on the cylinder?

(b) What is the smallest value of W allowable if the cylinder is not to tip over?

Ans.: $P = \dfrac{w(R-r)}{[R(2r-R)]^{\frac{1}{2}}}$;

(b) $W = 2w\left(1 - \dfrac{r}{R}\right)$.

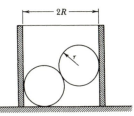

Exer. 3.4-10

3.4-11: Generalize part (b) of Exercise 3.4-10 to the case in which there are $2n$ spheres in the cylinder. Show that under these circumstances $W_{min} = 2nw(1 - r/R)$.

3.4-12: Solve Exercise 3.4-10 for a cylinder having a solid bottom. Now what is the pressure exerted by either of the spheres on the cylinder? Can the cylinder possibly tip?

3.4-13: A flat-bottomed cup of radius R and weight nW stands on a table. A uniform rod of length $2L$ and weight W rests over the rim of the cup and presses against its smooth vertical interior. (a) Find the angle θ for which rod and cup can be in equilibrium. (b) Show that if this angle is smaller than the angle whose cosine is $(n+2)R/L$ the cup will tip over.

Ans.: $\theta = \arccos\left(\dfrac{2R}{L}\right)^{\frac{1}{3}}$.

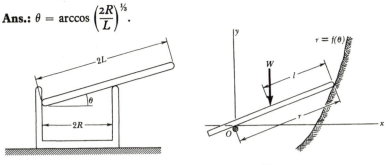

Exer. 3.4-13   Exer. 3.4-15

3.4-14: In the situation described in Exercise 3.4-13 find the length of the longest rod that can possibly be in equilibrium with one end pressing against the smooth vertical interior wall.

Ans.: $L = R(n+2)^{\frac{3}{2}}/\sqrt{2}$.

3.4-15: A rod of weight W rests on a smooth peg at O with one end pressing on a surface whose equation is $r = f(\theta)$. Show that the rod will be in equilibrium provided $dr/d\theta = -(r-l)\cot\theta$. Here l denotes the distance from one end of the rod to the center of gravity.

3.4-16: Use the results of Exercise 3.4-15 to show that, if the equation of the surface is $r = l + C\,\mathrm{cosec}\,\theta$, the rod will be in equilibrium for any angle θ. Here C is a constant of integration and may have any value. Sketch the shape of the curve for $\theta = 0°$ to $\theta = -90°$ when C is such that $r = 2l$ at $\theta = -90°$.

3.4-17: A uniform rod AB rests in equilibrium with A touching a smooth wall and B touching a smooth floor. B is attached by a cord to a point C directly below A at the intersection of wall and floor. The cord makes an angle of 60° with that intersection and with AB. A is held by a horizontal cord which is attached to the wall. Find the tensions in the two cords if the rod weighs 100 lb.

Ans.: 14.5 lb at A; 29 lb at B.

3.4-18: The figure shows plan and elevation of a shaft with two cranks. One crank is subjected at C to a single force of 112 lb directed as shown in the figure (note that neither view looks perpendicular to this force). The shaft is kept in equilibrium by a vertical force R applied at D, an end thrust T, and bearing reactions having components H_A, V_A, H_B, V_B as shown. Find their magnitudes.

Ans.: 12, 50, −100, 52, 13.5, 10 lb.

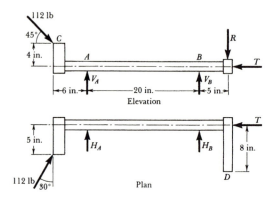

Exer. 3.4-18

CHAPTER **IV**

Equilibrium under

Distributed Forces

4.1 Introduction

The analysis of equilibrium of systems subjected to distributed loads parallels that of systems subjected to concentrated loads insofar as finding external constraint forces is concerned. The distributed load may be replaced by an equipollent resultant force set and the equations of equilibrium for the system can be written. If the problem is statically indeterminate, however, or if, for some reason, we need to know the details of the internal force distribution in a member that is statically determinate over-all, we must take into account the true external force distribution in determining the internal force distribution. A complete solution of such a problem—particularly statically indeterminate problems for deformable bodies—requires the introduction of displacement and strain analysis. The solution also requires that the strain quantities be related to the load quantities through the properties of the material of which the body is fashioned. We shall not be concerned with such questions here, but will consider only those that admit of a solution in terms of external load quantities alone, with a minimum assumption about deformation, such as the assumption of inextensibility for a cable.

In Section 2.7, the meaning of the equivalence of distributed loads to concentrated loads was discussed. For instance, the distribution of internal forces over the area of a beam cross-section was replaced by an equipollent resultant force, with axial and shear components, and a couple. In some problems, "partial" resultants may be the quantities of importance. For example, in the theory of thin plates, integrals of distributed forces per unit area are used as the fundamental quantities, much as in beam theory—but only integrals through the thickness of the plate, i.e., forces and moments per unit length of edge, not integrals over a whole area. Whatever type of internal force distribution is considered—the actual forces per unit area of some equipollent set of partial or total resultants—the equilibrium equations to be solved must generally be formulated as differential equations of equilibrium.

In this chapter, we shall treat a sequence of questions in which the internal forces to be found become successively of more complex types—although the problems solved may be more difficult for some of the simpler forces than for some of the more complex. First, the inextensible cable loaded along its length will be considered. The theory of the cable in a plane will be developed, and the problems of the suspension bridge cable, the belt on a smooth pulley, and the cable hanging under its own weight will be solved. Next, the distribution of internal shear force and bending moment in a statically determinate beam will be considered. The concept of pressure and the equations of fluid statics will lead to the general equilibrium equations governing internal force distributions and the concept of the stress tensor. Distributed forces of a frictional nature will be treated in Chapter V.

4.2 The Loaded Cable

The light cable treated so far has been considered to be a straight two-force member; the only forces of concern have been the tensile forces of equal magnitude applied to the cable at its ends, and hence applied by the cables to the bodies at either end. The cable supporting the deck of a suspension bridge, or a transmission line hanging under its own weight, assumes a sagging curved shape under the load distributed along the cable length and the tensile loads at the ends. In order to find the end tensions exerted by the cable on its supports, we must determine the equilibrium shape assumed by the cable. For a belt wrapped around a smooth pulley, we can often find the tensions at either end; however, to compute the pressure distribution between pulley and cable from the equilibrium equations, we must make use of the known shape of the belt.

The geometry of the curve assumed by the cable is, therefore, of fundamental importance. We shall treat only those problems in which the cable and all loads upon it can be considered to be in a plane; the geometry of space curves has more complexity than we care to introduce at this point. We shall begin by summarizing some results from the calculus about the properties of plane curves before formulating the problem of the loaded cable.

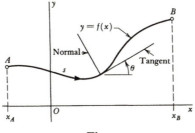

Fig. 4.2-1

Suppose a smooth curve between points A and B is given in a plane, and x-, y-axes are chosen so that the equation of the curve is $y = f(x)$. We suppose $f(x)$ to have at least two continuous derivatives. The tangent line to the curve at any point has *slope* $dy/dx = f'(x)$, with the *slope angle* θ from the x-axis to the tangent line determined by the relation

$$\frac{dy}{dx} = \tan \theta. \qquad \text{4.2-1}$$

The angle θ is, of course, a function of position. The *arclength* s along the curve is defined through the differential relation

$$(ds)^2 = (dx)^2 + (dy)^2, \qquad \text{4.2-2a}$$

or by

$$\left(\frac{ds}{dx}\right)^2 = 1 + \left(\frac{dy}{dx}\right)^2. \qquad \text{4.2-2b}$$

If we decide upon an orientation for the curve, i.e., decide in which sense along the curve the arclength increases, we may integrate

Eq. 4.2-2b for s as a function of x. Suppose that $s=0$ at A and increases as the curve is traversed toward B; then

$$s = \int_{x_A}^{x} \left[1 + \left(\frac{dy}{dx}\right)^2 \right]^{\frac{1}{2}} dx, \qquad\qquad 4.2\text{-}3$$

with the total length L of the curve being given by

$$L = \int_{x_A}^{x_B} \left[1 + \left(\frac{dy}{dx}\right)^2 \right]^{\frac{1}{2}} dx. \qquad\qquad 4.2\text{-}4$$

Once the arclength has been introduced, we may consider the equation of the curve to be given parametrically in the form $x=x(s), y=y(s)$. Then the slope angle $\theta = \theta(s)$ is determined by

$$\tan \theta = \frac{dy}{dx} = \frac{dy}{ds} \Big/ \frac{dx}{ds}, \qquad\qquad 4.2\text{-}5a$$

or by

$$\frac{dx}{ds} = \cos \theta, \qquad \frac{dy}{ds} = \sin \theta. \qquad\qquad 4.2\text{-}5b$$

The change in the slope angle with arclength can be determined by differentiating Eq. 4.2-5a:

$$\frac{d}{ds} (\tan \theta) = \sec^2 \theta \, \frac{d\theta}{ds} = \frac{d}{ds}\left(\frac{dy}{dx}\right) = \frac{d^2y}{dx^2}\frac{dx}{ds},$$

or

$$\frac{d\theta}{ds} = \cos^3 \theta \, \frac{d^2y}{dx^2}.$$

This is one form of the expression for the *curvature* of the curve; for, since $\cos \theta = 1/\sec \theta = \pm [1 + \tan^2 \theta]^{-\frac{1}{2}}$, we have

$$\frac{d\theta}{ds} = \frac{\pm \dfrac{d^2y}{dx^2}}{\left[1 + \left(\dfrac{dy}{dx}\right)^2\right]^{\frac{3}{2}}} = \pm \frac{1}{\rho} = \pm \kappa$$

where κ is the curvature and ρ is the radius of curvature. The choice of sign depends upon the sign convention selected for curvature. Since we shall not need to use the curvature formally, we shall not pursue this question further.

It will be helpful to establish formal coordinates tangent and normal to the curve. At any point on the curve, the direction tangent to the curve in the sense of increasing arclength is called the *tangential direction,* and a *unit tangent vector* $\mathbf{e}_t$ in that direction can be constructed. The direction 90° counterclockwise from the tangential direction is the *normal direction*, and the *unit normal vector* is denoted by $\mathbf{e}_n$. Since the slope angle θ is the angle between the x-axis and the tangential direction, the vectors $\mathbf{e}_t$ and $\mathbf{e}_n$ can be written down in cartesian components:

$$\mathbf{e}_t = \cos\theta\,\mathbf{i} + \sin\theta\,\mathbf{j}, \qquad \mathbf{e}_n = -\sin\theta\,\mathbf{i} + \cos\theta\,\mathbf{j}. \qquad \textbf{4.2-6}$$

These can be solved for $\mathbf{i}$ and $\mathbf{j}$:

$$\mathbf{i} = \cos\theta\,\mathbf{e}_t - \sin\theta\,\mathbf{e}_n, \qquad \mathbf{j} = \sin\theta\,\mathbf{e}_t + \cos\theta\,\mathbf{e}_n. \qquad \textbf{4.2-7}$$

The reader may compare these unit vector relations with those developed for polar coordinates in Section 1.5. The normal and tangential *intrinsic* coordinates are also of fundamental importance in the study of the motion of a particle along a curved path.

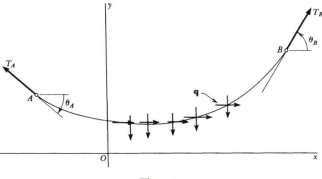

Fig. 4.2-2

We turn now to the formulation of the equilibrium equations for the loaded cable. Suppose a cable of length L is supported at two points, A and B, and subjected to a distributed load $\mathbf{q} = q_x\mathbf{i} + q_y\mathbf{j}$ (in units of force/length) in its plane. The load $\mathbf{q}$ may be regarded as known for the present, although there are questions where $\mathbf{q}$ is an unknown. A free-body diagram of the whole cable appears in Fig. 4.2-2, with the end reactions represented as tension forces of magnitudes T_A and T_B in directions determined by the slope angles θ_A and θ_B of the tangents to the curve at the endpoints. The problem is statically indeterminate, since there are four unknowns, T_A, T_B, θ_A, and θ_B, and only three non-trivial equilibrium equations. We expect, therefore, that we must add some statement about the cable geometry to the equations of equilibrium in order to obtain a solution. This statement will derive from the assumption of *inextensibility*;

i.e., the cable will be assumed to remain unchanged in length under load.

   In order to proceed further, we examine the relations between the changing shape of the cable and the changing tension along the cable under the applied loads.   Suppose we make a cut normal to the cable at some typical point $s = s_0$, and examine the distribution of

Fig. 4.2-3

force on the section (Fig. 4.2-3a).   The assumption of *perfect flexibility* implies that the cable cannot sustain forces tangent to the section, and that the normal forces are equipollent to a single force, without any resultant moment, at the centroid of the section.   This force is the tension force of magnitude $T(s)$ and direction tangent to the

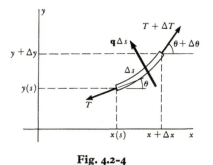

Fig. 4.2-4

center line of the cable.   We replace the cable, in fact, by its center line as our model for analysis.   The distributed load is thought of as applied at the center line, and will be considered either as a force $\mathbf{q}(s)$ per unit length of arc or a force $\mathbf{q}(x)$ or $\mathbf{q}(y)$ per unit x- or y-distance.

   Let us now isolate a small segment of cable of length Δs and draw a free-body diagram of it (Fig. 4.2-4).   The forces acting on the segment are three in number: the tension force of magnitude $T = T(s)$ in the negative tangential direction at the end $[x(s), y(s)]$, the tension force of magnitude $T(s+\Delta s) = T + \Delta T$ in the positive tangential direction at the end $(x+\Delta x, y+\Delta y)$, and the resultant* $\mathbf{q}(s) \Delta s$ of the distributed load. The distributed force $\mathbf{q}(s)$ is assumed to be given per unit of arclength. To write the force equilibrium equations for the cable segment, we can use either the fixed x-, y-directions or the intrinsic tangential and normal directions. We shall do both, for both forms of the equations are useful and important. We shall begin with the cartesian x-, y-form of the equations.

As shown in Fig. 4.2-4, we denote the slope angle at (x, y) by $\theta(s)$ and the slope angle at $(x+\Delta x, y+\Delta y)$ by $\theta + \Delta\theta$. The distributed force is written in cartesian components: $\mathbf{q}(s) = q_x(s)\mathbf{i} + q_y(s)\mathbf{j}$. The equations of force equilibrium are

$$\sum R_x = (T+\Delta T) \cos(\theta+\Delta\theta) - T \cos\theta + q_x \Delta s = 0,$$

$$\sum R_y = (T+\Delta T) \sin(\theta+\Delta\theta) - T \sin\theta + q_y \Delta s = 0. \qquad \textbf{4.2-8a}$$

Using the trigonometric formulas for the sine and cosine of the sum of two angles, we can rewrite these equations:

$$\sum R_x = \Delta T \cos(\theta+\Delta\theta)$$
$$- T \sin\theta \sin\Delta\theta + T \cos\theta(\cos\Delta\theta - 1) + q_x \Delta s = 0,$$

$$\sum R_y = \Delta T \sin(\theta+\Delta\theta)$$
$$+ T \cos\theta \sin\Delta\theta + T \sin\theta(\cos\Delta\theta - 1) + q_y \Delta s = 0. \qquad \textbf{4.2-8b}$$

---

* Properly we should consider the resultant $\int_s^{s+\Delta s} \mathbf{q}(\sigma) \, d\sigma$ of the distributed load on a finite segment $(s, s+\Delta s)$, as well as the tensions. We should also write a moment equilibrium equation, as well as the force equations, divide by Δs, and examine the limiting forms as $\Delta s \to 0$. When this is done, we find that the moment equation becomes trivial, i.e., of the $0 = 0$ form, and gives us no information. Moreover, in dealing with the force and moment resultants of the distributed forces, it is convenient to use the mean-value theorem of the integral calculus, which states that integrals such as $\int_s^{s+\Delta s} \mathbf{q}(\sigma) \, d\sigma$ can be replaced by $\mathbf{q}(s+\epsilon \Delta s) \Delta s, 0 \leq \epsilon \leq 1$, the number ϵ approaching zero as Δs approaches zero. It is implied in the main text development that all of these replacement and limiting operations can be performed rigorously. The model of the segment is, therefore, essentially a particle model for which force equations of equilibrium alone suffice.

We shall now divide by Δs and pass to the limit $\Delta s \to 0$. Of course, ΔT and $\Delta \theta$ approach zero with Δs. Also, we make use of the two fundamental limits

$$\lim_{\Delta\theta\to 0} \frac{\sin \Delta\theta}{\Delta\theta} = 1, \qquad \lim_{\Delta\theta\to 0} \frac{1-\cos \Delta\theta}{\Delta\theta} = 0 \qquad \textbf{4.2-9}$$

by multiplying and dividing the second and third terms in both of Eqs. 4.2-8b by $\Delta\theta$ before passing to the limit. We find

$$\frac{dT}{ds} \cos \theta - T \sin \theta \frac{d\theta}{ds} + q_x(s) = 0,$$

$$\frac{dT}{ds} \sin \theta + T \cos \theta \frac{d\theta}{ds} + q_y(s) = 0 \qquad \textbf{4.2-10a}$$

as our differential equations of equilibrium in the x- and y- directions. A more compact form can be given to each equation:

$$\frac{d}{ds}(T \cos \theta) + q_x(s) = 0,$$

$$\frac{d}{ds}(T \sin \theta) + q_y(s) = 0. \qquad \textbf{4.2-10b}$$

Equations 4.2-10b may be interpreted in the following way. Since $T \cos \theta$ and $T \sin \theta$ are the x- and y-components of the cable tension, respectively, the equations state that the rate of change with arclength of the cable tension component in the x- or y-direction is the negative of the applied load (per unit arclength) in that direction.

As we have mentioned, the intrinsic coordinate form of the equations is useful. Suppose we solve Eqs. 4.2-10a for dT/ds by multiplying the first equation by $\cos \theta$, the second by $\sin \theta$, and adding:

$$\frac{dT}{ds} + q_x \cos \theta + q_y \sin \theta = 0.$$

Now $q_x \cos \theta + q_y \sin \theta$ is the component of $\mathbf{q}$ in the tangential direction. From Eqs. 4.2-6,

$$q_t = \mathbf{q} \cdot \mathbf{e}_t = q_x \cos \theta + q_y \sin \theta$$

and

$$q_n = \mathbf{q} \cdot \mathbf{e}_n = -q_x \sin \theta + q_y \cos \theta \qquad \textbf{4.2-11}$$

are the tangential and normal components* of the distributed load at any point. Therefore, the equation for dT/ds can be written

$$\frac{dT}{ds} + q_t(s) = 0,$$

4.2-12a

and, similarly, an equation for the curvature $d\theta/ds$ can be found:

$$T\frac{d\theta}{ds} + q_n(s) = 0.$$

4.2-12b

From Eq. 4.2-12a, we conclude that the tension changes magnitude along the curve only if there is a load tangent to the curve.

Finally, if the load **q** is considered to be a function of some other position variable, say x, instead of arclength, then all the integrations and limiting processes should be performed with respect to that variable instead of s. The equilibrium equations 4.2-10 are replaced by similar ones with derivatives with respect to x replacing derivatives with respect to s:

$$\frac{d}{dx}(T\cos\theta) + q_x(x) = 0,$$

$$\frac{d}{dx}(T\sin\theta) + q_y(x) = 0.$$

4.2-13

We now have two differential equations in the two unknowns, T and θ. Our boundary conditions, however, are not on T and θ but rather on $y=f(x)$: $y_A=f(x_A)$, $y_B=f(x_B)$. Equation 4.2-1, $dy/dx=\tan\theta$ (or its equivalent), must be used. But now we have three equations and only two boundary conditions; the problem is still statically indeterminate. The third condition that shall be imposed,

---

* It should be emphasized that the sign conventions for positive q_t and q_n depend upon the choice we have made for the positive tangential and normal directions—that is, on the direction of increasing arclength and on the 90° counterclockwise rotation from that direction. In a particular problem, we may find it more convenient to use a 90° clockwise convention for the positive normal—in which case Eqs. 4.2-6, 11, and 12 should be suitably modified. Equation 4.2-12b, in particular, may cause difficulty; but an examination of the loading conditions and of the sense in which θ increases should let one decide in each case whether $T\,d\theta/ds$ equals $+q_n$ or $-q_n$. For examples of the direct derivation of normal and tangential equilibrium equations, the reader is referred to Section 5.4 on belt and cable friction.

as we have mentioned, is that of inextensibility: the length L of the cable is a constant, and the cable shape $y = f(x)$ must satisfy Eq. 4.2-4:

$$L = \int_{x_A}^{x_B} \left[1 + \left(\frac{dy}{dx} \right)^2 \right]^{\frac{1}{2}} dx.$$

Example 4.2-1

A belt passes around a smooth pulley of radius r, covering an arc of θ_0 radians (Fig. 4.2-5a). From equilibrium considerations, it is found that the tension in the belt is a constant T_0 at either end of the segment touching the pulley. What is the force exerted by the pulley on the belt?

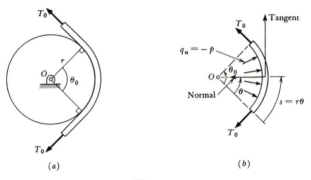

Fig. 4.2-5

Solution: The belt segment in contact with the pulley is circular in shape, so that here the equilibrium curve is known and the distributed loads are to be found. The free-body diagram of the segment is shown in Fig. 4.2-5b. The contact is smooth, so that no tangential loads q_t are exerted. The normal load q_n is of the form of a pressure per unit length, $q_n = -p(s)$. From Eq. 4.2-12a, we have

$$\frac{dT}{ds} + q_t = \frac{dT}{ds} = 0,$$

so that the tension is constant at any section: $T = T_0$. From the circular shape of the cable, we have $s = r\theta$, $d\theta/ds = 1/r$; the second of Eqs. 4.2-12 becomes

$$T \frac{d\theta}{ds} + q_n = \frac{T_0}{r} - p = 0,$$

and the pressure distribution is uniform in magnitude:

$$p = \frac{T_0}{r}.$$

Example **4.2-2**

A suspension bridge deck of weight W is supported by vertical stringers from a light cable of length L which is supported at its ends by towers of equal height (Fig. 4.2-6). Assuming that the deck weight can be considered to be distributed uniformly along the horizontal span of length l, what is the tension on the cable at either support? What is the maximum sag of the cable?

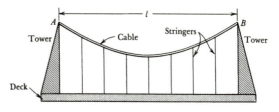

Fig. 4.2-6

Solution: The distributed load here is a constant $w = W/l$ in the vertically downward direction. The weight of the cable itself is neglected by comparison with this load. The uniformity of the load and the symmetry of the support conditions suggest that the equilibrium shape will be symmetrical, with a lowest point having horizontal tangent occurring at the midspan. A free-body diagram of the whole cable, with appropriate coordinates, is shown in Fig. 4.2-7. The sag is denoted by f; the supports are then at $x = \pm l/2$, $y = f$.

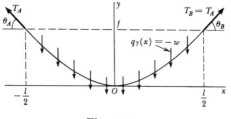

Fig. 4.2-7

Since the distributed load is given per unit horizontal span, Eqs. 4.2-13 with $q_x = 0$, $q_y = -w$ are used. The x-equation may be integrated immediately to obtain

$$T \cos \theta = H, \qquad \qquad \textbf{4.2-14}$$

where H is a constant of integration. Here it may be interpreted as the tension in the cable at the lowest point O, since the tension there is the same as its horizontal component $T \cos \theta$. Solving Eq. 4.2-14 for $T(x)$, we find

$$T(x) = H \sec \theta(x), \qquad \qquad \textbf{4.2-15}$$

which, when substituted in the second of Eqs. 4.2-13, leads to an equation for $\theta(x)$:

$$\frac{d}{dx}(H \tan \theta) - w = 0.$$

Since (Eq. 4.2-1) $\tan \theta = dy/dx$, we find that

$$\frac{d^2 y}{dx^2} = \frac{w}{H} \qquad \qquad \textbf{4.2-16}$$

is the differential equation of the cable curve. Integrating this under the conditions $y = dy/dx = 0$ at $x = 0$, we find the parabola

$$y = \frac{w}{2H} x^2 = \frac{W}{2Hl} x^2 \qquad \qquad \textbf{4.2-17}$$

as the shape of the curve.

In order to select the correct parabola, we must determine H or some equivalent parameter. One such parameter which determines the shape and which is also easily measured is the sag f. The relation between f and H is

$$f = \frac{w}{2H} \left(\frac{l}{2} \right)^2 = \frac{Wl}{8H}. \qquad \qquad \textbf{4.2-18}$$

To find H, we must use the inextensibility condition Eq. 4.2-4:

$$L = \int_{-l/2}^{l/2} \left[1 + \left(\frac{wx}{H} \right)^2 \right]^{\frac{1}{2}} dx = 2 \int_{0}^{l/2} \left[1 + \frac{w^2 x^2}{H^2} \right]^{\frac{1}{2}} dx.$$

This may be found in standard tables of integrals or integrated directly by use of the substitution

$$\frac{wx}{H} = \sinh u, \qquad \frac{w \, dx}{H} = \cosh u \, du.$$

Then

$$L = \frac{2H}{w} \int_{0}^{\text{arcsinh} \left(\frac{wl}{2H} \right)} \cosh^2 u \, du = \frac{H}{w} \int_{0}^{\text{arcsinh} \left(\frac{wl}{2H} \right)} (1 + \cosh 2u) \, du$$

$$= \frac{H}{w} \left[u + \frac{1}{2} \sinh 2u \right] \Big|_{0}^{\text{arcsinh} \left(\frac{wl}{2H} \right)}$$

$$= \frac{H}{w} [u + \sinh u \cosh u] \Big|_{0}^{\text{arcsinh} \left(\frac{wl}{2H} \right)}$$

or

$$L = \frac{H}{w} \left\{ \text{arcsinh} \left(\frac{wl}{2H} \right) + \frac{wl}{2H} \left[1 + \left(\frac{wl}{2H} \right)^2 \right]^{\frac{1}{2}} \right\}. \qquad \textbf{4.2-19}$$

This equation defines H in terms of the known quantities w, l, and L implicitly, and must be solved by numerical means. Dividing both sides of 4.2-19 by l, we may put the equation in a more convenient form in terms of non-dimensional parameters L/l (length-to-span ratio) and W/H:

$$\frac{2L}{l} = \left[1 + \left(\frac{W}{2H} \right)^2 \right]^{\frac{1}{2}} + \left(\frac{W}{2H} \right)^{-1} \text{arcsinh}\left(\frac{W}{2H} \right). \qquad \textbf{4.2-20}$$

Since the sag f is related to H through Eq. 4.2-18, we can also find the sag, or rather the ratio f/l of the sag to the span, directly from the length-to-span ratio:

$$\frac{2L}{l} = \left[1 + \left(\frac{4f}{l} \right)^2 \right]^{\frac{1}{2}} + \left(\frac{4f}{l} \right)^{-1} \text{arcsinh}\left(\frac{4f}{l} \right). \qquad \textbf{4.2-21}$$

When the span and cable length are of the same order of magnitude, the sag can be expected to be small. In this case, when f/l is small compared to unity, power series expansion of the right-hand side of Eq. 4.2-21 leads to an approximate result. Since

$$\left[1 + \left(\frac{4f}{l} \right)^2 \right]^{\frac{1}{2}} = 1 + \frac{1}{2}\left(\frac{4f}{l} \right)^2 - \frac{1}{8}\left(\frac{4f}{l} \right)^4 + \cdots,$$

$$\frac{\text{arcsinh}\,(4f/l)}{(4f/l)} = 1 - \frac{1}{6}\left(\frac{4f}{l} \right)^2 + \frac{3}{40}\left(\frac{4f}{l} \right)^4 + \cdots,$$

we have (neglecting terms of order four)

$$\frac{2L}{l} \simeq 2 + \frac{1}{3}\left(\frac{4f}{l} \right)^2.$$

Therefore, for small sag ratios, we have

$$\frac{4f}{l} \simeq \left[6\left(\frac{L}{l} - 1 \right) \right]^{\frac{1}{2}} \qquad \textbf{4.2-22a}$$

and

$$\frac{2H}{W} = \left(\frac{4f}{l} \right)^{-1} \simeq \left[6\left(\frac{L}{l} - 1 \right) \right]^{-\frac{1}{2}}. \qquad \textbf{4.2-22b}$$

Having found both exact and approximate expressions for determining the sag, we turn now to the other question asked: what is the tension $T_A = T_B$ at either support, which from Eq. 4.2-15 is the maximum tension in the parabolic cable? Since (from 4.2-17)

$$\left. \frac{dy}{dx} \right|_{x=l/2} = \tan \theta_B = \left(\frac{W}{Hl} \right)\left(\frac{l}{2} \right) = \frac{W}{2H},$$

we find

$$\sec \theta_B = [1 + \tan^2 \theta_B]^{\frac{1}{2}}$$

$$= \left[1 + \left(\frac{W}{2H} \right)^2 \right]^{\frac{1}{2}} = \left[1 + \left(\frac{4f}{l} \right)^2 \right]^{\frac{1}{2}}.$$

The maximum tension, therefore, is given by

$$\frac{2T_{max}}{W} = \frac{2H}{W} \sec \theta_B$$

$$= \frac{2H}{W} \left[1 + \left(\frac{W}{2H} \right)^2 \right]^{\frac{1}{2}} = \left[\left(\frac{2H}{W} \right)^2 + 1 \right]^{\frac{1}{2}} \qquad \textbf{4.2-23a}$$

$$= \left(\frac{4f}{l} \right)^{-1} \left[1 + \left(\frac{4f}{l} \right)^2 \right]^{\frac{1}{2}} = \left[\left(\frac{l}{4f} \right)^2 + 1 \right]^{\frac{1}{2}}. \qquad \textbf{4.2-23b}$$

Any number of special problems can now be solved. For instance, suppose we are given the span and weight to be carried and the maximum allowable cable tension, and we ask for the least length of cable necessary to support the weight. From 4.2-23b we can compute the allowable sag ratio and then compute the cable length needed from 4.2-21 or the approximate relation 4.2-22a.

Example 4.2-3

What shape does a uniform cable hanging under its own weight assume in equilibrium? The supports are at equal height.

Solution: From uniformity of load and symmetry of the external geometry of support, Fig. 4.2-7 can again serve as the free-body diagram of the whole cable, provided that we now take $q_y = -w = -W/L$ to be the distributed weight per unit length of cable as a function of the arclength s. The fundamental equations are now Eqs. 4.2-10, the first of which again integrates immediately to the result that the x-component of the tension is constant, or, paralleling Eq. 4.2-15,

$$T(s) = H \sec \theta(s). \qquad \textbf{4.2-24}$$

The second of Eqs. 4.2-10b becomes

$$\frac{d}{ds} (H \tan \theta) - w = 0,$$

so that

$$H \tan \theta = ws + V.$$

If we measure arclength s from the lowest point O of the curve, with increasing s agreeing with increasing x, the constant of integration $V=0$. Therefore,

$$\frac{dy}{dx} = \tan \theta = \frac{ws}{H}. \qquad \textbf{4.2-25}$$

To integrate this, we must use Eqs. 4.2-5b. Since $1 + \tan^2 \theta = \sec^2 \theta$ and $\sin \theta = \tan \theta \cos \theta$, we find

$$\frac{dx}{ds} = \cos \theta = [1 + \tan^2 \theta]^{-\frac{1}{2}} = \left[1 + \left(\frac{ws}{H} \right)^2 \right]^{-\frac{1}{2}},$$

$$\frac{dy}{ds} = \sin \theta = \frac{ws}{H} \left[1 + \left(\frac{ws}{H} \right)^2 \right]^{-\frac{1}{2}}. \qquad \textbf{4.2-26}$$

Again, we may refer to tables of elementary integrals or else integrate directly (with $x = y = 0$ at $s = 0$ as conditions) by substituting

$$\frac{ws}{H} = \sinh u, \qquad \frac{w\,ds}{H} = \cosh u\,du.$$

We obtain:

$$x = \int_0^s \left[1 + \left(\frac{ws}{H}\right)^2 \right]^{-\frac{1}{2}} ds = \frac{H}{w} \int_0^{\operatorname{arcsinh}\left(\frac{ws}{H}\right)} \frac{\cosh u}{\cosh u}\,du$$

$$= \frac{H}{w} \operatorname{arcsinh}\left(\frac{ws}{H}\right);$$ **4.2-27a**

$$y = \int_0^s \frac{ws}{H} \left[1 + \left(\frac{ws}{H}\right)^2 \right]^{-\frac{1}{2}} ds = \frac{H}{w} \int_0^{\operatorname{arcsinh}\left(\frac{ws}{H}\right)} \sinh u\,du$$

$$= \frac{H}{w} \left[\cosh u \right]_0^{\operatorname{arcsinh}\left(\frac{ws}{H}\right)} = \frac{H}{w} \left\{ \cosh \left[\operatorname{arcsinh}\left(\frac{ws}{H}\right) \right] - 1 \right\}.$$ **4.2-27b**

These are the parametric equations of the curve. Since s may be expressed as a function of x easily from 4.2-27a, we can find the $y = f(x)$ form of the curve equation:

$$y = \frac{H}{w} \left\{ \cosh \left(\frac{wx}{H}\right) - 1 \right\}.$$ **4.2-28**

Such a curve is called a *catenary*.

The condition that the curve pass through points A and B again relates the sag f to the constant H:

$$f = \frac{H}{w} \left\{ \cosh \left(\frac{wl}{2H}\right) - 1 \right\}.$$ **4.2-29**

The determination of H follows from the inextensibility condition. Rather than perform the integration, however, we can find the appropriate relation from Eq. 4.2-27a, since $s = L/2$ when $x = l/2$:

$$\frac{wL}{2H} = \sinh \left(\frac{wl}{2H}\right),$$ **4.2-30a**

or (with $wL = W$),

$$\frac{W}{2H} = \sinh \left[\frac{W}{2H} \left(\frac{l}{L}\right) \right].$$ **4.2-30b**

Again, this must be solved numerically in general. For small sag ratios ($f/l \ll 1$), i.e., for $l \simeq L$, approximations based on series expansions may be used in Eqs. 4.2-29 and 4.2-30. From Eq. 4.2-29, we have

$$\frac{2f}{l} = \frac{2H}{wl} \left[\cosh \left(\frac{wl}{2H}\right) - 1 \right]$$

$$\simeq \frac{1}{2} \left(\frac{wl}{2H}\right) = \frac{W}{4H} \left(\frac{l}{L}\right),$$ **4.2-31a**

neglecting terms of order $(wl/2H)^3$; using this in 4.2-30, we find

$$\frac{W}{2H} \cong \sinh\left(\frac{4f}{l}\right),$$

4.2-31b

or

$$\frac{L}{l} \cong \frac{\sinh\left(\dfrac{4f}{l}\right)}{\left(\dfrac{4f}{l}\right)} \cong 1 + \frac{1}{6}\left(\frac{4f}{l}\right)^2.$$

4.2-31c

Solving for the sag ratio, we find

$$\frac{4f}{l} \cong \left[6\left(\frac{L}{l} - 1\right)\right]^{\frac{1}{2}},$$

4.2-32

which should be compared with Eq. 4.2-22 for some idea of the degree to which the parabola approximates the catenary for small sag ratios. The minimum tension H, as determined from 4.2-31a or 4.2-22b, is somewhat different for the same l, L, f, and W.

Example 4.2-4

A cable of length L hangs under its own weight over a span l. For what range of vertical distance h between supports does the cable not sag below the lowest support?

Solution: The maximum vertical distance between supports possible for a given length L and given span l occurs, of course, when the cable is along the straight line joining the supports: $h_{\max}^2 = L^2 - l^2$. As h decreases from this value, the cable hanging under its own weight will assume the shape of a piece of a catenary until, when $h = 0$, the symmetrical catenary of the last example is attained. For some range $h^* \leqq h \leqq h_{\max}$, no lowest point with horizontal tangent will occur between the supports. The value h^* is determined by the condition that the slope of the curve is just horizontal at one support.

Under such conditions, when no external geometric symmetry is imposed, the x-, y-coordinates may as well be chosen with origin at one support, say the lower one. Indeed, in the previous two examples, where the distributed loads are constant in magnitude and direction, one need not use the full differential equations of equilibrium but can write algebraic equations for a finite segment of cable. Take a segment between the lowest point, where the tension is H, and a general point $x(s)$, $y(s)$, where the tension is T, and replace the distributed load $-w\mathbf{j}$ by an equipollent concentrated load $-wx\mathbf{j}$ or $-ws\mathbf{j}$ passing through the point of concurrence on the x-axis of $-H\mathbf{i}$ and $T\mathbf{e}_t$. This finite segment is then equivalent to the three-force member of Example 2.4-2, and we have effectively performed one integration of the equations of equilibrium.

In this example, however, or any example wherein geometrical symmetry cannot be assumed to occur, whether due to load distribution or external support geometry, we must use the full differential equations. The free-body diagram of the full cable appears in Fig. 4.2-8. Since $q_z=0$ again, the first of Eqs. 4.2-10b integrates to

$$T(s) = H \sec \theta(s)$$

as before (Eq. 4.2-24); and the second equation will integrate to

$$H \tan \theta = ws + V. \qquad \textbf{4.2-33}$$

Now, in general, V does not vanish when $s=0$ at support A, for $\theta = \theta_A$ is not zero. Since, however, we want to determine the limits on h so that no relative minimum point occurs in $0 \leq x \leq l$, we may take the limiting

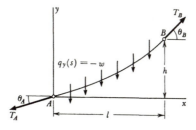

$q_y(s) = -w$

Fig. 4.2-8

condition $h=h^*$ for which $\theta_A=0$. Then $V=0$, and the integration of 4.2-33 proceeds as in the last example, with $x=y=s=0$, to give Eqs. 4.2-27 and 4.2-28. The solution to our problem, therefore, is that the cable with the just-horizontal tangent at the support looks like the right half of the cable of the last example; now the end condition at B, however, is $x=l$ and $y=h^*$ when $s=L$. That is, substituting in Eqs. 4.2-27a and 4.2-28, we find

$$\frac{wL}{H} = \sinh\left(\frac{wl}{H}\right) \qquad \textbf{4.2-34a}$$

and

$$h^* = \frac{H}{w}\left[\cosh\left(\frac{wl}{H}\right) - 1\right]. \qquad \textbf{4.2-34b}$$

These differ from 4.2-30a and 4.2-29 of the previous example. The first of 4.2-34 is to be solved (numerically) for H in terms of l, L, and $w=W/L$; the second then determines h^*.

Example 4.2-5

A 50 ft cable weighing 10 lb/ft hangs between two supports on the same level. The observed central sag is 15 ft. How far apart are the supports? What are the minimum and maximum tensions in the cable?

Solution: Here we show how numerical results are obtained from the equations that have been derived. The problem is a particular example of the symmetrical catenary for which the general solution was derived in Example 4.2-3. We are given $L = 50$ ft, $f = 15$ ft, $-q_y = w = 10$ lb/ft, $W = wL = 500$ lb; we wish to find l, H, and T_{max}, the tension at either support.

The first decision that should be made is whether the approximate or exact theory should be used. We note that, if the 50 ft cable were simply supporting a concentrated 500 lb load at its center, so that the two 25 ft halves were straight, then a 15 ft sag would give a half-span of $(25^2 - 15^2)^{1/2} = 20$ ft or a total span of $l = 40$ ft. Thus a first rough approximation to the span is forty feet, and whether or not an apparent $f/l \cong 3/8$ is "small" compared with unity, or a possible $l/L \cong 4/5$ is "close enough" to one, for the next refinement of approximate theory to be used depends upon judgment.

Let us suppose that it is proper to use the approximate theory. Then, from Eq. 4.2-31c or Eq. 4.2-32, we may write

$$\frac{L}{l} = 1 + \frac{1}{6}\left(\frac{4f}{l}\right)^2 = 1 + \frac{1}{6}\left(\frac{4f}{L}\right)^2\left(\frac{L}{l}\right)^2,$$

$$0.24\left(\frac{L}{l}\right)^2 - \frac{L}{l} + 1 = 0.$$

Solving for L/l, we find

$$\frac{L}{l} = \frac{1 \pm \sqrt{1 - 0.96}}{2(0.24)} = \frac{1 \pm 0.2}{0.48}. \qquad \text{4.2-35}$$

Both roots are larger than two in value, and certainly do not appear close to one. The approximate theory should, therefore, be discarded.

We turn to the exact equations 4.2-29 and 4.2-30a and form their ratio:

$$\frac{2f}{L} = \frac{\cosh\left(\dfrac{wl}{2H}\right) - 1}{\sinh\left(\dfrac{wl}{2H}\right)} = \tanh\left(\frac{wl}{4H}\right). \qquad \text{4.2-36}$$

This is a transcendental equation for the quantity $\alpha = wl/4H$. If α can be found, then we can compute H from Eq. 4.2-30a:

$$\frac{wL}{2H} = \frac{W}{2H} = \sinh 2\alpha, \qquad \text{4.2-37}$$

and then l from α:

$$l = \frac{4H\alpha}{w}. \qquad \text{4.2-38}$$

Our whole problem, therefore, rests on the solution of

$$\tanh \alpha = \frac{2f}{L} = 0.6. \qquad \text{4.2-39}$$

Reference to tables of the hyperbolic tangent shows that $0.69 < \alpha < 0.70$. If a more accurate answer is desired than can be obtained from such a table and if tables of natural logarithms are available, then the definition of the hyperbolic tangent proves useful:

$$\tanh \alpha = \frac{e^{\alpha} - e^{-\alpha}}{e^{\alpha} + e^{-\alpha}} = 0.6;$$

$$0.4e^{\alpha} = 1.6e^{-\alpha}, \qquad e^{2\alpha} = 4, \qquad e^{\alpha} = 2;$$

$$\alpha = \log 2 = 0.69315. \tag{4.2-40}$$

Therefore,

$$\sinh 2\alpha = (e^{2\alpha} - e^{-2\alpha})/2 = 15/8,$$

and

$$H = \frac{W}{2 \sinh 2\alpha} = \frac{400}{3} \text{ lb}, \tag{4.2-41a}$$

$$l = \frac{4H\alpha}{w} = 36.968 \text{ ft}; \tag{4.2-41b}$$

further, using Eqs. 4.2-24, 4.2-25, and 4.2-37, we find

$$T_{\max} = H \sec\left(\theta\left(\frac{L}{2}\right)\right) = H\left[1 + \left(\frac{wL}{2H}\right)^2\right]^{\frac{1}{2}}$$

$$= H \cosh 2\alpha = \left(\frac{400}{3}\right)\left(\frac{17}{8}\right)$$

$$= \frac{850}{3} \text{ lb} \cong 283 \text{ lb}. \tag{4.2-41c}$$

4.3   Statically Determinate Beams; Moment and Shear Diagrams

In Section 2.7, the beam subject to loads in a median plane was introduced and the resultant force set equipollent to the internal forces on a cross-section was defined (Fig. 2.7-4). In Example 2.7-5, the resultant set at one cross-section due to given loads on a particular simply-supported beam was found. Here we wish to consider the general problem of finding the resultant set at any cross-section by establishing the differential equations of equilibrium satisfied by the force and couple components of the resultant.

First, we describe the geometry of the straight beam of uniform cross-section and the coordinate system to be used (Fig. 4.3-1). The beam is a right cylinder of length l constructed on an area A, which we will take to be doubly symmetric. The center line of the beam passes through the centroids of the cross-sections, and will be taken as the x-axis; the y- and z-axes will be taken as shown at each section, along the lines of symmetry of the cross-section. (Removal of the symmetry condition will be discussed later.)

Let us isolate a length of the beam under load to the left of some cross-section normal to the center line, with the effect of the material removed being represented by a distribution of forces over the area

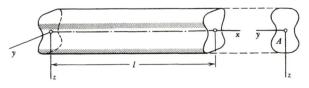

Fig. 4.3-1

of the cross-section (Fig. 4.3-2a). These are replaced by a resultant force **F** at the centroid of the section and a resultant couple **C**. The x-component of **F** is the axial force N of Section 2.7; the z-component is the shear force V of Section 2.7. The y-component of **F** is another shear force Q in the plane of the section; we shall consider for the present only those cases where we may set $Q=0$. Therefore, $\mathbf{F}=F_x\mathbf{i}+F_z\mathbf{k}=N\mathbf{i}+V\mathbf{k}$. The x-component of **C** is a torque about the axis of the cylinder and the z-component is a bending moment

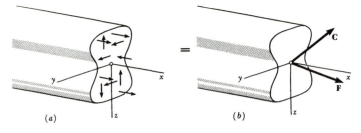

Fig. 4.3-2

about the z-axis; these torsional and bending moments will also be set equal to zero for the straight beam problem. The only resultant moment will then be the bending moment about the y-axis, which we shall denote by M as in Section 2.7: $\mathbf{C}=M\mathbf{j}$. If N and V are zero and only $M\neq0$ over a length of beam, that length is said to be in pure bending; if $N=0$ but M and $V\neq0$, the beam is in flexure with shear;

and, if M, V, and $N \neq 0$, we speak of a beam-column in combined flexure and tension or compression.

We shall now establish the differential equations of equilibrium of a beam-column subjected to distributed loads; the effect of concentrated loads will be considered later.   Since the bending moment, shear force, and axial load depend upon the value of x, that is, upon which cross-section is being considered, we may expect the equations

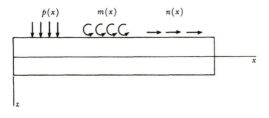

Fig. 4.3-3

to be ordinary differential equations with x as independent variable. The external loads will then be given as functions of the distance along the center line, i.e., will be forces and couples per unit length of beam. The distributed loads are considered to act in the xz-plane along the x-axis, and the distributed couples are due to forces in that plane such

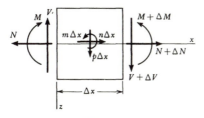

Fig. 4.3-4

that their resultant is a moment about the y-axis.   Let $\mathbf{q}(x) = n(x)\mathbf{i} + p(x)\mathbf{k}$ be the force per unit length and $\mathbf{c}(x) = m(x)\mathbf{j}$ be the moment per unit length (Fig. 4.3-3).   For the beam problem, we are, in technology, usually most concerned with the case $n(x) = m(x) = 0$.

Consider a small element of the beam of length Δx (Fig. 4.3-4). The forces and couples on the beam consist of (N, V, M) in the negative coordinate directions at the generic section x, $(N + \Delta N, V + \Delta V, M + \Delta M)$ in the positive senses at $x + \Delta x$, and the resultants

$(n\,\Delta x,\,p\,\Delta x,\,m\,\Delta x)$ of the distributed loads. The equilibrium equations for the segment are

$$\sum R_x = (N+\Delta N)-N+n(x)\,\Delta x = 0,$$

$$\sum R_z = (V+\Delta V)-V+p(x)\,\Delta x = 0, \qquad \text{4.3-1}$$

$$\sum M_y = (M+\Delta M)-M-(V+\Delta V)\,\Delta x+m(x)\,\Delta x = 0.$$

Simplifying these equations, we find

$$\Delta N+n(x)\,\Delta x = 0,$$

$$\Delta V+p(x)\,\Delta x = 0, \qquad \text{4.3-2}$$

$$\Delta M-V\,\Delta x+m\,\Delta x-\Delta V\,\Delta x = 0.$$

Dividing each of Eqs. 4.3-2 by Δx and passing to the limit as $\Delta x\to 0$, we find

$$\frac{dN}{dx}+n(x) = 0,$$

$$\frac{dV}{dx}+p(x) = 0, \qquad \text{4.3-3}$$

$$\frac{dM}{dx}-V+m(x) = 0$$

as the differential equations of equilibrium.

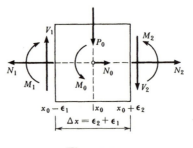

Fig. 4.3-5

Before discussing Eqs. 4.3-3, let us derive the "jump" conditions of equilibrium for concentrated loads. Suppose a concentrated load $N_0\mathbf{i}+P_0\mathbf{k}$ is applied at $x=x_0$, as is a concentrated couple $M_0\mathbf{j}$. Isolate a segment of the beam which includes that section $x=x_0$, and suppose no other external loads act on the segment except the given ones and the resultants on the cross-sections (Fig. 4.3-5). The equilibrium equations are

$$N_2 - N_1 + N_0 = 0,$$
$$V_2 - V_1 + P_0 = 0,$$
$$M_2 - M_1 + M_0 - \epsilon_1 V_1 - \epsilon_2 V_2 = 0;$$

letting ϵ_1 and ϵ_2 approach zero, so that the segment approaches the cross-section $x = x_0$, we see that equilibrium requires that

$$[N(x_0)] = -N_0, \qquad [V(x_0)] = -P_0, \qquad [M(x_0)] = -M_0, \qquad \textbf{4.3-4}$$

where the notation $[f(x_0)]$ denotes a discontinuity, or "jump," in the value of the function $f(x)$ at $x = x_0$. That is,

$$[f(x_0)] = \lim_{\substack{\varepsilon_1 \to 0 \\ \varepsilon_2 \to 0}} \{f(x_0 + \epsilon_2) - f(x_0 - \epsilon_1)\} = f(x_0^+) - f(x_0^-). \qquad \textbf{4.3-5}$$

The equations of equilibrium 4.3-3 state that the axial force N and shear force V decrease along the length of the beam at a rate equal to the applied load per unit length in the axial or transverse directions. The bending moment M decreases with the applied couple and increases at a rate equal to the shear force on the section. The jump conditions 4.3-4 are equivalents to 4.3-3 when the applied loads are concentrated; note that the shear force effect does not appear in the jump condition for "concentrated" moment.

The shear force may be eliminated from the last two of Eqs. 4.3-3, resulting in a second-order differential equation for $M(x)$ in terms of the loads. Differentiate the moment equation and substitute for dV/dx from the shear equation:

$$\frac{d^2M}{dx^2} - \frac{dV}{dx} + \frac{dm}{dx} = 0,$$

so that

$$\frac{d^2M}{dx^2} + \frac{dm}{dx} + p(x) = 0. \qquad \textbf{4.3-6}$$

Let us now turn to the solution of the equations. We shall consider only transverse loadings $p(x)$ or P_0 in the z-direction, taking $m(x) = n(x) = N_0 = M_0 = 0$ and, in fact, $N(x) \equiv 0$. These effects are easily accounted for if necessary. To integrate the equations, we must adjoin boundary conditions. The boundary or support conditions for simple beams are usually taken to be one of the standard three: free, clamped, or simply-supported. Other end conditions are possible—an elastic constraint, for instance, that relates an end moment linearly to the slope of the (deformed) center line. Generally, the conditions set at the ends of the beam involve both applied forces and moments and possible deflections and slopes. The general beam problem is statically indeterminate, and, to the equilibrium equations, we must adjoin relations between the displacements and strains in the deformed beam and relations between the internal stress resultants, such as bending moment, and the strains, such as center-line curvature. The deformable beam is not of concern to us here; we shall consider only those statically determinate problems where the equations of equilibrium can be solved for the shear force and bending moment distributions in a rigid beam.

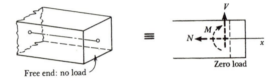

Free end: no load   Zero load

Fig. 4.3-6

The conditions at a *free* end of a beam express the fact that no load is applied: the bending moment and shear force (and axial load if considered) on any internal section must vanish as the free end is approached (Fig. 4.3-6). A simply-supported end may be pinned

Simply-supported (pinned) end

Fig. 4.3-7

or on rollers; the conditions are that the center line has zero deflection in the z-direction transverse to the center line and that no moment can be transmitted. In Fig. 4.3-7, the representation of the effect of such a support is shown: an unknown z-reaction Q, zero moment, and no axial load. (Strictly, there will be no axial load only at the smooth roller type of simple support, but the possibility of one at a pinned joint should be considered.) The clamped or built-in joint was mentioned in Chapter II; the conditions at such a point are the vanishing of deflection and slope of the center line, so that unknown moment and transverse force reactions (and, again, an axial reaction in general) are exerted (Fig. 4.3-8). It should be noted that in the free-body diagrams of Figs. 4.3-7 and 4.3-8, the sign conventions for positive shear force and bending moment have been followed in drawing the reactions; these reactions are shown applied to a cross-section of the beam that has outward normal in the negative x-direction. The end conditions on M and V as we approach such a section from the positive x-direction are then $M = 0$, $V = Q$ for the simple support and $M = C$, $V = Q$ for the clamped end.

Fig. 4.3-8

Combinations of these end conditions, one at $x = 0$ and the other at $x = l$, then give the boundary conditions for integrating the equations. Some combinations lead to indeterminate problems. All of our examples will be determinate problems, for which the external reactions can be found from consideration of equilibrium of the beam as a whole.

A useful device for visualizing the shear and moment distribution along the beam is the construction of the *shear and bending moment diagrams*. These are simply plots of V and M against x as abscissa. Since (from Eqs. 4.3-3) the slope of the V-x curve is the negative of the applied load $p(x)$ and that of the M-x curve is V, the graphical construction of solutions for V, and then M, is not difficult. In the study of advanced beam theory, such diagrams are helpful in identifying critical sections subject to extremal values of bending moment.

We have developed the theory for doubly-symmetric cross-sections with loads in one principal plane of bending. But we have not explained why such planes are important. This is a task for deformable-body mechanics, where it is shown that, without this symmetry, loads in the xz-plane tend to produce torsion of the beam in addition to flexure.

Example 4.3-1

Find the bending moment and shear force distributions in a uniform beam of length l due to its own weight W distributed along the beam if the beam is (a) simply-supported at both ends, and (b) cantilevered at the left end.

Solution: The weight force is taken as a constant $w = W/l$ distributed along the center line. In Fig. 4.3-9, the simply-supported beam and the cantilevered beam together with complete free-body diagrams are shown. The support reactions are easily found in each case, with the distributed load replaced by the equipollent concentrated load W at the midpoint.

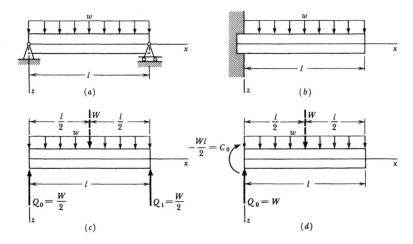

Fig. 4.3-9

The support reactions for the simply-supported beam are $Q_0 = Q_1 = W/2$, while the reactions at the built-in end of the cantilevered beam are $Q_0 = W$, $C_0 = -Wl/2$ (note the sign convention for the assumed direction of the couple in the free-body diagram 4.3-9d). In both cases, the transverse load $p(x)$ is the constant w. Let us integrate Eqs. 4.3-3, which here take the form

$$\frac{dV}{dx} + \frac{W}{l} = \frac{dV}{dx} + w = 0, \qquad \frac{dM}{dx} - V = 0.$$

In both cases, the shear force decreases linearly with x:

$$V = -wx + V_0; \qquad\qquad\qquad \textbf{4.3-7}$$

the bending moment depends quadratically on x:

$$M = -\frac{wx^2}{2} + V_0 x + M_0. \qquad\qquad \textbf{4.3-8}$$

The two cases differ in the evaluation of the constants of integration $V_0 = V(0)$ and $M_0 = M(0)$.

For the simply-supported beam, $V_0 = Q_0 = W/2$, $M_0 = 0$; therefore,

$$V(x) = -wx + \frac{W}{2} = \frac{W}{2}\left[1 - \frac{2x}{l}\right],$$

$$M(x) = -\frac{wx^2}{2} + \frac{Wx}{2} = \frac{Wx}{2}\left[1 - \frac{x}{l}\right]. \qquad \textbf{4.3-9}$$

Note that $V(x)$ vanishes at the midsection of the beam and equals $-W/2$ $= -Q_1$ at $x = l$, as it should; $M(x)$ vanishes at both ends and reaches a

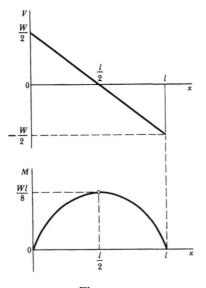

Fig. 4.3-10

maximum of $Wl/8$ at the midsection. The shear and bending moment diagrams are shown in Fig. 4.3-10.

For the cantilevered beam, $V_0 = Q_0 = W$, $M_0 = C_0 = -Wl/2$. Therefore,

$$V(x) = -wx + W = W\left(1 - \frac{x}{l}\right),$$

$$M(x) = -\frac{wx^2}{2} + Wx - \frac{Wl}{2} \qquad \textbf{4.3-10}$$

$$= -\frac{Wl}{2}\left(1 - \frac{x}{l}\right)^2.$$

Both $V(x)$ and $M(x)$ vanish, as they should, at the free end $x=l$. The shear and bending moment diagrams appear in Fig. 4.3-11.

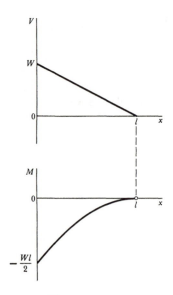

Fig. 4.3-11

It is worthwhile comparing these solutions for a uniformly distributed load to the solutions for a concentrated load W at the midpoint. Now $p(x)=0$ for $0 \leqq x < \dfrac{l}{2}, \dfrac{l}{2} < x \leqq l$, and the equations integrate to a constant V and linearly varying M for each half of the beam. Across the midsection $x=l/2$, we must use the concentrated load conditions (Eqs. 4.3-4) with $P_0 = W$, M_0 (and N_0) $=0$. Therefore,

$$\left[V\!\left(\frac{l}{2}\right) \right] = -W, \qquad \left[M\!\left(\frac{l}{2}\right) \right] = 0. \qquad\qquad \textbf{4.3-11}$$

There is a discontinuity in V of magnitude W, while M is continuous. The analytical and graphical solutions follow.

For the simply-supported beam (Fig. 4.3-12),

$$V(x) = W/2, \qquad 0 \leqq x < l/2;$$
$$ = -W/2, \qquad l/2 < x \leqq l;$$
$$M(x) = Wx/2, \qquad 0 \leqq x \leqq l/2;$$
$$ = W(l-x)/2, \qquad l/2 \leqq x \leqq l.$$

4.3-12

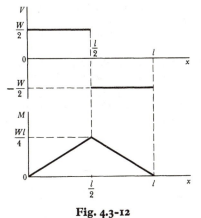

Fig. 4.3-12

For the cantilevered beam (Fig. 4.3-13),

$$V(x) = W, \qquad 0 \leqq x < l/2;$$
$$= 0, \qquad l/2 < x \leqq l;$$
$$M(x) = -\frac{Wl}{2}\left(1 - \frac{2x}{l}\right), \qquad 0 \leqq x \leqq l/2;$$
$$= 0, \qquad l/2 \leqq x \leqq l.$$

4.3-13

By comparing Eqs. 4.3-9 and 4.3-12 (or Figs. 4.3-10 and 4.3-12) for the simply-supported beam, and Eqs. 4.3-10 and 4.3-13 (Figs. 4.3-11 and 4.3-13) for the cantilever, we see that the concentrated load leads to a bending moment distribution that is algebraically larger than the bending

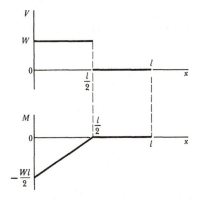

Fig. 4.3-13

moment for the uniformly distributed load. Considering absolute values of the bending moment, however, we find a difference in behavior for the two different support conditions. The moment distribution for the simply-supported beam under concentrated load is never less than that for the same beam under the distributed load, whereas, for the cantilevered beam, the reverse is true for the absolute value of M. The maximum absolute value $|M|_{max}$ in both cases is at least as great for the concentrated load as for the distributed load. The proof of this for general distributions is important for the development of a collapse design criterion for the safe loads that a beam may carry.

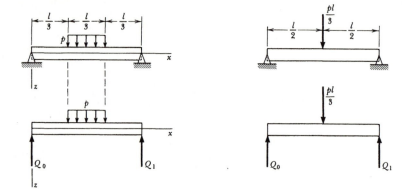

Fig. 4.3-14

Example **4.3-2**

A simply-supported beam carries a uniform distributed load p over the middle third of its span. Find the shear and bending moment distributions. Compare with the distributions for an equipollent concentrated load. Neglect the weight of the beam by comparison with the applied load.

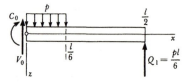

Fig. 4.3-15

Solution: In Fig. 4.3-14, the beam and its free-body diagram are shown for the two loading conditions. The beam under concentrated load has been solved already in the last example; simply replace W by $pl/3$ in Eqs. 4.3-12 and Fig. 4.3-12.

The beam under distributed load must be treated in three sections, $0 \leq x < l/3$, $l/3 < x < 2l/3$, and $2l/3 < x \leq l$. From the symmetry of both geometry and loading, we see that $Q_0 = Q_1 = pl/6$. Moreover, the symmetry permits the solution of the problem through consideration of half the beam only. Shift the origin of coordinates to the midsection. The free-body diagram of the right half of the beam is given in Fig. 4.3-15. The bending moment and shear force at the center are determined from equilibrium of the half beam:

$$R_z = \frac{pl}{6} - Q_1 - V_0 = -V_0 = 0, \qquad V_0 = 0;$$

$$M_y = \frac{l}{2}\,Q_1 - C_0 - \int_0^{l/6} xp\,dx = \frac{pl^2}{12} - C_0 - \frac{pl^2}{72} = 0, \qquad C_0 = \frac{5pl^2}{72}. \qquad \textbf{4.3-14}$$

The differential equation for the shear force is

$$\frac{dV}{dx} = -p, \qquad 0 \leq x < l/6;$$

$$= 0, \qquad l/6 < x \leq l/2. \qquad \textbf{4.3-15}$$

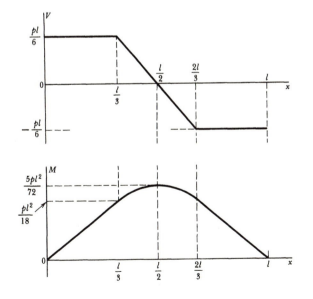

Fig. 4.3-16

Integrating subject to the conditions that $V(0)=0$ and $[V(l/6)]=0$ (since there is no concentrated load at $x=l/6$), we find

$$V(x) = -px, \qquad 0 \leqq x \leqq l/6;$$

$$= -pl/6, \qquad l/6 \leqq x \leqq l/2. \qquad \textbf{4.3-16}$$

From

$$dM/dx = V, \qquad M(0) = C_0, \qquad \left[M\!\left(\frac{l}{6}\right)\right] = 0,$$

we find

$$M(x) = -\frac{px^2}{2}+C_0 = \frac{pl^2}{72}\left[5-36\!\left(\frac{x}{l}\right)^{\!2}\right], \qquad 0 \leqq x \leqq l/6;$$

$$= -\frac{pl}{6}\left(x-\frac{l}{6}\right)+M\!\left(\frac{l}{6}\right)$$

$$= -\frac{plx}{6}+\frac{pl^2}{36}+\frac{pl^2}{18} = \frac{pl^2}{12}\left(1-\frac{2x}{l}\right), \qquad l/6 \leqq x \leqq l/2. \quad \textbf{4.3-17}$$

Remember that these hold for the right half of the beam only; they can be continued to the left half by recognizing that $V(-x) = -V(x)$, $M(-x)= M(x)$. The shear and bending moment diagrams are given in Fig. 4.3-16, with the origin shifted back to the left end for direct comparison with Fig. 4.3-12.

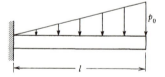

Fig. 4.3-17

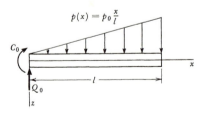

Fig. 4.3-18

Example **4·3-3**

A cantilevered beam (*Fig.* 4.3-17) *is subjected to a triangular load distribution. Find the shear and bending moment distributions.*

Solution: The free-body diagram of the whole beam is shown in Fig. 4.3-18. The applied load is $p(x) = p_0 x/l$. Equilibrium equations for the whole beam determine C_0 and Q_0:

$$R_z = -Q_0 + \int_0^l p(x)\,dx = -Q_0 + \frac{p_0 l}{2} = 0, \qquad Q_0 = \frac{p_0 l}{2};$$

$$M_y = -C_0 - \int_0^l x p(x)\,dx = -C_0 - \frac{p_0 l^2}{3} = 0, \qquad C_0 = -\frac{p_0 l^2}{3}.$$

4·3-18

Therefore, we wish to integrate

$$\frac{dV}{dx} = -\frac{p_0 x}{l}, \qquad V(0) = Q_0 = \frac{p_0 l}{2};$$

$$\frac{dM}{dx} = V, \qquad M(0) = C_0 = -\frac{p_0 l^2}{3}.$$

4·3-19

The solutions are

$$V(x) = \frac{p_0 l}{2}\left[1 - \left(\frac{x}{l}\right)^2\right];$$

$$M(x) = \frac{p_0 l}{2}\left[x - \frac{x^3}{3l^2}\right] - \frac{p_0 l^2}{3}$$

$$= -\frac{p_0 l^2}{6}\left(1 - \frac{x}{l}\right)^2\left(2 + \frac{x}{l}\right).$$

4·3-20

The shear and bending moment diagrams appear in Fig. 4.3-19.

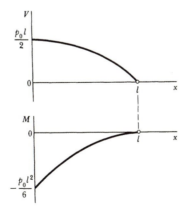

Fig. 4·3-19

Example **4.3-4**

Suppose a beam is pinned at one end and supported on rollers at its middle, and a tip load of P is applied to the free end (Fig. 4.3-20). What are the shear and moment distributions?

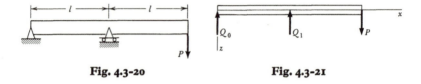

Fig. 4.3-20   **Fig. 4.3-21**

Solution: The whole beam is statically determinate. The free-body diagram appears in Fig. 4.3-21, and the equations of equilibrium are

$$Q_0 + Q_1 - P = 0,$$

$$lQ_1 - 2lP = 0. \qquad\qquad \textbf{4.3-21}$$

Therefore, the reactions are

$$Q_1 = 2P, \qquad Q_0 = -P. \qquad\qquad \textbf{4.3-22}$$

The application of the load P at the otherwise unconstrained end $x = 2l$ changes the free end boundary conditions $M = 0$, $V = 0$ to $M = 0$, $V = P$. The internal support at $x = l$ may be treated by noting the equivalence of the reaction Q_1 to a concentrated load at the point. The "jump" conditions of equilibrium 4.3-4 are appropriately used here. Besides the vanishing deflection condition that is present at any simple support (and which does not concern us in the present analysis), we have a condition of continuity of bending moment

$$[M(l)] = 0, \qquad\qquad \textbf{4.3-23}$$

rather than the vanishing of M as we do at $x = 0$. The jump condition on the shear force is

$$[V(l)] = Q_1 = 2P, \qquad\qquad \textbf{4.3-24}$$

since Q_1 is in the negative z-direction.

With no distributed load, $V(x)$ must be constant in each unloaded segment:

$$V(x) = Q_0 = -P, \qquad 0 \leqq x < l;$$

$$= Q_0 + [V(l)] = +P, \qquad l < x \leqq 2l. \qquad \textbf{4.3-25}$$

The bending moment is

$$M(x) = -Px, \qquad 0 \leq x \leq l;$$
$$= P(x-2l), \qquad l \leq x \leq 2l. \qquad \textbf{4.3-26}$$

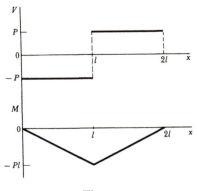

Fig. 4.3-22

Example **4·3-5**

A curved beam is formed in the shape of a circular quadrant of radius r, with cross-sections normal to the circular center line being rectangular in shape (Fig. 4.3-23). The beam is cantilevered at one end and subjected to a downward load P at the centroid of the other. The weight of the beam may be neglected. Find the reactions at the built-in end and the resultant force set at any section.

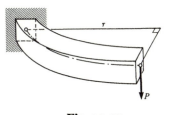

Fig. 4.3-23

Solution: Such curved beam problems can be solved in a manner analogous to that used for the straight beam. The analysis is generally more difficult, especially if the problem is statically indeterminate and the deformations of the beam must be taken into account. However, by considering the arclength along the beam and appropriate tangential and normal directions in the plane of the centerline (as well as a z-direction normal to the plane of the centerline), we can establish differential equations of equilibrium for the resultant force and moment components on any normal cross-section in terms of the distributed loads on the beam. Here, we may solve the problem without using differential equations.

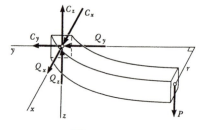

Fig. 4.3-24

In Fig. 4.3-24, a free-body diagram of the whole beam is shown. (Q_x, Q_y, Q_z) are the components of the resultant reaction force on the beam at the built-in end; (C_x, C_y, C_z) are the components of the reactive couple. The force equilibrium equations, when solved, result in

$$Q_x = Q_y = 0, \qquad Q_z = P. \qquad\qquad 4.3\text{-}27$$

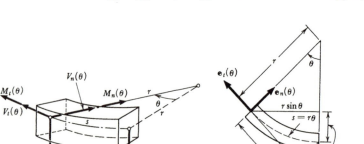

Fig. 4.3-25

The moment equation about the centroid of the built-in end is

$$C_x\mathbf{i} + C_y\mathbf{j} + C_z\mathbf{k} + (-r\mathbf{j} + r\mathbf{i}) \times P\mathbf{k} = \mathbf{0},$$

from which we find

$$C_x = rP, \qquad C_y = rP, \qquad C_z = 0. \qquad\qquad 4.3\text{-}28$$

The C_y-component is a bending moment such as that found for the cantilevered straight beam; in addition, a torsional moment C_x about the center line is found.

A section of the beam from the loaded end to a general section is pictured in the free-body diagram of Fig. 4.3-25a. The angle θ or the arclength $s = r\theta$ may be used as the independent variable denoting the section. The force and moment components shown are in directions of positive tangential (agreeing with increasing θ and s), normal, and z-directions; these directions form a right-handed system at each section. A top view, looking in the positive z-direction, is given in Fig. 4.3-25b.

The equilibrium equations are:

$$P\mathbf{k} + [V_t(\theta)\mathbf{e}_t(\theta) + V_n(\theta)\mathbf{e}_n(\theta) + V_z(\theta)\mathbf{k}] = \mathbf{0},$$

$$[-r\sin\theta\mathbf{e}_t + r(\mathrm{I} - \cos\theta)\mathbf{e}_n] \times P\mathbf{k} + [M_t(\theta)\mathbf{e}_t + M_n(\theta)\mathbf{e}_n + M_z(\theta)\mathbf{k}] = \mathbf{0}.$$

$$\text{4.3-29}$$

Therefore,

$$V_t = V_n = \mathrm{o}, \qquad V_z(\theta) = -P \qquad \text{4.3-30}$$

and

$$M_t(\theta) = -rP(\mathrm{I} - \cos\theta), \qquad M_n(\theta) = -rP\sin\theta, \qquad M_z = \mathrm{o}. \quad \text{4.3-31}$$

As $\theta \to \pi/2$, i.e., as we approach the built-in end, we find $M_t(\pi/2) = M_n(\pi/2) = -rP$. At the built-in end, the tangential direction in the sense of increasing arclength is the negative of the x-direction of Fig. 4.3-24, and the normal direction there is the negative of the y-direction. Thus we have the proper agreement of these end values with the reactions 4.3-28: $M_t = -C_x$, $M_n = -C_y$ at $\theta = \pi/2$.

4.4 Hydrostatics and Aerostatics

So far, we have been replacing distributed forces over internal areas by equipollent concentrated resultants. We now turn to the study of the equilibrium of fluids, wherein we must examine the behavior of the distributed force itself.

A fluid can be differentiated mechanically from a solid by the statement that a tangential load increment, however small, applied to the surface of a fluid will cause it to move indefinitely as long as the load is applied. A solid, on the other hand, will have at most a perfectly definite additional deformation for a given additional increment of load. Fluids are classified as liquids or gases. A given mass of liquid occupies a definite volume under given conditions of temperature and external load; the boundary of the volume, however, is not of any definite shape but is formed by the solid container of the liquid. A given mass of gas does not have a definite volume, but will expand to fill its whole container. The equilibrium of liquids forms the study known as *hydrostatics*; the study of the equilibrium of gases is called *aerostatics*. Only a few of the principles and results of fluid statics will be treated here. The reader should note that there are no formal numbered "examples" in this section (or the next); the applications of the principles are worked into the general line of the textual development. The results on centers of pressure, the incompressible fluid under gravity, the isothermal atmosphere, and the hydraulic press may all be viewed as examples following from the theory that we now develop.

We proceed with the characterization of fluids. In Section 2.7, the replacement of the internal forces between parts of a body by an equipollent force set was discussed. Suppose we have a body,

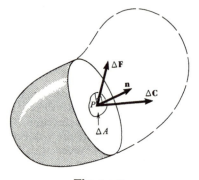

Fig. 4.4-1

either fluid or solid, and pick a point, P, in its interior (Fig. 4.4-1); as in Section 2.7, we pass a plane through P and replace one part of the body by the forces it exerts on the rest. These are distributed over the whole plane area. This time, rather than replace these forces by their full resultant set, we divide the total area into a number of parts and replace the forces on each part by their resultant. In particular, suppose that an element of area ΔA containing point P is chosen, and that the forces on ΔA are replaced by a resultant force $\Delta \mathbf{F}$ at P and a couple $\Delta \mathbf{C}$ (Fig. 4.4-1). We also draw the unit vector $\mathbf{n}$ at P perpendicular to the area ΔA and pointing away from the material we have retained as our system and into the part we have replaced. The average force per unit area, or *average traction*, at P on an area with normal $\mathbf{n}$ is then $\Delta \mathbf{F}/\Delta A$; the average moment is $\Delta \mathbf{C}/\Delta A$. The limiting process we are about to perform should now be obvious from our choice of notation: we let ΔA shrink to the point P and examine the limits of the average traction and average moment vectors. We suppose these limits to exist and to be independent of the shape of ΔA and of the way in which $\Delta A \to 0$. The limit of the average traction is the *traction vector* $\mathbf{T}^{(n)}$ at P associated with a unit plane area through P with normal $\mathbf{n}$:*

$$\mathbf{T}^{(n)} = \lim_{\Delta A \to 0} \frac{\Delta \mathbf{F}}{\Delta A}. \qquad \textbf{4.4-1}$$

Note that the normal $\mathbf{n}$ is held constant in the process. It is usually assumed that $|\Delta \mathbf{C}|$ is of the same order of magnitude as $|\Delta \mathbf{F}|$ multiplied by a moment arm whose length lies within the shrinking region ΔA. Therefore, even though $\Delta \mathbf{F}/\Delta A$ does not vanish as $\Delta A \to 0$,

$$\lim_{\Delta A \to 0} \frac{\Delta \mathbf{C}}{\Delta A} = 0. \qquad \textbf{4.4-2}$$

We shall pursue the consequences of these concepts further in the next section. Now we characterize fluids in terms of $\mathbf{T}^{(n)}$. The traction vector can be divided into two components, the *normal* or *direct stress* and the *tangential* or *shearing stress*. The normal stress component is

$$\mathbf{T}_\sigma^{(n)} = (\mathbf{T}^{(n)} \cdot \mathbf{n})\mathbf{n} = \sigma_n \mathbf{n}; \qquad \textbf{4.4-3}$$

the shearing stress component is

$$\mathbf{T}_\tau^{(n)} = \mathbf{T}^{(n)} - \mathbf{T}_\sigma^{(n)}. \qquad \textbf{4.4-4}$$

On the basis of experimental observations, all fluids at rest are characterized by the vanishing of shear stress, so that only a normal stress is exerted on any interior plane drawn in the fluid. Fluids in motion may exhibit a shear stress between neighboring layers; such fluids are said to be *viscous*. The assumption of vanishing shear stress is often made, whether or not the fluid is in motion; such fluids are termed *inviscid* or "*perfect*" fluids. Further, the normal stress in fluids is compressive in nature, resisting the interpenetration of the parts; for a fluid at rest,

$$\mathbf{T}^{(n)} = \mathbf{T}_\sigma^{(n)} = \sigma_n \mathbf{n} = -p\mathbf{n}, \qquad \textbf{4.4-5}$$

---

* The traction vector is often called the stress vector.

where $p = p(x, y, z)$ is the *pressure** at point P. One of the first theorems that we shall prove is that we can indeed speak of the pressure at a point, independently of our choice of the normal $\mathbf{n}$ to the area we imagine at P.

Let us now consider the equilibrium of a small mass of fluid surrounding the point P. We first introduce the density, ρ, of the fluid at any point, as we have done before, as the limit of the ratio of the mass, Δm, in a volume, ΔV, around the point as the volume approaches zero. If ρ remains a constant, whatever loads are applied, then the material is *incompressible*; if ρ changes with load, the material is *compressible*. Liquids have relatively low compressibility, and are often assumed to be strictly incompressible fluids; the compressibility of gases is an essential feature of their behavior.

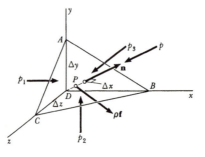

Fig. 4.4-2

Consider a small tetrahedron $ABCD$ of fluid surrounding point P, which may be taken at the centroid (Fig. 4.4-2); three faces of the tetrahedron are perpendicular to axes chosen as shown, while the fourth face ABC has outer normal $\mathbf{n} = n_x\mathbf{i} + n_y\mathbf{j} + n_z\mathbf{k}$. The outer normals to faces ACD, BCD, and ABD are $-\mathbf{i}$, $-\mathbf{j}$, and $-\mathbf{k}$, respectively. Let ΔA be the area of the slant face ABC; then the usual projection property enables us to write the areas of the other faces in terms of ΔA:

$$ACD: \frac{1}{2}\Delta y \, \Delta z = n_x \, \Delta A,$$

$$BCD: \frac{1}{2}\Delta x \, \Delta z = n_y \, \Delta A, \qquad\qquad \textbf{4.4-6}$$

$$ABD: \frac{1}{2}\Delta x \, \Delta y = n_z \, \Delta A.$$

---

* A liquid carefully freed from dissolved gases can stand appreciable tension. In practice, however, p is normally a positive quantity.

The volume of the tetrahedron is

$$\Delta V = \frac{1}{6} \Delta x \, \Delta y \, \Delta z = \frac{1}{3} \Delta h \, \Delta A, \qquad \textbf{4·4-7}$$

where Δh is the length of the altitude dropped from vertex D on the slant face.

The forces acting on the tetrahedron are the pressure forces normal to the faces and body forces, such as gravity, acting on the volume. The pressure forces have resultant on each face equal to the integral of the pressure over the face; by the mean-value theorem, we replace this by the product of the pressure at some point on the face and the area of the face. Similarly, the body forces $\mathbf{f}$, which we take to be per unit mass, have force resultant equal to the integral over the volume of the product of the density ρ and $\mathbf{f}$. This we replace by the product of the volume and the value of $\rho\mathbf{f}$ at some interior point. The force equilibrium equation is

$$p_1(\tfrac{1}{2}\Delta y\,\Delta z)\mathbf{i} + p_2(\tfrac{1}{2}\Delta x\,\Delta z)\mathbf{j} + p_3(\tfrac{1}{2}\Delta x\,\Delta y)\mathbf{k} - p\,\Delta A\mathbf{n} + \rho\mathbf{f}\,\Delta V = \mathbf{0},$$
$$\textbf{4·4-8}$$

where p_1, p_2, p_3, and p are the pressures on the faces as shown in Fig. 4.4-2. Now, dividing by ΔA and using Eqs. 4.4-6, 4.4-7, we find

$$(p_1 - p)n_x\mathbf{i} + (p_2 - p)n_y\mathbf{j} + (p_3 - p)n_z\mathbf{k} + \rho\mathbf{f}\,\Delta h = 0. \qquad \textbf{4·4-9}$$

If the tetrahedron volume is now allowed to shrink to a point, so that $\Delta h \to 0$, the body force term vanishes; and, since the direction cosines (n_x, n_y, n_z) or the orientation $\mathbf{n}$ of ΔA was arbitrarily chosen, the only way the equilibrium equation can be satisfied is to have $p_1 = p_2 = p_3 = p$. Therefore, in a fluid at rest, the pressure on all planes through a point must be the same.

The same result holds for an inviscid fluid in motion. There are no tangential stresses on the faces in such a fluid, so that the surface force terms in Eq. 4.4-8 or 4.4-9 do not change. If the mass center of the tetrahedron has acceleration $\mathbf{a}^*$, then Newton's second law of motion requires a term $\rho\mathbf{a}^* \, \Delta V$ on the right side of 4.4-8 and hence a term $\rho\mathbf{a}^* \, \Delta h$ on the right side of 4.4-9. This term vanishes, like the body-force term, in the limiting process. For a viscous fluid, tangential stresses are present, and we cannot come to the same conclusion. However, in the usual theory, we still speak of the pressure at a point as the average of the normal compressive stresses over any three orthogonal planes through the point.

In order to find out how the pressure varies from point to point in a fluid, we must include the next higher order of pressure variation term in the equation of equilibrium 4.4-8. That is, we have established the equality of the pressure on any plane through the point P by keeping only those average pressure terms over the face areas that lead to forces proportional in magnitude to the area.   To include the effect of the body forces, we must find the variation of pressure in the volume.   It is easiest to do this if we drop our tetrahedron "particle" model and choose instead a rectangular parallelepiped "particle" (Fig. 4.4-3).

If we let $p = p(x, y, z)$ be the pressure at the center point P, then the pressures on the various faces are (approximately) as indicated in the figure; the partial derivatives are evaluated at the center P.   Then the equation of equilibrium becomes (to within the first-order approximation ordinarily used)

$$\left[-p\left(x + \frac{\Delta x}{2}, y, z\right) \Delta y\, \Delta z + p\left(x - \frac{\Delta x}{2}, y, z\right) \Delta y\, \Delta z \right] \mathbf{i}$$

$$+ \left[-p\left(x, y + \frac{\Delta y}{2}, z\right) \Delta x\, \Delta z + p\left(x, y - \frac{\Delta y}{2}, z\right) \Delta x\, \Delta z \right] \mathbf{j}$$

$$+ \left[-p\left(x, y, z + \frac{\Delta z}{2}\right) \Delta x\, \Delta y + p\left(x, y, z - \frac{\Delta z}{2}\right) \Delta x\, \Delta y \right] \mathbf{k}$$

$$+ \rho \mathbf{f}\, \Delta x\, \Delta y\, \Delta z$$

$$= -\frac{\partial p}{\partial x} \Delta x\, \Delta y\, \Delta z\mathbf{i} - \frac{\partial p}{\partial y} \Delta x\, \Delta y\, \Delta z\mathbf{j} - \frac{\partial p}{\partial z} \Delta x\, \Delta y\, \Delta z\mathbf{k} + \rho \mathbf{f}\, \Delta x\, \Delta y\, \Delta z = \mathbf{0}.$$

Therefore, dividing by the volume and passing to the limit $\Delta V \to 0$, we find the vector equation of equilibrium[*]

$$\frac{\partial p}{\partial x} \mathbf{i} + \frac{\partial p}{\partial y} \mathbf{j} + \frac{\partial p}{\partial z} \mathbf{k} = \rho \mathbf{f}. \qquad \textbf{4.4-10}$$

If we neglect all body forces ($\mathbf{f} = \mathbf{0}$), then the pressure must be constant everywhere in a fluid in equilibrium.   The most common non-zero body force considered in fluid statics is the gravitational force.   Since the gravitational force per unit mass is simply the local gravitational acceleration, $\mathbf{f}$ is either $-g\mathbf{e}$, where $\mathbf{e}$ is the local upward

---

[*] The vector $\dfrac{\partial p}{\partial x} \mathbf{i} + \dfrac{\partial p}{\partial y} \mathbf{j} + \dfrac{\partial p}{\partial z} \mathbf{k}$ derived from the scalar field $p(x, y, z)$ is known as the *gradient* of p; Eq. 4.4-10 may be written grad $p = \rho \mathbf{f}$ or $\nabla p = \rho \mathbf{f}$, the *vector differential operator* $\mathbf{i} \dfrac{\partial}{\partial x} + \mathbf{j} \dfrac{\partial}{\partial y} + \mathbf{k} \dfrac{\partial}{\partial z}$ (in rectangular coordinates) being denoted by "grad" or ∇ (read "del" or "nabla").

vertical unit vector and g ($\cong 32.2$ ft/sec^2) is the constant gravitational acceleration near the earth, or $(-GM/r^2)\mathbf{e}$, from the full gravitational law. The latter expression is of use only if we wish to consider changes in pressure through large vertical distances in the earth's atmosphere.

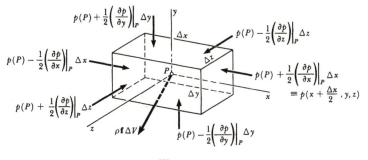

Fig. 4.4-3

The fundamental equilibrium problem of fluid statics is the solution of Eq. 4.4-10. Boundary conditions of an appropriate type must be adjoined. A distinction between liquid and gas, or at least between incompressible and compressible fluid, must be made in order to treat the relation between p and ρ properly. However, before examining some solutions, let us suppose that the pressure distribution $p(x, y, z)$ has been found in a fluid and let us then examine the resultant force due to fluid pressure on surfaces, real or imagined, acted upon by the fluid.

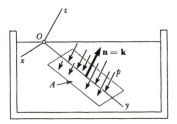

Fig. 4.4-4

Suppose that a plane area A bounds the fluid, or is imagined to be drawn in the fluid (Fig. 4.4-4). The fluid pressing on the plane boundary, or on one side of the imagined surface, gives rise to a pressure distribution of constant direction, even if of varying magnitude p, since the normal to the plane at every point is the same. Choose (x, y) axes in the plane and z in the positive normal direction. The resultant force is then

$$\mathbf{P} = \int\int_A (-p\mathbf{k})\, dA = \left(-\int\int_A p\, dx\, dy\right)\mathbf{k}. \qquad \textbf{4.4-11}$$

The resultant moment of the pressure distribution about the fixed origin of coordinates is

$$\mathbf{M}_O = \int\int_A (x\mathbf{i} + y\mathbf{j}) \times (-p\mathbf{k})\, dA$$

$$= \int\int_A [xp\mathbf{j} - yp\mathbf{i}]\, dA$$

$$= \left(\int\int_A xp\, dx\, dy\right)\mathbf{j} - \left(\int\int_A yp\, dx\, dy\right)\mathbf{i}. \qquad \textbf{4.4-12}$$

Since, after all, the pressure distribution is a parallel force system, we obtain the unsurprising result that the pressure distribution on a plane area may be replaced by a single force $\mathbf{P} = -P\mathbf{k}$ located at a point in the plane—the *center of pressure*—the coordinates $(\hat{x}, \hat{y})$ of which are given by

$$(\hat{x}\mathbf{i} + \hat{y}\mathbf{j}) \times (-P\mathbf{k}) = \mathbf{M}_O$$

or

$$\hat{x} = \frac{\int\int_A xp\, dx\, dy}{\int\int_A p\, dx\, dy}, \qquad \hat{y} = \frac{\int\int_A yp\, dx\, dy}{\int\int_A p\, dx\, dy}. \qquad \textbf{4.4-13}$$

We do not always find it convenient, of course, to choose such axes in and normal to the plane. Since the pressure distribution is a parallel force distribution $-p\mathbf{n}$, with $\mathbf{n}$ a constant unit vector, the result holds in any coordinate system. We find (with $\mathbf{r} = x\mathbf{i} + y\mathbf{j} + z\mathbf{k}$ the position vector from O to a general point in the plane)

$$\mathbf{P} = \int\int_A (-p\mathbf{n})\, dA = \left(-\int\int_A p\, dA\right)\mathbf{n},$$

$$\mathbf{M}_O = \int\int_A [\mathbf{r} \times (-p\mathbf{n})]\, dA = \left(-\int\int_A p\mathbf{r}\, dA\right) \times \mathbf{n},$$

so that the center of pressure is located at $\hat{\mathbf{r}}$ given by

$$\hat{\mathbf{r}} \times \mathbf{P} = \left(-\int\int_A p\, dA\right)(\hat{\mathbf{r}} \times \mathbf{n})$$

$$= \left(-\int\int_A p\mathbf{r}\, dA\right) \times \mathbf{n} = \mathbf{M}_O, \qquad \textbf{4.4-14}$$

or at $(\hat{x}, \hat{y}, \hat{z})$ given by

$$\hat{x} = \frac{\iint_A xp \, dA}{\iint_A p \, dA}, \qquad \hat{y} = \frac{\iint_A yp \, dA}{\iint_A p \, dA}, \qquad \hat{z} = \frac{\iint_A zp \, dA}{\iint_A p \, dA}. \qquad \textbf{4.4-15}$$

That the vector $\hat{\mathbf{r}}$ with these components satisfies Eq. 4.4-14 may be verified by substitution in that equation. As a final note on centers of pressure for the present, we see that Eqs. 4.4-13 and 4.4-15 follow the pattern for center of mass, centroid of area, and such other averaged quantities involving the first moment of a scalar. Indeed, if the pressure is constant, the center of pressure and centroid of area will coincide.

Suppose now that the surface we are considering is not plane but curved. We can no longer conclude in general that a resultant force set is a single force at a center of pressure. The force-and-couple resultant set must be used. Under certain circumstances, when there exists a plane of vertical symmetry of the surface and the pressure varies linearly with depth, a single resultant force may be found. Otherwise, the surface integrals—where $\mathbf{n}$ cannot now be removed from the integrand—for force and moment must both be evaluated and an appropriate couple kept, together with the force:

$$\mathbf{P} = -\iint_S p\mathbf{n} \, dS, \qquad \mathbf{C} = -\iint_S p(\mathbf{r} \times \mathbf{n}) \, dS.$$

The line of action of the force is, of course, through the point from which $\mathbf{r}$ is measured.

Returning now to the solution of the equilibrium equation 4.4-10, let us see what the difficulties are. If $\mathbf{f} = \mathbf{0}$, then the pressure is constant—but what constant? If $\mathbf{f}$ is not zero, indeed if it is but a constant non-zero vector, then the density function must be prescribed. Conservation of mass in the equilibrium case tells us only that ρ is time-independent, as are all of our quantities. In order to solve our problem, we need an additional hypothesis about our fluid material that relates pressure and density. This is, in general, a *thermodynamic equation of state*

$$F(p, \rho, T) = 0 \qquad \textbf{4.4-16}$$

which also involves the temperature $T(x, y, z)$ of the fluid. This, in turn, requires that we know the temperature or that we can write an equation of thermal energy balance that will determine it. We shall not treat this question, for it is not our purpose to discuss the laws of thermodynamics in any detail. For *isothermal* processes, wherein the temperature at all points is the same, Eq. 4.4-16 becomes an equation in p and ρ alone.

For incompressible fluids, the equation of state is $\rho =$ constant. This we will take as characteristic of liquids for a wide range of temperature and pressure conditions. Suppose the body force per unit mass is due to gravity, and that the z-direction is the upward vertical. Then $\mathbf{f} = -g\mathbf{k}$, and Eq. 4.4-10 becomes

$$\frac{\partial p}{\partial x} = \frac{\partial p}{\partial y} = 0, \qquad \frac{\partial p}{\partial z} = -\rho g. \qquad \textbf{4.4-17}$$

The first two of Eqs. 4.4-17 say that the pressure is independent of the horizontal coordinates, and so must be the same at any given horizontal level; the last integrates to

$$p = p_0 - \rho g z. \qquad \textbf{4.4-18}$$

The pressure in an incompressible fluid subject to gravity decreases linearly with height. The pressure at all points in the same horizontal level is the same. The constant, p_0, is the pressure at the level $z = 0$. If $z = 0$ is taken at a free surface between air and the liquid, then p_0 is atmospheric pressure, with the fluid occupying a region in which $z < 0$. If we write $p_0 = \rho g z_0$, then 4.4-18 becomes $p = -\rho g(z - z_0)$; the level $z = z_0$ at which the pressure vanishes is the *effective surface* of the fluid, and the pressure due to a *head* of h feet means the pressure at a distance of h feet below the effective surface: $z = z_0 - h$, $p = +\rho g h$.

For gases, the assumption of incompressibility is generally no longer tenable. Aerostatics is based upon the additional assumption of the gas laws, usually those of the perfect gas. Boyle's law states that, if the temperature of a given mass of gas is constant, the pressure varies inversely as the volume:

$$pV = \text{const.}; \qquad \textbf{4.4-19}$$

and Charles' law states that, if the pressure in a given mass of gas is constant, the volume of the gas increases linearly with the temperature:

$$V = V_0(1 + \alpha T_c). \qquad \textbf{4.4-20}$$

Here V_0 is the volume at the zero point on the temperature scale. If T_c is measured in degrees Centigrade, then the coefficient of volume expansion α is approximately $1/273$ for many gases—leading to an "absolute zero" of $-273°C$ at which V vanishes. Introducing the density by the conservation of mass equation $\rho V =$ constant, we may combine the laws into one of the two equivalent forms

$$p = R\rho T \qquad\qquad \textbf{4.4-21a}$$

or

$$pV = nRT \qquad\qquad \textbf{4.4-21b}$$

where R is a constant characteristic of the gas, $n = \rho V$ is the total mass, and T is the absolute temperature.

The gas laws, as stated, are for a total mass of gas in which the pressure is everywhere the same, so that $\mathbf{f}$ must be the zero vector in 4.4-10. If we wish to have $\mathbf{f} \neq \mathbf{0}$, say again the gravitational force $-g\mathbf{k}$ per unit mass, then we must modify these laws to ones valid point by point in the fluid. This is easily done by reinterpreting V as a specific volume, or volume per unit mass $(V = 1/\rho)$. In particular, 4.4-21a leads to a linear relation between p and ρ for an isothermal process:

$$\frac{p}{p_0} = \frac{\rho}{\rho_0} \qquad\qquad \textbf{4.4-22}$$

where (p_0, ρ_0) are the pressure and density at some particular point. Equilibrium of an isothermal gas under gravity is then governed by 4.4-10 and 4.4-22, resulting in the pressure equilibrium equation

$$\frac{\partial p}{\partial x}\mathbf{i} + \frac{\partial p}{\partial y}\mathbf{j} + \frac{\partial p}{\partial z}\mathbf{k} = -\frac{\rho_0}{p_0}\,pg\mathbf{k}. \qquad\qquad \textbf{4.4-23}$$

Again, p is independent of the horizontal coordinates x and y: $p = p(z)$, and 4.4-23 becomes the ordinary differential equation

$$\frac{dp}{dz} + \frac{\rho_0}{p_0}gp = 0. \qquad\qquad \textbf{4.4-24}$$

The solution is

$$p = p_0 e^{(-\rho_0/p_0)gz}, \qquad\qquad \textbf{4.4-25}$$

if we interpret p_0 and ρ_0 to be the pressure and density at $z = 0$. Thus, in an isothermal equilibrium atmosphere, the pressure (and density) decrease exponentially with altitude.

One important theorem that forms the basis for the design of the hydraulic press may be deduced from the general equilibrium equation 4.4-10. Suppose that we have found a solution $p = p_1(x, y, z)$ to the equation for a given $\mathbf{f}$, that is, for a given set of external forces. Then $p = p_1 + p_0$ is also a solution, where p_0 is an arbitrary constant that is the same at every point in the fluid. If this change of pressure

p_0 is somehow caused at any one point, it is immediately transmitted to every other point. In Fig. 4.4-5, the application of this to the hydraulic press is sketched. If the fluid in the press is in equilibrium, and a force F_1 is applied to the piston of area A_1, a constant pressure $p_0 = F_1/A_1$ is transmitted through the fluid to the piston of area A_2. The equilibrating force $F_2 = A_2 p_0 = A_2 F_1/A_1$ then acts to press body B between the piston and the upper wall of the container. The so-called "hydrostatic paradox" finds its expression here: by adjusting

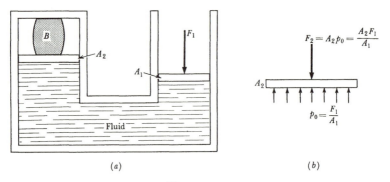

(a)   (b)

Fig. 4.4-5

the piston areas, any load F_2, however large, may be balanced by as small a load F_1 as desired.

Another class of equilibrium problems involving fluids are those involving floating or submerged bodies. Solution of such problems depends upon *Archimedes' principle*:

A body wholly or partially immersed in a fluid at rest is subject to a resultant force due to fluid pressure which is the negative of the weight force vector of the displaced fluid and which has the same line of action, through the center of gravity of the displaced fluid.

This force is called the *buoyant force*, and the center of gravity of the displaced fluid is the *center of buoyancy*. To prove Archimedes' principle, we first note that the resultant force, being the integral of the pressure distribution over the surface between body and fluid, depends only on the shape of that surface and its location in the fluid, and not on the material of which the body is made. Imagine, then, that the body is removed and that the space which it occupied is filled with a mass of the same fluid that surrounds the body. This additional mass of fluid—the "displaced mass"—is in equilibrium under its own weight and the resultant pressure of the surrounding fluid; if it were not, flow would start in it and in the surrounding fluid under an equilibrium pressure distribution. This cannot be. Thus the resultant of the pressure force, i.e., the buoyancy force, must be the negative of the weight of the displaced fluid and must act through its center of gravity.

There are many more problems of interest in aerostatics, hydrostatics, and the equilibrium of floating bodies. Among them may be mentioned capillarity and the hypothesis of *surface tension* to explain meniscus effects in tubes of small bore containing fluid; partial pressures in mixtures of fluids; and the equilibrium of stratified layers of immiscible fluids. For details of these and many others, reference should be made to a work on hydrostatics.*

4.5 The Stress Tensor; Equilibrium Equations for the Continuum

In the last section we defined the traction vector, its normal stress component, and its shearing stress component (Eqs. 4.4-2, 4.4-3, and 4.4-4) acting at a point in a continuous body on a unit area through the point having normal **n**. By assuming that the shearing stress on any plane vanished, we were able to derive the conditions of equilibrium at any point of the first two orders: that the normal stress—the pressure—on any plane through the point is the same, and that the rate of change of the pressure with position is proportional to the component of the body force per unit mass in the direction of the position coordinate. These two results were derived by consideration of the behavior of forces that vanished proportionally to area and volume, respectively.

For the general continuous distribution of matter, or *continuum*, corresponding results can be derived. Now we assume that the material can sustain shearing stress. The first results that we derive are *Cauchy's relations*; these express the traction vector on any plane through a point in terms of the traction vectors on three perpendicular planes through the point. The fact that the traction vector on any plane can be found from the three other vectors when the orientation of the plane is given leads us to the concept of the *state of stress* at a point and its representation by a mathematical entity, the *stress tensor*, that generalizes the vector concept. Our final result will be the derivation of the equations of equilibrium for a continuum in terms of body forces and the scalar components of the stress tensor in a rectangular cartesian coordinate system.

---

* For example, A. S. Ramsey, *Hydrostatics* (New York: Cambridge University Press, 1946).

As in Fig. 4.4-2, let us surround the point P by a tetrahedron $ABCD$ of material (Fig. 4.5-1), with three faces in the coordinate planes and the fourth slant face having unit normal vector $\mathbf{n} = n_x\mathbf{i} + n_y\mathbf{j} + n_z\mathbf{k}$. A body force $\mathbf{f}$ per unit mass is assumed to act. The traction vector on each face can have any direction; the notation given in the figure will be used. The area of the slant face will be denoted by ΔA once again, and the altitude

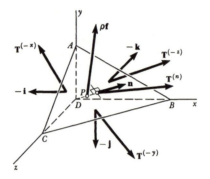

Fig. 4.5-1

from D to the face by Δh. With $BD = \Delta x$, $AD = \Delta y$, and $CD = \Delta z$, we have the relations 4.4-6 and 4.4-7 once again for the areas of the other faces and the volume of the tetrahedron:

$$\frac{1}{2}\Delta y\,\Delta z = n_x\,\Delta A, \qquad \frac{1}{2}\Delta x\,\Delta z = n_y\,\Delta A, \qquad \frac{1}{2}\Delta x\,\Delta y = n_z\,\Delta A,$$

$$\Delta V = \frac{1}{6}\Delta x\,\Delta y\,\Delta z = \frac{1}{3}\Delta h\,\Delta A. \qquad \text{4.5-1}$$

The average forces on each face and on the volume lead to the equilibrium equation

$$\mathbf{T}^{(-x)}n_x\,\Delta A + \mathbf{T}^{(-y)}n_y\,\Delta A + \mathbf{T}^{(-z)}n_z\,\Delta A + \mathbf{T}^{(n)}\,\Delta A + \rho\mathbf{f}\,\Delta V = \mathbf{0},$$

which becomes, after division by ΔA,

$$\mathbf{T}^{(-x)}n_x + \mathbf{T}^{(-y)}n_y + \mathbf{T}^{(-z)}n_z + \mathbf{T}^{(n)} + \rho\mathbf{f}\,\Delta h = \mathbf{0}. \qquad \text{4.5-2}$$

Before letting the volume shrink to the point P, we replace $\mathbf{T}^{(-x)}$ by $-\mathbf{T}^{(x)}$, etc.; that is, the traction vector on each coordinate plane through D with normal in a negative coordinate direction must, by the third law of motion, be the negative of the traction vector on the same plane with positive orientation. Then, if the tetrahedron shrinks to the point P so that Δh approaches zero, the body force term vanishes and the traction vectors approach their values on the appropriate planes through P. Cauchy's relations in vector form are

$$\mathbf{T}^{(n)} = \mathbf{T}^{(x)}n_x + \mathbf{T}^{(y)}n_y + \mathbf{T}^{(z)}n_z. \qquad \text{4.5-3}$$

Equation 4.5-3 is valid in the dynamic case as well. Setting the forces equal to the mass of the tetrahedron multiplied by the acceleration of its mass center, $\rho \, \Delta V \, \mathbf{a}^*$, we see that this term also goes to zero with Δh if the acceleration is to remain finite.

Cauchy's relations may be written in scalar form by introducing the (x, y, z) components of the four vectors that occur in 4.5-3. A consistent notation is helpful here. Let us consider the vector $\mathbf{T}^{(x)}$, the force at P per unit area with normal in the positive x-direction. The normal stress is the component of $\mathbf{T}^{(x)}$ in the x-direction. Let us denote this component by σ_{xx}—the scalar component on a plane with normal in the positive x-direction (the first subscript, corresponding to the superscript x on $\mathbf{T}^{(x)}$) of the vector $\mathbf{T}^{(x)}$ in that normal x-direction (the second subscript). The normal component of the traction vector is, therefore, $\sigma_{xx}\mathbf{i}$. The remainder of $\mathbf{T}^{(x)}$ is the shear stress on that plane: $\mathbf{T}^{(x)} - \sigma_{xx}\mathbf{i}$. This shear stress can be resolved into y- and z-components, which we will denote by σ_{xy} and σ_{xz}—the components in the y- and z-directions (second subscripts) of the traction vector on a plane with normal in the positive x-direction (first subscripts). Thus

$$\mathbf{T}^{(x)} = \sigma_{xx}\mathbf{i} + \sigma_{xy}\mathbf{j} + \sigma_{xz}\mathbf{k}; \qquad \text{4.5-4a}$$

similarly,

$$\mathbf{T}^{(y)} = \sigma_{yx}\mathbf{i} + \sigma_{yy}\mathbf{j} + \sigma_{yz}\mathbf{k}, \qquad \text{4.5-4b}$$

$$\mathbf{T}^{(z)} = \sigma_{zx}\mathbf{i} + \sigma_{zy}\mathbf{j} + \sigma_{zz}\mathbf{k}. \qquad \text{4.5-4c}$$

The sign convention for positive scalar stress components is the same as for the components of any other vector; in particular, the normal stresses $(\sigma_{xx}, \sigma_{yy}, \sigma_{zz})$ are counted as positive if they are tensile. For the fluid at rest, we found that $\sigma_{xx} = \sigma_{yy} = \sigma_{zz} = -p$; all the shearing stresses were assumed to be zero.

The Cauchy relations 4.5-3 in scalar form are:

$$T_x^{(n)} = n_x\sigma_{xx} + n_y\sigma_{yx} + n_z\sigma_{zx},$$

$$T_y^{(n)} = n_x\sigma_{xy} + n_y\sigma_{yy} + n_z\sigma_{zy}, \qquad \text{4.5-5}$$

$$T_z^{(n)} = n_x\sigma_{xz} + n_y\sigma_{yz} + n_z\sigma_{zz}.$$

The nine scalar quantities $\sigma_{xx}, \sigma_{xy}, \ldots \sigma_{zz}$ are called the *components of stress*. They have the dimensions of force per unit area. In general, they are functions of position (x, y, z) and time.

The traction components on any plane through P are thus determined by the nine stress components $\sigma_{xx}, \sigma_{xy}, \ldots \sigma_{zz}$. We can therefore speak of the *stress at a point* in a continuum, and say that this state of stress is known once the nine components of stress relative to any given set of cartesian axes are known. A very similar situation occurs with vector quantities: once the cartesian components (A_x, A_y, A_z) of a vector $\mathbf{A}$ are known with respect to one coordinate system, the component of the vector in any direction is known. If the arbitrary direction is prescribed by the unit vector $\mathbf{n} = n_x\mathbf{i} + n_y\mathbf{j} + n_z\mathbf{k}$, then the scalar component $A^{(n)}$ of $\mathbf{A}$ in the direction of $\mathbf{n}$ is computed by

$$A^{(n)} = \mathbf{A}\cdot\mathbf{n} = n_x A_x + n_y A_y + n_z A_z. \qquad \textbf{4·5-6}$$

The analogy between Eq. 4.5-6 and any one of the three Eqs. 4.5-5 is no coincidence. Just as we speak of a vector entity $\mathbf{A}$ and its representation (A_x, A_y, A_z) in some particular coordinate system, so may we speak of a new entity $\tilde{B}$ and its representation

$$\begin{pmatrix} B_{xx} & B_{xy} & B_{xz} \\ B_{yx} & B_{yy} & B_{yz} \\ B_{zx} & B_{zy} & B_{zz} \end{pmatrix} \qquad \textbf{4·5-7}$$

in a particular coordinate system. This new entity obeys certain rules of transformation (which can be derived from Eqs. 4.5-5) that relate its

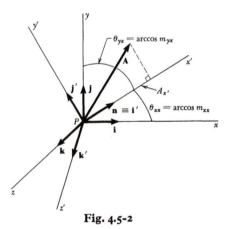

Fig. 4·5-2

components in any pair of coordinate systems. These rules of transformation are generalizations of those relating the scalar components of a vector in two different coordinate systems. Indeed, vectors and scalars are special cases of a more general concept, the *tensor*. In particular, the state of stress at a point is represented by a *second-order tensor*, each component of which depends upon two directions in space. Vectors are *first-order tensors*, involving one direction in space; and scalars, independent of spatial direction, are *zero-order tensors*. Higher-order tensors can be defined and are useful in various branches of mechanics and physics in general as well as being a study in themselves in mathematics. Besides the stress tensor, two other second-order tensors should be mentioned: the strain tensor, encountered in the study of deformable bodies, and the moment of inertia tensor, encountered in the dynamics of rigid bodies in three dimensions. We shall now derive the transformation rules for the components of a second-order tensor in two different coordinate systems and compare them to the rules for a vector. We are concerned with right-handed rectangular cartesian systems only, and with the transformation rules from one such system to another. The tensor quantities we define are not the most general possible but are known as cartesian tensors and, indeed, as oriented cartesian tensors.

In Fig. 4.5-2, two rectangular cartesian coordinate systems (x, y, z) and (x', y', z') with origin at P are shown. The unit vector $\mathbf{n}$, and the direction of that vector, have now been called $\mathbf{i}'$ and x'—one of our new coordinate directions. The other new directions are chosen perpendicular to x' in such a way that the primed coordinate system is right-handed. Rather than call the components of $\mathbf{n} \equiv \mathbf{i}'$ in the (x, y, z) system (n_x, n_y, n_z) as before, we introduce a new notation. Let $(\theta_{xx}, \theta_{yx}, \theta_{zx})$ be the direction angles between the x-axis and the x'-axis, y-axis and x'-axis, and z-axis and x'-axis, respectively. The *direction cosines* of x' with respect to (x, y, z) are, in fact, the components of the unit vector $\mathbf{i}'$ in the (x, y, z) system. We introduce the following notation for these direction cosines:

$$m_{xx} = \cos \theta_{xx}, \quad m_{yx} = \cos \theta_{yx}, \quad m_{zx} = \cos \theta_{zx}; \qquad \textbf{4.5-8a}$$

the first subscript refers to an axis of the (x, y, z) set, and the second to the x'-axis of the second set. Similarly, the direction cosines of the other primed axes are introduced:

$$m_{xy} = \cos \theta_{xy}, \quad m_{yy} = \cos \theta_{yy}, \quad m_{zy} = \cos \theta_{zy}; \qquad \textbf{4.5-8b}$$

$$m_{xz} = \cos \theta_{xz}, \quad m_{yz} = \cos \theta_{yz}, \quad m_{zz} = \cos \theta_{zz}. \qquad \textbf{4.5-8c}$$

It is worth noting that, although a double subscript notation has been used for the direction cosines, the array of nine direction cosines does *not* constitute a tensor. The subscripts refer to different axis systems, not to the same axis system as in the stress components of 4.5-5 or the components of $\tilde{B}$ in 4.5-7. The unit vectors $(\mathbf{i}', \mathbf{j}', \mathbf{k}')$ can be written in terms of the direction cosines and the vectors $(\mathbf{i}, \mathbf{j}, \mathbf{k})$:

$$\begin{aligned}
\mathbf{i}' &= m_{xx}\mathbf{i} + m_{yx}\mathbf{j} + m_{zx}\mathbf{k}, \\
\mathbf{j}' &= m_{xy}\mathbf{i} + m_{yy}\mathbf{j} + m_{zy}\mathbf{k}, \qquad\qquad \textbf{4.5-9} \\
\mathbf{k}' &= m_{xz}\mathbf{i} + m_{yz}\mathbf{j} + m_{zz}\mathbf{k}.
\end{aligned}$$

With these results, we proceed to the transformation rules for vectors and second-order tensors. Suppose a vector $\mathbf{A}$ at point P is given (Fig. 4.5-2), with representations

$$\mathbf{A} = A_x\mathbf{i} + A_y\mathbf{j} + A_z\mathbf{k} = A_{x'}\mathbf{i}' + A_{y'}\mathbf{j}' + A_{z'}\mathbf{k}' \qquad \textbf{4.5-10}$$

in the two coordinate systems.   We wish to express $(A_{x'}, A_{y'}, A_{z'})$ in terms of (A_x, A_y, A_z).   We have already found $A_{x'}$ in Eq. 4.5-6; replacing $\mathbf{n}$ by $\mathbf{i}'$ and $A^{(n)}$ by $A_{x'}$, we have

$$A_{x'} = \mathbf{A} \cdot \mathbf{i}' = A_x \mathbf{i} \cdot \mathbf{i}' + A_y \mathbf{j} \cdot \mathbf{i}' + A_z \mathbf{k} \cdot \mathbf{i}'$$
$$= m_{xx} A_x + m_{yx} A_y + m_{zx} A_z; \qquad \text{4.5-11a}$$

similarly,

$$A_{y'} = \mathbf{A} \cdot \mathbf{j}' = A_x \mathbf{i} \cdot \mathbf{j}' + A_y \mathbf{j} \cdot \mathbf{j}' + A_z \mathbf{k} \cdot \mathbf{j}'$$
$$= m_{xy} A_x + m_{yy} A_y + m_{zy} A_z, \qquad \text{4.5-11b}$$

and

$$A_{z'} = \mathbf{A} \cdot \mathbf{k}' = A_x \mathbf{i} \cdot \mathbf{k}' + A_y \mathbf{j} \cdot \mathbf{k}' + A_z \mathbf{k} \cdot \mathbf{k}'$$
$$= m_{xz} A_x + m_{yz} A_y + m_{zz} A_z. \qquad \text{4.5-11c}$$

Equations 4.5-11 are the transformation rules for vector components under the coordinate rotation 4.5-9.

The tensor transformation rules for any $\tilde{B}$ can be obtained from the rules for the special tensor $\bar{\sigma}$ by computing (from Eqs. 4.5-5) the normal and shear stress components on the plane with normal $\mathbf{n}$—or, in our new notation, the normal stress $\sigma_{x'x'}$ and the shearing stresses $\sigma_{x'y'}$ and $\sigma_{x'z'}$ on the plane through P with normal in the x'-direction.   In the new notation, the components of $\mathbf{n}$ in 4.5-5 are replaced by the appropriate direction cosines and the components in the (x, y, z) directions of the traction vector $\mathbf{T}^{(n)}$ are replaced by the (x, y, z) components of the traction vector $\mathbf{T}^{(x')}$. Rewriting Eqs. 4.5-5 in the new notation, we have:

$$T_x^{(x')} = m_{xx} \sigma_{xx} + m_{yx} \sigma_{yx} + m_{zx} \sigma_{zx},$$
$$T_y^{(x')} = m_{xx} \sigma_{xy} + m_{yx} \sigma_{yy} + m_{zx} \sigma_{zy}, \qquad \text{4.5-12}$$
$$T_z^{(x')} = m_{xx} \sigma_{xz} + m_{yx} \sigma_{yz} + m_{zx} \sigma_{zz}.$$

The normal stress $\sigma_{x'x'}$ is the component $T_x^{(x')}$ of the traction vector $\mathbf{T}^{(x')}$ in the direction of the normal to the plane:

$$\sigma_{x'x'} = T_x^{(x')} = \mathbf{T}^{(x')} \cdot \mathbf{i}' = T_x^{(x')} \mathbf{i} \cdot \mathbf{i}' + T_y^{(x')} \mathbf{j} \cdot \mathbf{i}' + T_z^{(x')} \mathbf{k} \cdot \mathbf{i}'$$
$$= m_{xx} T_x^{(x')} + m_{yx} T_y^{(x')} + m_{zx} T_z^{(x')}. \qquad \text{4.5-13}$$

Therefore, $\sigma_{x'x'}$ can be written in terms of $(\sigma_{xx}, \sigma_{xy}, \ldots \sigma_{zz})$; substituting 4.5-12 into 4.5-13, we obtain

$$\sigma_{x'x'} = m_{xx} m_{xx} \sigma_{xx} + m_{xx} m_{yx} \sigma_{xy} + m_{xx} m_{zx} \sigma_{xz}$$
$$+ m_{yx} m_{xx} \sigma_{yx} + m_{yx} m_{yx} \sigma_{yy} + m_{yx} m_{zx} \sigma_{yz} \qquad \text{4.5-14a}$$
$$+ m_{zx} m_{xx} \sigma_{zx} + m_{zx} m_{yx} \sigma_{zy} + m_{zx} m_{zx} \sigma_{zz}.$$

The shear stress component in the y'-direction on the plane with normal x' is

$$\sigma_{x'y'} = \mathbf{T}^{(x')} \cdot \mathbf{j}' = m_{xy} T_x^{(x')} + m_{yy} T_y^{(x')} + m_{zy} T_z^{(x')},$$

or, if we use 4.5-12 again,

$$\sigma_{x'y'} = m_{xx}m_{xy}\sigma_{xx} + m_{xx}m_{yy}\sigma_{xy} + m_{xx}m_{zy}\sigma_{xz}$$
$$+ m_{yx}m_{xy}\sigma_{yx} + m_{yx}m_{yy}\sigma_{yy} + m_{yx}m_{zy}\sigma_{yz} \qquad \textbf{4.5-14b}$$
$$+ m_{zx}m_{xy}\sigma_{zx} + m_{zx}m_{yy}\sigma_{zy} + m_{zx}m_{zy}\sigma_{zz}.$$

The equations 4.5-14a and 4.5-14b are typical of the transformation rule for computing any component in the (x', y', z') system of a second-order tensor from the components in the (x, y, z) system. Let us state in words what the rule is. The typical term on the right is the product of two direction cosines and an (x, y, z) component of the tensor. The direction cosines are arranged so that the second subscripts, *taken in order*, are the same as the (x', y', z') subscripts on the component we are computing—(x', x') in 4.5-14a, (x', y') in 4.5-14b. The first subscripts on the direction cosines, again taken in order, are the same as those of the (x, y, z) component that the direction cosines multiply. Add up all nine possible terms in the (x, y, z) components; this sum is equal to the (x', y', z') component given by the second subscripts on the direction cosines. Following this rule, we can compute the other seven components in the (x', y', z') axes. We note that the vector component transformations 4.5-11 follow the same rule, with but one direction cosine involved in each term in the sum; and higher-order tensor transformations follow the same pattern, with direction cosines equal in number to the order in each term. Also, a much more efficient notation can be developed to make the expressions more compact and useful for theoretical and computational purposes.

Let us now derive the force equilibrium equations of the order of volume terms; we expect these to involve the partial derivatives of the stress components. The fluid equilibrium equation 4.4-10 is a special case of these general equations. Now, rather than replace the stress distribution on each face of the tetrahedron or parallelepiped by a mean stress over the face, we take into account the first variation of the stress

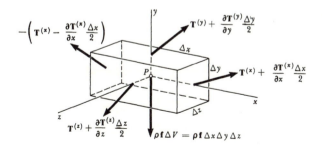

Fig. 4·5-3

components with distance. We take the parallelepiped of Fig. 4.4-3, with center at the point P and edges Δx, Δy, Δz, as shown here in Fig. 4.5-3. The body force per unit mass is again denoted $\mathbf{f} = f_x\mathbf{i} + f_y\mathbf{j} + f_z\mathbf{k}$. The stresses on the faces are taken to be the stresses at P plus the changes in stress, to first-order terms in the edge lengths, due to the distance of the face from P. Again, the variation in these stresses over the face is "averaged out," so that the stress is the same at each point of a given face. In Fig. 4.5-3, the traction vectors on the three faces with normals in the positive coordinate directions are shown, as is the body force $\rho\mathbf{f}\,\Delta V$. There are stresses on the other three faces as well. Only those on the face with unit normal $-\mathbf{i}$ are shown. Note the minus sign that appears for that traction vector (and for the other two similar traction vectors also); this sign occurs since $\mathbf{T}^{(x)}$ is the traction vector on a plane with unit normal $+\mathbf{i}$. The vectors $\mathbf{T}^{(x)}$, $\mathbf{T}^{(y)}$, $\mathbf{T}^{(z)}$ have cartesian components $(\sigma_{xx}, \sigma_{xy}, \sigma_{xz})$, $(\sigma_{yx}, \sigma_{yy}, \sigma_{yz})$, and $(\sigma_{zx}, \sigma_{zy}, \sigma_{zz})$, respectively; a partial derivative, such as $\partial\mathbf{T}^{(x)}/\partial x$, of a vector means the vector whose components are the partial derivatives of the components of the initial vector.

The force equilibrium equations are

$$\left[\left(\mathbf{T}^{(x)} + \frac{\partial\mathbf{T}^{(x)}}{\partial x}\frac{\Delta x}{2}\right) - \left(\mathbf{T}^{(x)} - \frac{\partial\mathbf{T}^{(x)}}{\partial x}\frac{\Delta x}{2}\right)\right]\Delta y\,\Delta z$$

$$+\left[\left(\mathbf{T}^{(y)} + \frac{\partial\mathbf{T}^{(y)}}{\partial y}\frac{\Delta y}{2}\right) - \left(\mathbf{T}^{(y)} - \frac{\partial\mathbf{T}^{(y)}}{\partial y}\frac{\Delta y}{2}\right)\right]\Delta x\,\Delta z$$

$$+\left[\left(\mathbf{T}^{(z)} + \frac{\partial\mathbf{T}^{(z)}}{\partial z}\frac{\Delta z}{2}\right) - \left(\mathbf{T}^{(z)} - \frac{\partial\mathbf{T}^{(z)}}{\partial z}\frac{\Delta z}{2}\right)\right]\Delta x\,\Delta y$$

$$+\rho\mathbf{f}\,\Delta x\,\Delta y\,\Delta z = \mathbf{0},$$

or (after division by $\Delta V = \Delta x\,\Delta y\,\Delta z$),

$$\frac{\partial\mathbf{T}^{(x)}}{\partial x} + \frac{\partial\mathbf{T}^{(y)}}{\partial y} + \frac{\partial\mathbf{T}^{(z)}}{\partial z} + \rho\mathbf{f} = \mathbf{0}. \qquad\qquad \textbf{4.5-15}$$

The "shrinking of the parallelepiped to the limit point P" has been carried out implicitly in this argument, with higher-order terms involving second derivatives of the stresses vanishing in the limit process.

If we now replace the traction vectors $\mathbf{T}^{(x)}$, $\mathbf{T}^{(y)}$, and $\mathbf{T}^{(z)}$ by their cartesian component representations, the vector equation 4.5-15 may be written in equivalent scalar form:

$$\frac{\partial\sigma_{xx}}{\partial x} + \frac{\partial\sigma_{yx}}{\partial y} + \frac{\partial\sigma_{zx}}{\partial z} + \rho f_x = 0,$$

$$\frac{\partial\sigma_{xy}}{\partial x} + \frac{\partial\sigma_{yy}}{\partial y} + \frac{\partial\sigma_{zy}}{\partial z} + \rho f_y = 0, \qquad\qquad \textbf{4.5-16}$$

$$\frac{\partial\sigma_{xz}}{\partial x} + \frac{\partial\sigma_{yz}}{\partial y} + \frac{\partial\sigma_{zz}}{\partial z} + \rho f_z = 0.$$

These are known as Navier's equations of equilibrium. They play a basic role in stress analysis.

To these three partial differential equations representing force equilibrium per unit volume we must adjoin moment equilibrium equations. Under the assumption that there are no surface or body moments, i.e., no couples per unit area or volume, the moment equations express a condition on the stress components themselves and not on the spatial derivatives of the stresses. The vectorial moment equilibrium equation about point P for the forces on the parallelepiped of Fig. 4.5-3 is:

$$\left(\frac{\Delta x}{2}\mathbf{i}\right) \times \left[\left(\mathbf{T}^{(x)} + \frac{\partial \mathbf{T}^{(x)}}{\partial x}\frac{\Delta x}{2}\right) \Delta y\,\Delta z\right]$$

$$+ \left(-\frac{\Delta x}{2}\mathbf{i}\right) \times \left[-\left(\mathbf{T}^{(x)} - \frac{\partial \mathbf{T}^{(x)}}{\partial x}\frac{\Delta x}{2}\right) \Delta y\,\Delta z\right]$$

$$+ \left(\frac{\Delta y}{2}\mathbf{j}\right) \times \left[\left(\mathbf{T}^{(y)} + \frac{\partial \mathbf{T}^{(y)}}{\partial y}\frac{\Delta y}{2}\right) \Delta x\,\Delta z\right]$$

$$+ \left(-\frac{\Delta y}{2}\mathbf{j}\right) \times \left[-\left(\mathbf{T}^{(y)} - \frac{\partial \mathbf{T}^{(y)}}{\partial y}\frac{\Delta y}{2}\right) \Delta x\,\Delta z\right]$$

$$+ \left(\frac{\Delta z}{2}\mathbf{k}\right) \times \left[\left(\mathbf{T}^{(z)} + \frac{\partial \mathbf{T}^{(z)}}{\partial z}\frac{\Delta z}{2}\right) \Delta x\,\Delta y\right]$$

$$+ \left(-\frac{\Delta z}{2}\mathbf{k}\right) \times \left[-\left(\mathbf{T}^{(z)} - \frac{\partial \mathbf{T}^{(z)}}{\partial z}\frac{\Delta z}{2}\right) \Delta x\,\Delta y\right] = \mathbf{0},$$

which reduces to

$$(\Delta x\mathbf{i}) \times (\mathbf{T}^{(x)}\,\Delta y\,\Delta z) + (\Delta y\mathbf{j}) \times (\mathbf{T}^{(y)}\,\Delta x\,\Delta z) + (\Delta z\mathbf{k}) \times (\mathbf{T}^{(z)}\,\Delta x\,\Delta y) = \mathbf{0}.$$
$$\textbf{4.5-17}$$

Note that the body force $\rho\mathbf{f}\,\Delta V$ has been considered to act at P; if it does not, its moment will be of higher order in the edge lengths than those of the stress vectors and will vanish in the limiting process $\Delta V \to 0$. Dividing Eq. 4.5-17 by the volume $\Delta V = \Delta x\,\Delta y\,\Delta z$ of the parallelepiped and performing the limit process $\Delta V \to 0$, we find

$$\mathbf{i} \times \mathbf{T}^{(x)} + \mathbf{j} \times \mathbf{T}^{(y)} + \mathbf{k} \times \mathbf{T}^{(z)} = \mathbf{0} \qquad \textbf{4.5-18}$$

as the vector moment equilibrium equation. From this, we obtain

$$\mathbf{i} \times (\sigma_{xx}\mathbf{i} + \sigma_{xy}\mathbf{j} + \sigma_{xz}\mathbf{k}) + \mathbf{j} \times (\sigma_{yx}\mathbf{i} + \sigma_{yy}\mathbf{j} + \sigma_{yz}\mathbf{k}) + \mathbf{k} \times (\sigma_{zx}\mathbf{i} + \sigma_{zy}\mathbf{j} + \sigma_{zz}\mathbf{k}) = \mathbf{0},$$

or

$$(\sigma_{yz} - \sigma_{zy})\mathbf{i} + (\sigma_{zx} - \sigma_{xz})\mathbf{j} + (\sigma_{xy} - \sigma_{yx})\mathbf{k} = \mathbf{0}. \qquad \textbf{4.5-19}$$

Therefore, for moment equilibrium with no body or surface moments, we must have the shear stresses on orthogonal planes at a point equal in pairs:

$$\sigma_{xy} = \sigma_{yx}, \qquad \sigma_{yz} = \sigma_{zy}, \qquad \sigma_{zx} = \sigma_{xz}. \qquad \textbf{4.5-20}$$

The six equations of equilibrium 4.5-16 and 4.5-20 in nine unknowns—or the three equations in six unknowns resulting from the substitution of

4.5-20 in 4.5-16—are the only independent ones.   Problems requiring the solution of these equations for the stress distribution are, therefore, statically indeterminate.

The dynamical equations of motion corresponding to 4.5-16 and 4.5-20 are simple to state, although we shall not derive them.   Equations 4.5-16 are modified by the replacement of the right-hand side by the product of the density, ρ, and the appropriate component of the acceleration vector at point P.   These equations are derived from the dynamical principle of motion of the mass center.   The dynamical principle of moment of momentum, or angular momentum, leads to Eqs. 4.5-20 once again if there are no surface or body moments.

The relations 4.5-20 state that the stress tensor $\tilde{\sigma}$ is *symmetric*.   That is, the component representation (in any rectangular cartesian coordinate system)

$$\tilde{\sigma} = \begin{pmatrix} \sigma_{xx} & \sigma_{xy} & \sigma_{xz} \\ \sigma_{yx} & \sigma_{yy} & \sigma_{yz} \\ \sigma_{zx} & \sigma_{zy} & \sigma_{zz} \end{pmatrix}$$

has equal-valued components in positions symmetrically placed with respect to the main diagonal formed by the normal stress components.   This symmetry has important consequences for the development of the theory of stress analysis, perhaps the primary one being the existence at any point of three perpendicular planes on which there are no shearing stresses. Similar results hold for the symmetric strain tensor and the symmetric moment of inertia tensor.

Exercises

4.2-1: A cable in uniform tension fits tightly against a cam of arbitrary shape (except that there are no sharp corners).   What is the tangential load, q_t, exerted by the cam on the cable ?

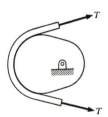

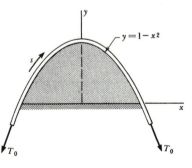

Exer. 4.2-1   Exer. 4.2-2

4.2-2: A cable subject to uniform tension T_0 lb passes around a parabolic cam, the equation of the boundary being $y = 1 - x^2$ (foot units are used). Measure arclength s in the sense shown; the slope angle θ is given by $\tan \theta = dy/dx$. By first finding ds/dx and $d\theta/ds = (d^2y/dx^2)/(ds/dx)^3$, find the normal load distribution q_n between cam and cable.

Ans.: $q_n(x) = 2T_0/[1 + 4x^2]^{3/2}$ lbs/ft (of arclength).

4.2-3: A cable weighing 5 lb/ft is 100 ft long. How far apart should the supports be (on the same horizontal level) if the cable is not to sag more than 12 ft at its center? What is the minimum tension H?

Ans.: Exact: $l = 96.12$ ft, $H = 490.8$ lb; Approx.: $l = 96$ ft, $H = 480$ lb.

4.2-4: Suppose a 100 ft cable is supporting a 5 lb load per foot of horizontal span, the weight of the cable being neglected. How far apart must the supports be if the cable is not to sag more than 12 ft? What are the maximum and minimum tensions in the cable? What is the total load carried by the cable? Compare the answer for the total span with the answer to the previous exercise.

Ans.: Approx: $l = 96$ ft, $H = W = 480$ lb, $T_{max} = 537$ lb.

4.2-5: The maximum tension permitted in a cable before it will snap is 1500 lb. A uniform load of 20 lb per foot of horizontal span is to be carried over a 50 ft span. What is the minimum length of cable permissible? What will be the sag?

Ans.: $L = 51.0$ ft, $f = 4.42$ ft.

4.2-6: The maximum tension permitted in a cable before it will snap is T^* lb. It hangs under its own weight over a span of l ft. Find the relation between the length L of cable needed and its total weight W so that the maximum tension condition will be just satisfied.

Ans.: $\dfrac{L\sqrt{4T^{*2} - W^2}}{W} \operatorname{arcsinh}\left(\dfrac{W}{\sqrt{4T^{*2} - W^2}}\right) = l.$

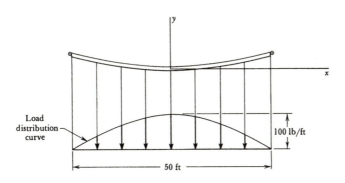

Exer. 4.2-7

4.2-7: A cable of negligible weight is supported with its ends at the same level 50 ft apart. A load is distributed along the span sinusoidally, vanishing at the ends and reaching a peak value of 100 lb/ft at the center. What is the shape of the cable if the maximum tension is $(15{,}000/\pi)$ lb?

Ans.: $y = \dfrac{25\sqrt{2}}{2\pi}\left[1 - \cos\left(\dfrac{\pi x}{50}\right)\right]$ ft.

4.2-8: A cable subject to uniform tension T_0 passes over half of a fixed elliptical bar as shown. The equation of the bar surface is

$$\frac{x^2}{a^2} + \frac{y^2}{b^2} = 1,$$

with $a > b$. Find the normal pressure $q_n(x, y)$ between bar and cable and determine the maximum value of the pressure.

Ans.: $q_n(x, y) = \dfrac{a^4 b^4 T_0}{(b^4 x^2 + a^4 y^2)^{3/2}}$ (per unit arclength).

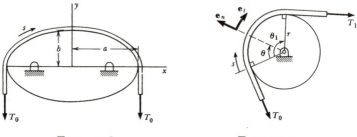

Exer. 4.2-8   Exer. 4.2-9

4.2-9: A cable passes around a fixed circular rough surface in such a way that the ratio of the tangential load (friction force) to the normal load is a constant:

$$\frac{q_t(s)}{q_n(s)} = -\mu.$$

If the value of the tension at the point from which s and θ are measured is T_0, what will be the tension, $T(\theta)$, at a general point along the cable? What will be the tension T_1 at $\theta = \theta_1$, where the cable leaves the surface? Note that, for the positive normal direction shown, $-T\,d\theta/ds + q_n = 0$.

Ans.: $T(\theta) = T_0\, e^{\mu\theta}$.

4.3-1: The simply-supported beam of Exercise 2.7-17 is shown here again. It is subjected to a triangular load compared to which the weight of the beam may be ignored. Find the shear force and bending moment distributions and draw the diagrams for them.

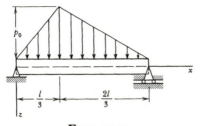

Exer. 4.3-1

Ans.: $M(x) = \dfrac{p_0 x}{18l}(5l^2 - 9x^2)$, $0 \leq x \leq l/3$;

$$M(x) = \dfrac{p_0}{36l}(x-l)(9x^2 - 18lx + l^2),\ l/3 \leq x \leq l.$$

4.3-2: Suppose the beam of Exercise 4.3-1 under the same load is clamped at the left and free at the right; what are the shear and moment distributions now? Draw the shear and bending moment diagrams.

Ans.: $M(x) = -\dfrac{p_0}{18l}(4l^3 - 9l^2 x + 9x^3)$, $0 \leq x \leq l/3$;

$$M(x) = -\dfrac{p_0}{4l}(l-x)^3,\ l/3 \leq x \leq l.$$

4.3-3: Suppose the beam of Exercise 4.3-1 under the same load is free at the left and clamped at the right; now what are the distributions of, and diagrams for, the shear force and bending moment?

Ans.: $M(x) = -\dfrac{p_0 x^3}{2l}$, $0 \leq x \leq l/3$;

$$M(x) = -\dfrac{p_0}{36l}(l^3 - 9l^2 x + 27lx^2 - 9x^3),\ l/3 \leq x \leq l.$$

4.3-4: The simply-supported beam of Exercise 2.7-16 is shown here again. It is subjected to the load $p(x) = p_0 \sin(\pi x/l)$ compared to which the weight

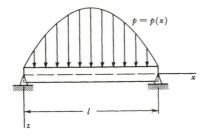

Exer. 4.3-4

of the beam may be neglected. Find the shear force and bending moment distributions and draw the corresponding diagrams.

Ans.: $V(x) = \dfrac{p_0 l}{\pi} \cos\left(\dfrac{\pi x}{l}\right); \; M(x) = \dfrac{p_0 l^2}{\pi^2} \sin\left(\dfrac{\pi x}{l}\right).$

4.3-5: (a) Suppose the beam of Exercise 4.3-4 under the same load is built-in at the left end and free at the right. Solve for the shear and bending moment distributions. (b) What changes occur if the free and fixed ends are interchanged?

Ans.: (a) $V(x) = \dfrac{p_0 l}{\pi}\left[\cos\left(\dfrac{\pi x}{l}\right) + 1\right]; \; M(x) = \dfrac{p_0 l^2}{\pi^2}\left[\sin\left(\dfrac{\pi x}{l}\right) - \pi\left(1 - \dfrac{x}{l}\right)\right].$

(b) $V(x) = \dfrac{p_0 l}{\pi}\left[\cos\left(\dfrac{\pi x}{l}\right) - 1\right]; \; M(x) = \dfrac{p_0 l^2}{\pi^2}\left[\sin\left(\dfrac{\pi x}{l}\right) - \dfrac{\pi x}{l}\right].$

4.3-6: A cantilevered beam of length l is loaded as shown with loads of magnitude p_0 lb/ft. Find the shear force and bending moment distributions.

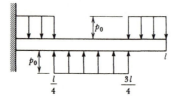

Exer. 4.3-6

Ans.: $M(x) = -\dfrac{p_0 x^2}{2}, \quad 0 \leq x \leq l/4;$

$M(x) = \dfrac{p_0}{16}(8x^2 - 8lx + l^2), \, l/4 \leq x \leq 3l/4;$

$M(x) = -\dfrac{p_0}{2}(l - x)^2, \, 3l/4 \leq x \leq l.$

4.3-7: Solve Exercise 4.3-1 if the two supports are not at the ends of the beam but are moved in to the points $x = l/3, \, x = 2l/3$.

Ans.: $M(x) = -\dfrac{p_0 l^2}{2}\left(\dfrac{x}{l}\right)^3, \, 0 \leq x \leq l/3;$

$M(x) = -\dfrac{p_0 l^2}{36}\left[9\left(1 - \dfrac{x}{l}\right)^3 - 6\left(1 - \dfrac{x}{l}\right) + 2\right], \, l/3 \leq x \leq 2l/3;$

$M(x) = -\dfrac{p_0 l^2}{4}\left(1 - \dfrac{x}{l}\right)^3, \, 2l/3 \leq x \leq l.$

4.3-8: Solve Exercise 4.3-4 if the right-hand roller support is moved in to the center of the beam.

Ans.: $M(x) = \dfrac{p_0 l^2}{\pi^2}\left[\sin\left(\dfrac{\pi x}{l}\right) - \dfrac{\pi x}{l}\right], 0 \leq x \leq l/2;$

$M(x) = \dfrac{p_0 l^2}{\pi^2}\left[\sin\left(\dfrac{\pi x}{l}\right) - \pi\left(1 - \dfrac{x}{l}\right)\right], l/2 \leq x \leq l.$

4.3-9: An L-section is clamped at one end and loaded at the other as shown. By replacing the given load by an equipollent force set at the point Q, compute the axial force, shear force, and bending moment in the beam-column OQ.

Ans.: $N(x) = -P,\ V(x) = 0,\ M(x) = aP.$

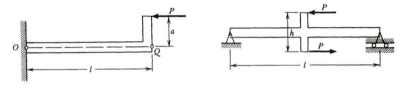

Exer. 4.3-9 **Exer. 4.3-10**

4.3-10: A simply-supported beam is loaded at its midsection effectively by a concentrated couple as shown. Find the shear and bending moment distributions. Solve the problem again with the beam cantilevered at the left.

Ans.: $V(x) = \dfrac{Ph}{l};\ M(x) = \dfrac{Phx}{l}, 0 \leq x < \dfrac{l}{2};\ M(x) = \dfrac{Ph}{l}(x - l), \dfrac{l}{2} < x \leq l.$

4.3-11: A uniform pole of length l ft and weight W lb is stuck in the ground at an angle θ to the vertical. Find the axial force, shear force, and bending moment distributions along the pole due to the distributed weight force.

Ans.: $N(x) = -W\cos\theta\left(1 - \dfrac{x}{l}\right);\ V(x) = W\sin\theta\left(1 - \dfrac{x}{l}\right);$

$M(x) = -\dfrac{Wl\sin\theta}{2}\left(1 - \dfrac{x}{l}\right)^2.$

4.3-12: If a circular cantilevered beam like that of Example 4.3-5 subtends an angle ϕ ($0 \leq \theta \leq \phi$) at the center, instead of 90°, what are the internal force set at any section and the reactions at the built-in end due to a downward load P at the free end? What happens as $\phi \to 360°$, i.e., as the beam approaches a split circular ring with one end clamped and the other free (except for the applied load P)?

Ans.: Forces and moments at any section θ from the loaded end are the same as in Example 4.3-5.

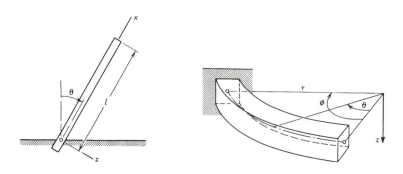

Exer. 4.3-11   **Exer. 4.3-12**

4.4-1: Show that, in a fluid at rest under gravity, horizontal planes are the surfaces of equal density as well as equal pressure.

4.4-2: Suppose that two fluids of different densities are poured in a container and come to rest under gravity. Suppose, further, that they do not mix. Using the result of the previous exercise, show that the surface separating the two fluids is a horizontal plane.

4.4-3: A plane area is immersed in an incompressible fluid at rest under gravity. Without changing the orientation of the area relative to fixed axes, it is lowered to greater and greater depths. Show that the center of pressure moves closer to the centroid of area as the depth is increased.

4.4-4: (a) An equilateral triangle of side s is submerged vertically in a fluid of constant density, with one vertex at depth h below the effective surface and the opposite (and lower) side horizontal. Find the distance from the effective surface to the center of pressure. (b) Reverse the triangle, putting the horizontal side at depth h and the vertex below it. Find the depth of the center of pressure. (c) Let h become large in each case and check the theorem of the last exercise.

Ans.: (a) $h+s\left(\dfrac{8\sqrt{3}h+9s}{24h+8\sqrt{3}s}\right)$; (b) $h+s\left(\dfrac{4\sqrt{3}h+3s}{24h+4\sqrt{3}s}\right)$.

4.4-5: A dam of concrete is built in the shape of an isosceles trapezoid, with the larger base up as shown. The wall is vertical, with the bases horizontal. Find the pressure distribution on the wall resulting only from water weighing 62.4 lb/ft³ completely filling the space behind the dam to the full height of 80 ft. Where is the center of pressure?

Ans.: $p = 62.4z$ lb/ft²; $\hat{z} = 50$ ft.

4.4-6: If the dam of the previous exercise makes an angle of 5° with the vertical, sloping out as shown, what is the pressure distribution? Where is the center of pressure?

Ans.: $p = 62.4z$ lb/ft^2; $\hat{x} = 50 \sin 5°$ ft; $\hat{z} = 50 \cos 5°$ ft.

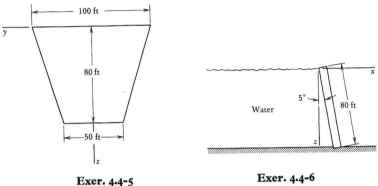

Exer. 4.4-5   Exer. 4.4-6

4.4-7: Suppose we must take into account the variation of gravity with altitude in establishing the pressure distribution in an isothermal atmosphere. Let p_0 and ρ_0 be the values of pressure and density at the earth's surface, z be the altitude measured outward from the earth's surface, R_E the radius of the earth, and g the gravitational acceleration at the surface. The gravitational force per unit mass then has magnitude $gR_E^2/(R_E+z)^2$; the equation of state is $p/p_0 = \rho/\rho_0$. Find $p(z)$.

Ans.: $p(z) = p_0 e^{-\left(\frac{\rho_0 g R_E}{p_0}\right)\left(\frac{z}{R_E+z}\right)}$.

4.4-8: A process in which no heat exchange occurs is termed *adiabatic*, with equation of state $p/p_0 = (\rho/\rho_0)^\gamma$, where γ is the ratio of the specific heat at constant pressure to that at constant volume. Find the pressure and temperature distributions (under constant gravity) in an atmosphere subject to the adiabatic equation of state—an atmosphere in *convective equilibrium*.

Ans.: $\dfrac{p}{p_0} = \left[1 - \dfrac{(\gamma-1)\rho_0 g}{\gamma p_0} z\right]^{\frac{\gamma}{\gamma-1}}$; $\dfrac{T}{T_0} = 1 - \dfrac{(\gamma-1)\rho_0 g}{\gamma p_0} z$,

where $p_0 = R\rho_0 T_0$ at $z = 0$.

CHAPTER V

Frictional Effects

in Statics

5.1 Sources of Friction in Technology

It will have been apparent to the reader that the mechanical analysis of structures and machine elements carried out in previous chapters has been greatly simplified by the assumption that the surfaces in contact were smooth. While the hypothesis of smoothness is closely approximated in a large number of engineering situations, and analyses are, in fact, often carried out on this basis, there are also many cases in which the presence of friction plays an essential role. Since frictional effects are always present to some extent whenever there is a tendency for one surface to move relative to another with which it is in contact, mechanical systems in which the presence of friction is taken into account are sometimes called "real" systems; those in which friction is neglected are termed "ideal" systems. The presence of friction in a machine entails an inevitable conversion of mechanical energy into heat. Although this energy "loss" can be considerable—amounting in the case of an automobile, for example, to about twenty per cent of the energy that would be available were there no friction in the moving parts—it is not the principal engineering objection to friction.   The wear, rough operation, and premature failure of elements is ordinarily far more important.   On the other hand, it must be recognized that were it not for the friction between tire and road the automobile would be unable to function at all, and that most holding and fastening devices rely on friction for their effectiveness.

Much progress has been made in recent years toward an understanding of frictional phenomena.   In the first place it is necessary to distinguish between the friction of dry rubbing solids, rolling friction, and the friction of fully lubricated bearings.   It is the first of these with which we shall be concerned.   If a block of material in the form of a rectangular parallelepiped is placed on a horizontal surface, a horizontal force is needed to make the block move along the

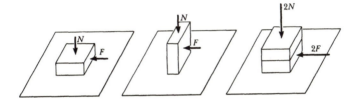

In each case the weight of a single block is N. The horizontal force shown is the minimum horizontal force which will produce sliding.

Fig. 5.1-1

surface.   The magnitude of this force is independent of the surface area of apparent contact; if we turn the block on end the force required to initiate motion is not affected.   Furthermore, the magnitude of the force is proportional to the weight of the block; when a second identical block is placed on top of the first, the horizontal force required to initiate motion is doubled.   These facts, illustrated schematically in Fig. 5.1-1, appear to have been known to Leonardo da Vinci.   They lay buried in his notebooks for many years.   In 1699 they were independently rediscovered by the French engineer Amontons, who verified them experimentally.   Coulomb in 1781 again independently discovered the law of rubbing friction.   He further distinguished between the force necessary to initiate motion and the force needed to maintain motion, once begun.   The latter, he found, was lower than the former and was, over a large speed range, independent of the relative velocities of the two surfaces.   In statics we are interested primarily in the situation that exists before motion actually occurs and shall not be much concerned with this kinetic friction.

The ratio between the force component parallel to the contact surface (F in Fig. 5.1-1) and the component normal to the contact surface (N in Fig. 5.1-1) *when motion impends* is known as the coefficient of limiting friction, f, or simply as the *coefficient of friction*. According to the Amontons-Coulomb friction law, f is supposed to depend only on the smoothness and material of which the surfaces are made and not at all on the nominal area of the surfaces or on the normal force pressing them together. This friction law can be explained if we realize that at a microscopic level no surface is truly smooth and that contact between two bodies actually takes place at the tips of minute "high spots" or asperities. The normal load is carried on the tips of these asperities. The real area of contact is much smaller than the nominal or "over-all" surface area. Under these circumstances the local stress at the tip of an asperity is very high and the material, if a ductile metal, is soon stressed to its plastic yield stress, p. If the normal load is doubled the area of real contact must also double, because the stress in a ductile material cannot exceed the yield value. This is true no matter how many points are actually in contact. Suppose the individual contact areas are $a_1, a_2, \ldots a_n$; the load carried by each will be $N_1 = pa_1$, $N_2 = pa_2, \ldots N_n = pa_n$ and therefore

$$N = N_1 + N_2 + \ldots N_n = p(a_1 + a_2 + \ldots a_n) = pA. \quad \textbf{5.1-1}$$

The area of real contact, A, is therefore proportional to the load, N, normal to the surface, and is independent of the shape, size, or nominal surface area of the bodies. Furthermore, on account of the high local pressures at the points of real contact, the bodies adhere at these locations; in metals an actual cold welding takes place. This adhesion is not ordinarily apparent upon removal of the normal load, N, because the release of the spring-like elastic stresses underlying the small plastic zone breaks the bond. But if, instead of removing the normal load, a tangential force, F, is applied so as to make the surfaces move relative to one another, the bonds will be sheared and we may write $F = As$, where s denotes the mean shear strength of the sheared material. Then

$$f = \frac{F}{N} = \frac{As}{Ap} = \frac{s}{p} \quad \textbf{5.1-2}$$

and the coefficient of limiting friction is seen to be essentially a material property, independent of the normal load and the nominal surface area, as required by the Amontons-Coulomb friction law.   Formulas such as 5.1-2, however, cannot be used to predict the magnitude of the coefficient of friction.   They represent an oversimplification of the actual situation.   The theory of plasticity shows that normal and tangential stresses cannot be treated as independent and that plastic flow is due to a combination of the two.

The foregoing explanation of the nature of the friction of rubbing solids is a restatement of the one advanced by Bowden and Tabor,[*] to whose work the student should refer for details of the supporting evidence.   They give the representative values shown in Table 5.1 for the coefficient of limiting friction, determined by careful experiment on small specimens.

It may be seen from this table that the "constant", f, is one to which it is difficult for the designer to assign a precise value in advance of actual test.   In cases of doubt, resort should be made to test results that duplicate the situation in question.   Certain generalizations may, however, be made concerning the range of validity of the Amontons-Coulomb friction law.   In the case of metals the presence of an oxide film is of prime importance.   The extremely high values of f quoted for pure metals thoroughly denuded of surface films refer to the results of laboratory experiments performed in vacuum, where no oxide film can form.   As has been noted, even a small tangential force produces a growth in junction area due to plastic yielding.   For ordinary metals in air this surface oxide film shears easily, little junction growth takes place, and the elementary theory is satisfactory.   But for laboratory-clean metals in vacuum, junction growth dominates; the real contact area bears no direct relation to the normal load and extremely high coefficients of friction are observed.   On the other hand, in the case of materials that have a relatively large range of elastic behavior, such as diamond or rubber, the deformation is primarily elastic and the coefficient of friction decreases as the load is increased.   This trend is also observed in plastics that deform viscoelastically.

Although we shall not be further concerned with them in the material of this text, it should be said that rolling friction and fully lubricated bearing friction differ fundamentally from the friction of

---

[*] F. P. Bowden and D. Tabor, *The Friction and Lubrication of Solids* (Oxford: The Clarendon Press, 1954).   The student will find an excellent introduction to these ideas in the monograph *Friction and Lubrication* (London: Methuen; New York: Wiley, 1956) by the same authors.

Table 5-1: VALUES OF THE COEFFICIENT OF LIMITING STATIC FRICTION

Metals, pure, rubbing on themselves:

Thoroughly denuded of surface films	> 100
In air, unlubricated	1
Lubricated, mineral oil	0.2–0.4
Lubricated, animal, vegetable oil	0.1

Alloys rubbing on steel:

Copper-lead, unlubricated	0.2
Copper-lead, lubricated, mineral oil	0.1
White metal, Wood's alloy, unlubricated	0.7
White metal, Wood's alloy, lubricated, mineral oil	0.1
Phosphor bronze, brass, unlubricated	0.35
Phosphor bronze, brass, lub., mineral oil	0.15–0.2
Cast iron, unlubricated	0.4
Cast iron, lubricated, mineral oil	0.1–0.2

Hard steel surfaces with various lubricants:

Unlubricated	0.6
Mineral oils	0.14–0.2
Molybdenum disulphide	0.1

Non-metals:

Glass on glass, clean	1.0
Ice on self, below $-50°C$	0.5
Ice on self, between $0°$ and $-20°C$	0.05–0.1
Diamond on self, in air, clean	0.1
Rock salt on self, clean	0.8
Tungsten carbide on steel, clean	0.4–0.6
Nylon on self	0.5
Polytetrafluoroethylene (Teflon) on self	0.04–0.1
Polytetrafluoroethylene (Teflon) on steel	0.04–0.1
Brake material on cast iron, clean	0.4
Brake material on cast iron, wet	0.2
Brake material on cast iron, greasy	0.1
Leather on metal, clean, dry	0.6
Wood on self, clean, dry	0.25–0.5
Wood on metals, clean, dry	0.2–0.6

rubbing solids.   In the case of the wheel of a heavy vehicle, large
tangential forces are indeed involved, but when a ball or disk rolls
freely on a horizontal surface the resistance to rolling corresponds
to a value of f of the order of 10^{-3}.   To understand why there should,
in fact, be any resistance at all to rolling, we need to realize that all
materials are deformable to some extent.   The rolling ball is in
contact with the bearing plate over a finite region called the contact
area, as pictured schematically in Fig. 5.1-2.   Over the trailing part

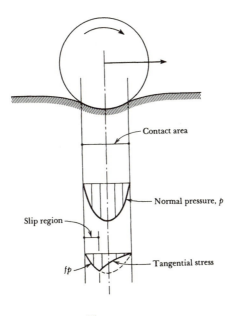

Fig. 5.1-2

of this region there is a local slip; the particle velocities of ball and
track differ and rubbing friction takes place.   The slip region and the
relative motions are, however, small compared with what they would
be were the sphere in translational rather than in rolling motion.
There is a further source of energy loss arising from what is known as
"*internal friction*" or "*hysteresis*," a name given to imperfect elasticity
in the material of ball and track.   Which of the two influences, surface
or internal friction, is the more important would appear to depend
upon particular circumstances.   The internal effect might be expected to dominate in the case of a rubber wheel on a flat track,
whereas surface effects would be more important in the case of a steel

ball rolling in a deep groove.   In fully lubricated bearings, on the
other hand, the surfaces in relative motion are separated by a continuous film of fluid and the resistance to motion arises from the viscosity of the fluid.   A load-carrying shaft in a circular bearing will
take up an eccentric position in the bearing.   When the oil, dragged
by the rotating shaft, is squeezed through the narrowest part of the
gap between shaft and bearing, pressure builds up in the fluid.   This
pressure keeps the surfaces separated.   A full examination of the
situation from the point of view of the dynamics of viscous fluids is
outside the scope of this text, but it may be said that such an analysis
shows the resisting torque to be proportional to the rotation rate of the
shaft and to the viscosity of the oil.

5.2 Elementary Frictional Analysis

When a body is pressed obliquely against a rough surface, the
surface exerts on the body a force N normal to the surface and a
frictional force F tangential to the surface.   The magnitude and
direction of the frictional component are governed by the following
rules due to Amontons and to Coulomb, the physical basis of which
has been described in the previous section.

1. The direction of the friction force is opposite to the direction
   in which the body tends to move.
2. The magnitude of the friction force is just sufficient to prevent
   motion.
3. The magnitude of the friction force cannot be larger than a
   certain value called the limiting friction.   This value is given
   by the product fN, where f is the so-called coefficient of
   friction.   The limiting friction is independent of the area of
   the body in contact with the rough surface.

These rules are supplemented by the statement that, when there is
relative motion between the surfaces, the magnitude of the friction
force is independent of the velocity, its direction opposes the relative
motion, and its magnitude is $f'N$, where f', the so-called coefficient
of kinetic friction, is, in general, less than f.   As has been mentioned,
we shall be concerned with the static situation and shall not make
use of this supplement.

It is important to appreciate that the frictional force actually
exerted is that required to satisfy the newtonian equations of equilibrium.   Only when motion impends is the frictional force equal in
magnitude to the product of the coefficient of friction and the normal
component of force.   Friction is one of that class of forces known as

resistances; it operates to oppose motion, never to produce it.   These points may be clarified by means of a very simple example.   Consider a block resting on a plane inclined to the horizontal, as shown in Fig. 5.2-1.   To fix ideas we may suppose that the block has a mass m (and therefore weight mg), that the plane is inclined at an angle θ to the horizontal, and that the coefficient of friction is f; we ask to know the magnitude of the friction force and whether or not the block will move.   The analysis begins by supposing that the block does *not* move, but rests in equilibrium.   Its free-body diagram is then as pictured in Fig. 5.2-1.   Of course the forces F and N represent components of the resultant of distributed pressures and shears all along the base of the block (see Section 2.7).   The exact distribution of these tractions cannot be found from rigid-body statics so that the

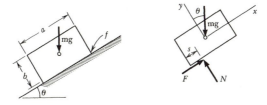

Fig. 5.2-1

point of application of the resultant pressure along the base must, temporarily, be regarded as unknown.   This makes it impossible to begin by writing the moment equation of equilibrium.   The other two equilibrium equations, however, take the form

$$\sum F_x = F - mg \sin \theta = 0, \qquad \sum F_y = N - mg \cos \theta = 0. \qquad \textbf{5.2-1}$$

We may at once conclude that $N = mg \cos \theta$ and that $F = mg \sin \theta$. This answers the question asked, on the supposition that the block does not move.   No use has yet been made of the coefficient of friction.   To see whether the block does or does not actually slide, we note that to maintain equilibrium we must have $F/N = \tan \theta$.   Clearly, if f is greater than $\tan \theta$, we will be able to muster a friction force equal to $N \tan \theta$, but if f is less than $\tan \theta$ the block will slide downhill.   In principle, f can be determined by increasing the angle θ until it reaches a critical value θ^* at which the block begins to slide. Then $f = \tan \theta^*$.   The angle θ^* is known as the *angle of friction*.

Knowing the normal and frictional components of the reaction of the plane on the block, it is simple to find the line of action of this resultant force. Of course the dimensions of the block (or at least the location of the mass center relative to the base) need to be known. Suppose that the block is homogeneous, that its breadth and depth are a and b, and that the reaction of the plane on the block intersects the base at a distance s from the lower end, as shown in Fig. 5.2-1. Then, taking moments about an axis through the mass center, we have

$$N\left(s - \frac{1}{2}a\right) + F\left(\frac{1}{2}b\right) = 0, \quad \text{or} \quad s = \frac{1}{2}(a - b \tan \theta). \qquad \textbf{5.2-2}$$

In the last step, use has been made of Eqs. 5.2-1 for F and N. Equation 5.2-2 locates the resultant pressure on the base. If θ is a small angle, this point on the line of action lies between the center of the base and the toe. But if θ is large enough the value of s given by Eq. 5.2-2 will be negative. This would imply that the reaction must

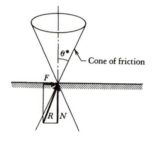

Fig. 5.2-2

fall outside the base in order to preserve rotational equilibrium. Since this is not possible, we conclude that tipping will occur if $\tan \theta$ exceeds a/b. If we think of θ as being increased gradually from zero, sliding will occur when $\theta = \arctan f$, tipping when $\theta = \arctan (a/b)$. If f is smaller than a/b sliding will occur first, otherwise tipping will occur first.

The idea of the angle of friction, θ^*, provides a geometric interpretation of the range of the frictional reaction which is often helpful. If we refer to Fig. 5.2-2 it may be seen that $F < N \tan \theta^*$ provided that the total reaction, R, of which F and N are components, falls within a cone of semivertical angle θ^*. If R were to fall outside this cone, F would have to exceed fN and sliding must ensue. This cone is sometimes termed the *cone of friction*.

A feature of interest in connection with the presence of friction is the possibility it offers the designer of producing what is known as *friction lock.* This condition arises when increase in load is automatically accompanied by increase in frictional capacity, so that a state of impending motion cannot be produced. Again, a simple example in the form of the block on an inclined plane may serve to illustrate the point. What horizontal force, for instance, will cause the block of Fig. 5.2-3a to slide up the plane? If we suppose that such a force exists, then, when it is applied and motion up the plane impends, the free-body diagram of the block is as shown in Fig. 5.2-3b. Since

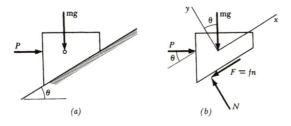

Fig. 5.2-3

motion of the block is supposed to impend, the frictional force on the block must, in view of rules 1 and 3, be directed down the plane and must be equal in magnitude to fN. When compared with Fig. 5.2-1, this free-body diagram has the same number of unknown elements, F being replaced by P as an unknown. Now the equations of translational equilibrium are

$$\sum F_x = P \cos \theta - mg \sin \theta - fN = 0,$$

and

$$\sum F_y = -P \sin \theta - mg \cos \theta + N = 0. \qquad \textbf{5.2-3}$$

On eliminating N and solving for P, we have

$$P = mg \, \frac{\tan \theta + f}{1 - f \tan \theta}. \qquad \textbf{5.2-4}$$

It follows from this result that as f or θ increases until $f \tan \theta = 1$, the force P required to produce impending motion uphill will increase without bound. For values of $f \tan \theta \geq 1$, no horizontal force to the right, however great, will push the block uphill; the situation of impending uphill motion originally hypothesized cannot, in fact, be created. What is happening is that the horizontal force P has two components, one normal to the friction surface and one tangential to it.   When P is increased, both these components increase.   If the normal component, multiplied by f, grows as rapidly as the tangential component, increasing P will not produce motion.   Friction locking may be a desirable effect, as in the case of many gripping and lifting devices, or it may be quite undesirable, as in the case of a poorly designed drawer that will not open.   It may be noted, finally, that expressions such as Eq. 5.2-4 may be put in a compact form through the introduction of the friction angle, $\theta^* = \arctan f$.   Since $\tan (A + B)$ $= (\tan A + \tan B)/(1 - \tan A \tan B)$, Eq. 5.2-4 may be rewritten as

$$P = mg \tan (\theta + \theta^*). \qquad\qquad \textbf{5.2-5}$$

As $\theta + \theta^*$ approaches $\pi/2$, the force P grows without limit.

The introduction of frictional considerations into statics makes it necessary to consider the various ways in which mechanical systems can move.   Only on those surfaces where motion is actually impending is the magnitude of the friction force equal to its maximum value, fN.   Furthermore, the direction of the friction force is then opposite to the direction of the impending motion.   In elementary mechanical analysis this direction is usually obvious, but there are cases where it is not.   These cases are treated in Section 5.3.   The student must beware of the false assumption that the friction force is equal in magnitude to fN.   The friction law is essentially an inequality: $F \leqq fN$, and only when words such as "impending motion," "just moving," "just slips," or "just slides" are appropriate does the equality sign hold.

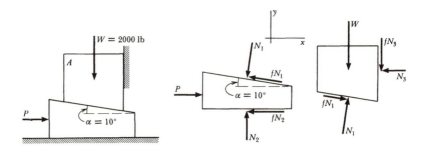

Fig. 5.2-4   Fig. 5.2-5

Example **5.2-1**

A load of 2000 *lb is to be raised by means of a wedge, as shown in Fig.* 5.2-4.   *Ignoring the weight of the wedge, and assuming that on all contact* surfaces the coefficient of friction is $f=0.2$, what is the minimum force, P, needed to lift the load? If P is removed will the weight remain in its raised position?

   Solution: We picture the situation when P is just large enough to cause motion of the wedge to the right to impend. But if motion of the wedge to the right impends, it follows logically that upward motion of the block A also impends. The free-body diagrams of wedge and block at this instant are shown in Fig. 5.2-5. The student should note the *directions* of the frictional forces. The frictional force exerted by the wedge on the heavy block, for example, is directed to the right because the block is about to move to the left with respect to the wedge. Notice that the forces N_1 and fN_1 which appear in both diagrams conform to Newton's third law. There are four unknowns, N_1, N_2, N_3, and P, so that the equations of equilibrium will enable us to find them all.

   For the wedge,

$$\sum F_x = P - fN_1 \cos \alpha - N_1 \sin \alpha - fN_2 = 0,$$

$$\sum F_y = -N_1 \cos \alpha + fN_1 \sin \alpha + N_2 = 0.$$

For the block,

$$\sum F_x = N_1 \sin \alpha + fN_1 \cos \alpha - N_3 = 0,$$

$$\sum F_y = N_1 \cos \alpha - fN_1 \sin \alpha - fN_3 - W = 0.$$

To solve these equations, multiply the second by f and add it to the first to get

$$P = N_1[2f \cos \alpha + (1 - f^2) \sin \alpha].$$

Multiply the third equation by $-f$ and add it to the fourth to get

$$W = N_1[(1 - f^2) \cos \alpha - 2f \sin \alpha].$$

The ratio of these expressions gives

$$P = W \frac{2f \cos \alpha + (1 - f^2) \sin \alpha}{(1 - f^2) \cos \alpha - 2f \sin \alpha}.$$

This answers the first part of the question asked. The student may check that this expression can be put in the form

$$P = W \tan (2\theta^* + \alpha).$$

With $W = 2000$ lb, $f = 0.2$, and $\alpha = 10°$, these expressions indicate that the required value of P is

$$P = 2000 \frac{(0.4)(0.985) + (0.96)(0.174)}{(0.96)(0.985) - (0.4)(0.174)} = 1280 \text{ lb}.$$

Although it is counsel of perfection, we should now check to see that the corresponding values of N_1, N_2, and N_3 are positive, since, presumably, the surfaces in contact must be in compression, as drawn in the free-body diagram. In the present case we find

$$N_1 = W/[(1-f^2)\cos\alpha - 2f\sin\alpha] = 2280\ \text{lb},$$
$$N_2 = N_1(\cos\alpha - f\sin\alpha) = 2170\ \text{lb},$$
$$N_3 = N_1(\sin\alpha + f\cos\alpha) = 850\ \text{lb}.$$

What will happen if the force P is now removed? Two possibilities present themselves: either the system moves or it does not. If motion ensues, the question ceases to be one in statics; on the other hand, if the system remains in equilibrium there is no reason for assuming that motion will impend. The frictional forces fN_1, fN_2, and fN_3 become simply unknowns F_1, F_2, and F_3. But this means that even with $P=0$ there will

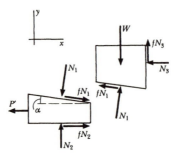

Fig. 5.2-6

be six unknowns in the free-body diagrams and only four equations of equilibrium. The system is statically indeterminate. To avoid this dilemma we suppose that a force, P', to the left, is required in an amount just sufficient to produce motion of the block A down the wedge. Then the free-body diagram is as shown in Fig. 5.2-6. If, on analysis, P' turns out to be positive, we can conclude that in the absence of the force P', motion to the left will not occur. The free-body diagrams now are as shown in Fig. 5.2-6. The equations of equilibrium are

$$-P' + fN_1\cos\alpha - N_1\sin\alpha + fN_2 = 0$$
$$-N_1\cos\alpha - fN_1\sin\alpha + N_2 = 0$$
$$N_1\sin\alpha - fN_1\cos\alpha - N_3 = 0$$
$$N_1\cos\alpha + fN_1\sin\alpha + fN_3 - W = 0.$$

We solve for P', exactly as before, to find

$$P' = W\frac{2f\cos\alpha - (1-f^2)\sin\alpha}{2f\sin\alpha + (1-f^2)\cos\alpha} = W\tan(2\theta^* - \alpha) = 450\ \text{lb}.$$

Our conclusion, therefore, is that, in order to make the wedge slide to the left, a leftward force of 450 lb must be applied to the wedge; in the absence of such a force the wedge and block will remain stationary.

Another way of regarding the situation would be to say that, as regards vertical load alone, the arrangement of wedge and block is self-locking. This is manifestly desirable from a practical engineering point of view; one would not want the raised weight to be continually slipping back. So that this self-locking effect will be achieved, the numerator of the expression for P' must be positive. That is, we must have

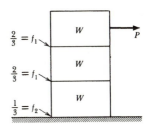

$$\frac{f}{1-f^2} > \frac{1}{2}\tan\alpha, \quad \text{or} \quad \theta^* > \frac{1}{2}\alpha. \qquad \textbf{Fig. 5.2-7}$$

Example 5.2-2

Three boxes, each of weight W, are stacked on the floor, as shown in Fig. 5.2-7. The coefficient of friction between boxes is $f_1 = \frac{2}{3}$ and between box and floor is $f_2 = \frac{1}{3}$. A horizontal force P is applied to the top box and gradually increased. What is the maximum P? How will equilibrium be broken?

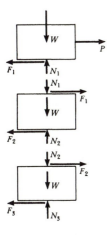

Fig. 5.2-8

Solution: The directions of the friction forces at each contact are known from equilibrium considerations; the only real problem is to find where slip will start. In Fig. 5.2-8 the free-body diagrams of the three boxes are given. Notice that the direction of the friction force acting on the uppermost block must be to the left; hence, from the third law, the friction force exerted by the uppermost block on the middle one must act to the right. It follows that the friction force exerted on the middle block by the lowest block must act to the left so that the equilibrium of the middle block will be maintained, and so on. Since the friction forces on the middle and bottom boxes must be opposite in direction, neither the middle nor the bottom box can possibly slip out by itself from the other two. Equilibrium can only be broken if the top box slides off the others or the whole group slides on the floor. The equilibrium equations are

$$\begin{aligned} F_1 &= P, & N_1 &= W, \\ F_2 &= F_1, & N_2 &= N_1 + W, \\ F_3 &= F_2, & N_3 &= N_2 + W, \end{aligned}$$

whence

$$F_1 = F_2 = F_3 = P \quad \text{and} \quad N_1 = W, \quad N_2 = 2W, \quad N_3 = 3W.$$

The system will be in equilibrium for P small enough; the limiting inequalities are

$$\frac{F_1}{N_1} = \frac{P}{W} \leqq f_1 = \frac{2}{3}, \qquad \frac{F_2}{N_2} = \frac{P}{2W} \leqq f_1 = \frac{2}{3}, \qquad \frac{F_3}{N_3} = \frac{P}{3W} \leqq f_2 = \frac{1}{3}.$$

Clearly the second inequality will always be satisfied if the first is. Slip will always occur at the bottom surface of the uppermost box before it occurs at the bottom surface of the middle box. The limiting value of P is therefore either $f_1 W$ or $3f_2 W$. In the present case, $f_1 W = (\frac{2}{3})W$, whereas $3f_2 W = W$, so that equilibrium will be broken when the top box slides over the other two as soon as $P = (\frac{2}{3})W$.

The student may wish to work this example for the case in which the force P is applied to the middle box.

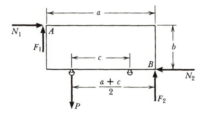

Fig. 5.2-9

Example 5.2-3

A drawer of width a and depth b has two symmetrically placed handles distant c apart. If the coefficient of friction of the drawer against the sides of the case is f, what should the dimensions be so that the drawer will not jam when it is pulled out by one handle?

Solution: Suppose that the drawer is pulled out by a force P exerted on the left-hand handle, as shown in Fig. 5.2-9. It will come in contact with the case at points A and B. We see that $N_1 = N_2$. Taking moments about B, we have

$$-aF_1 - bN_1 + \frac{1}{2} P(a+c) = 0,$$

and if there is to be motion in the direction of P we must have

$$P \geqq F_1 + F_2.$$

Now the maximum value of F_1 is fN_1 and the maximum value of F_2 is fN_2. To assure outward motion of the drawer we must have

$$P \geqq f(N_1 + N_2)$$

or, in view of the equality of N_1 and N_2,

$$P \geqq 2fN_1.$$

But with $F_1 = fN_1$ the moment equation yields

$$N_1 = P \frac{a+c}{2(fa+b)}.$$

We must therefore have

$$P \geqq 2fP \frac{a+c}{2(fa+b)}.$$

This inequality will be satisfied, for any P, provided

$$fa + b \geqq f(a+c)$$

or

$$b \geqq fc.$$

Example 5.2-4

A cylinder of radius a and weight W is wedged between a vertical wall and a light hinged bar, as shown in Fig. 5.2-10. The coefficient of friction between cylinder and wall is 0.2, and between bar and cylinder is 0.4. What

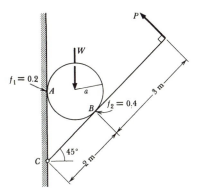

Fig. 5.2-10

is the force, P, which, when applied at the end of the bar and at right angles to the bar, is just sufficient to cause the cylinder to slip upward? The dimensions of the bar are given in the figure.

Solution: The essential point of this example is that if motion impends at the point of contact with the wall, *A*, it does not necessarily impend at the point of contact with the bar, *B*, and vice versa. This example therefore differs fundamentally from Example 5.2-1 in which, from the geometry of the question, motion had to impend simultaneously at all surfaces. Another way of putting this is to say that, in the present case, motion can begin with slipping at either *A* or *B* and rolling (without slip) at the other contact. We must therefore begin by making an assumption about the nature of the impending motion. Analysis will show whether or not this assumption is correct. Suppose, then, that when motion impends, slipping occurs at *B* and not at *A*. An elementary analysis of the equilibrium of the bar, taking moments about *C*, shows that the normal force exerted by the bar on the cylinder must be $(\frac{5}{2})P$. The free-body diagram of the cylinder is shown in Fig. 5.2-11. Note that the direction and magnitude of the

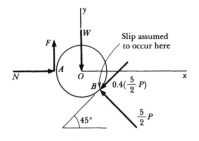

Fig. 5.2-11

friction force at *A* are arbitrary. Since no slip is assumed to impend at *A*, we rely on the equations of equilibrium to determine *F* completely. These equations are:

$$\sum F_x = N - P\left(\frac{\sqrt{2}}{2}\right) - \left(\frac{5}{2}P\right)\left(\frac{\sqrt{2}}{2}\right) = 0,$$

$$\sum F_y = F - P\left(\frac{\sqrt{2}}{2}\right) + \left(\frac{5}{2}P\right)\left(\frac{\sqrt{2}}{2}\right) - W = 0,$$

$$\sum M_O = -aF - aP = 0.$$

We see from these that

$$F = -16.5W, \quad N = 40.8W, \quad \text{and} \quad P = 16.5W.$$

How shall we interpret these results? The negative sign on *F* means simply that the actual direction of the friction force acting on the cylinder at *A* is downward, not upward. There is no inconsistency in this, and an experienced hand would have foreseen from the outset that *F* must in fact be downward. But is our assumption that there is no slip at *A* a valid one? We find $|F|/N = 16.5/40.8 = 0.40$, whereas the maximum F that can be mustered is $0.2N$ since the coefficient of friction at A is 0.2. We conclude that slip will occur at A before it occurs at B.

It is now necessary to repeat the analysis using this alternative assumption. The free-body diagram is as shown in Fig. 5.2-12. Note that the

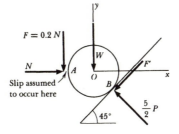

Fig. 5.2-12

friction force at B is now completely unknown and that the friction force at A must be drawn downward since upward slip of the disk impends at A. The equations of equilibrium now read:

$$\sum F_x = N - \left(\frac{5}{2}P\right)\left(\frac{\sqrt{2}}{2}\right) - F'\left(\frac{\sqrt{2}}{2}\right) = 0,$$

$$\sum F_y = -0.2N - W - F'\left(\frac{\sqrt{2}}{2}\right) + \left(\frac{5}{2}P\right)\left(\frac{\sqrt{2}}{2}\right) = 0,$$

$$\sum M_O = 0.2Na - F'a = 0$$

whence we conclude that

$$F' = 0.39W, \quad N = 1.94W, \quad \text{and} \quad P = 0.94W.$$

This answers the question raised; a force $P = 0.94W$ will cause the cylinder to slip upwards. We see that the ratio of frictional to normal force at point B, where motion was assumed not to impend, is

$$\frac{F'}{\left(\frac{5}{2}P\right)} = \frac{0.39}{(2.5)(0.94)} = 0.17,$$

and this is certainly less than the limiting value of 0.4; thus we have checked our assumption that motion would not impend at B.

There is an alternative way of looking at questions of this sort that is often instructive. Suppose we draw the limiting cones of friction at A and B, as shown in Fig. 5.2-13. These have semivertical angles of $11°$ and $22°$, respectively. The reactions at A and B must lie within

these cones.   But the cylinder is acted on by three forces, W and the
forces at A and B.   These must intersect at some point if there is to
be equilibrium, and that point must lie on the line of action of W, i.e., on
the vertical through the center of the cylinder.   If slipping impends at B,

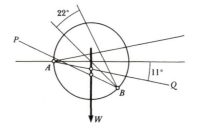

Fig. 5.2-13

the reaction at B will lie along the edge of the cone BP.   But this intersects the line of action of W at a point outside the friction cone at A.
Hence it is not possible to have motion impend at B.   On the other hand,
if upward motion impends at A, the reaction at A will lie along the line
AQ.   This line intersects the vertical through the mass center at a point
within the friction cone at B, so that motion can impend at A without
violating the friction laws at A.   The intersection of AQ and the vertical
also locates the line of action of the force at B and makes it simple to complete the analysis graphically.

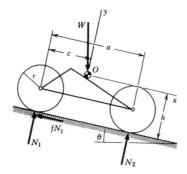

Fig. 5.2-14

Example 5.2-5

What is the steepest slope on which a bicycle can rest without slipping? Assume that only the uphill wheel is braked, that $f = 0.25$, that the axles of the bicycle are 48 in. apart, and that its mass center is 20 in. from the uphill wheel and 32 in. from the ground.

Solution: The free-body diagram of the bicycle, when slipping downhill impends, is shown in Fig. 5.2-14 together with a suggested nomenclature for the dimensions. Notice that there is no frictional force at the lower wheel because that wheel is neither powered nor braked in this example. The equations of equilibrium are

$$\sum F_x = W \sin \theta - fN_1 = 0,$$

$$\sum F_y = N_1 + N_2 - W \cos \theta = 0,$$

$$\sum M_O = N_2(a - c) - fN_1 h - N_1 c = 0.$$

We solve the first two of these equations for the unknowns N_1 and N_2, then substitute in the third equation to find the last unknown, θ.

$$N_1 = \left(\frac{W}{f}\right) \sin \theta, \qquad N_2 = W \cos \theta - \left(\frac{W}{f}\right) \sin \theta;$$

$$\left(\cos \theta - \frac{1}{f} \sin \theta\right)(a - c) - (fh + c)\left(\frac{1}{f} \sin \theta\right) = 0,$$

$$\tan \theta = \frac{f(a - c)}{a + fh}.$$

For $f = \frac{1}{4}$, $a = 48$ in., $c = 20$ in., $h = 32$ in., we obtain

$$\theta = \arctan\left(\frac{\frac{1}{4}(48 - 20)}{48 + 8}\right) = \arctan\left(\frac{1}{8}\right) = 7.125°.$$

The student may find it interesting to consider whether it would be a better design to have the brake on the other wheel.

Example 5.2-6

A simple friction grip has the general shape and dimensions shown in Fig. 5.2-15a. The designer wishes it to be used to pick up boxes of various sizes, with coefficients of friction that may be as low as 0.2. How should the dimensions a and c be chosen if the parcels are not to slip?

Solution: The free-body diagram of one of the two members that compose the grip is shown in Fig. 5.2-15b. The fact that the device is symmetrical and that the vertical and horizontal forces must add to zero implies that the vertical forces will each be $W/2$, half the weight of the box, and that the horizontal forces, N, will be equal.   To find N, take moments about the origin of coordinates:

$$\sum M_O = Nc - \frac{1}{2} Wb - \frac{1}{2} Wa = 0,$$

$$N = \frac{1}{2} W\left(\frac{a+b}{c}\right).$$

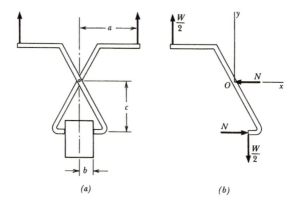

(a)   (b)

Fig. 5.2-15

It follows that the ratio of frictional to normal force at the contact between grip and box is $c/(a+b)$.   If the coefficient of friction may be as low as 0.2 we must have $c/(a+b) < 0.2$ in order for the device to function with complete effectiveness.   If the boxes are of various sizes, b may be small. For complete safety c should be kept to no more than one-fifth of a.

5.3 Further Aspects of Frictional Analysis

The logical structure of mechanics is so highly developed that additions to it, even when apparently well justified by experiment and physical theory, require careful consideration.   In the case of the friction law, no conflict with the principles of mechanics arises provided the deformability of the parts in contact is taken into account where necessary.★   We can see that deformability will affect matters if we

---

★ If deformability is completely excluded, pathological situations can arise in dynamics when friction enters. [For further references see E. T. Whittaker, *Analytical Dynamics*, 4th ed. (Cambridge: Cambridge University Press, 1952), p. 227.]   These are not of importance in engineering because no engineer would imagine his materials to be perfectly rigid.

consider the case of a body resting on a rough horizontal plane, on several supports. The loads, W_1, W_2,... W_n, at each support may be regarded as known. Suppose a horizontal force, **P**, or a couple, is applied to the body at a point A. As this force is gradually increased in magnitude, the friction at the point of support nearest A—say, F_s—will gradually increase, so as to balance **P**. When the magnitude of the friction at this point has reached its limiting value, fW_s, the next point of support will begin to develop a frictional reaction, and so on. Another way of looking at the situation is to say that from the point of view of rigid-body mechanics there are $2n$ unknown frictional reaction components and only three equations of equilibrium in the plane. The situation that presents itself when the applied load **P** is less than that required to produce over-all motion is therefore statically indeterminate. All we can say about the frictional reactions under these circumstances, without a knowledge of the elastic properties of the body, is that they are less in magnitude than the limiting value and that there are many possible different arrangements that satisfy both the friction law and the equations of over-all statical equilibrium. The analysis of such situations— which arise, as has been mentioned, in connection with the contact stresses in ball and roller bearings—requires the resources of continuum mechanics, including the theories of elasticity and plasticity.

On the other hand, when the force **P** is great enough to make motion impend, a more complete analysis can be made within the framework of rigid-body mechanics. Unlike the frictional analysis of Section 5.2, however, the directions of the n frictional forces are now unknown. In order to reduce the number of unknowns to three, we note that the body described in the previous paragraph as resting on supports on a rough horizontal plane will, when it begins to move, turn about some axis perpendicular to the plane. This follows from the geometrical theorem that when a lamina is moved from one position to another in its own plane there is always one point, C, rigidly connected to the lamina, whose position in space is unchanged. The lamina can therefore be brought from its initial to its final position by a rotation about C through some appropriate angle. To prove this theorem we need only suppose that A, B are two points fixed on the lamina and that they move to A', B', as shown in Fig. 5.3-1. If A, B can be brought to A', B' by rotation about some point, the whole lamina will be in its proper final position. The perpendicular bisectors of AA' and BB' intersect at a point C. Since $AC = A'C$ and $BC = B'C$ (and, of course, $AB = A'B'$) it follows that the triangles ACB and $A'CB'$ are congruent. The angles subtended at C by AB and $A'B'$

are therefore equal. If we rotate AB about C through the angle ϕ shown, A will be carried to A' and B will also be carried to B'. It follows that there is always—and therefore initially—a unique point C about which the body may be considered to be rotating. In

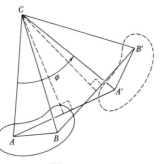

Fig. 5.3-1

dynamics the angle ϕ is taken vanishingly small and C is called the instantaneous center of zero velocity. Of course, if the lamina is in pure translation, C is infinitely distant. With the knowledge that there is a point, C, about which the body starts to rotate when motion impends, the analysis is reduced to finding the two coordinates of this point. These, together with the unknown limiting force magnitude, P, form the three unknowns of the analysis. Once C is found, the direction of the frictional force at any point of support must be at right angles to the line connecting that point to C and the magnitude of the frictional force at any point of support must be the normal load carried there multiplied by the coefficient of limiting friction. An exceptional case arises when C coincides with one of the points of support; this point cannot then be said to be on the verge of motion. In that special case the coordinates of C are replaced as unknowns by the components of the frictional reaction at C. These are subject to the limitation that their resultant is less than the limiting value.

Example 5.3-1

A small block of weight W rests on a rough plane which is inclined to the horizontal at an angle θ; a taut cord parallel to the plane passes over a pulley and supports the weight P, as shown in Fig. 5.3-2. The coefficient of limiting friction is f. Discuss equilibrium of the block. What weight P^* will cause it to slide?

Solution: In this case the body resting on the plane is a particle; it has, therefore, only one "point of contact" and the situation is statically determinate. Since the system is in equilibrium the cord tension will be

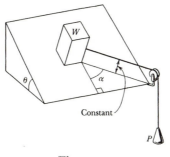

Fig. 5.3-2

P in magnitude. A free-body diagram (Fig. 5.3-3) of the block shows P, the weight W, the normal force N, and a friction force having two components, F_x and F_y, tangent to the plane. The equilibrium equations are

$$\sum R_x = P \cos \alpha + W \sin \theta - F_x = 0,$$
$$\sum R_y = P \sin \alpha - F_y = 0,$$
$$\sum R_z = N - W \cos \theta = 0.$$

We conclude that

$$N = W \cos \theta,$$
$$F_x = W \sin \theta + P \cos \alpha,$$
$$F_y = P \sin \alpha.$$

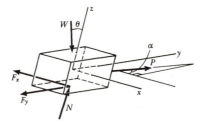

Fig. 5.3-3

Notice that the direction of the resultant friction force is neither along the cord nor straight down the plane. The magnitude of the total friction force is

$$F = (F_x^2 + F_y^2)^{1/2} = (W^2 \sin^2 \theta + 2WP \sin \theta \cos \alpha + P^2)^{1/2}.$$

Since equilibrium is only possible when $F/N \leqq f$ it follows that, for equilibrium,

$$(W^2 \sin^2 \theta + 2WP \sin \theta \cos \alpha + P^2)^{1/2} \leqq fW \cos \theta.$$

We can square both sides of this inequality, since both sides are positive, and put it in the form

$$W^2 \sin^2 \theta + 2WP \sin \theta \cos \alpha + P^2 \leqq f^2 W^2 \cos^2 \theta,$$

or

$$\left(\frac{P}{W}\right)^2 + 2 \sin \theta \cos \alpha \left(\frac{P}{W}\right) + \sin^2 \theta - f^2 \cos^2 \theta \leqq 0.$$

The limiting conditions for slip occur when equality first becomes possible. Then $P = P^*$ and we have a quadratic equation for P^*/W. The roots of this equation are

$$\frac{P^*}{W} = -\sin \theta \cos \alpha \pm (\sin^2 \theta \cos^2 \alpha - \sin^2 \theta + f^2 \cos^2 \theta)^{1/2}.$$

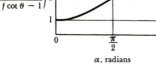

Fig. 5.3-4

The minus sign before the radical would make P^* negative, a physically meaningless result. We conclude that

$$\frac{P^*}{W} = (\sin^2 \theta \cos^2 \alpha - \sin^2 \theta + f^2 \cos^2 \theta)^{1/2} - \sin \theta \cos \alpha$$

$$= (f^2 \cos^2 \theta - \sin^2 \theta \sin^2 \alpha)^{1/2} - \sin \theta \cos \alpha.$$

P^* has its smallest value when $\alpha = 0$ (the cord straight down the plane). This value, denoted $P^*_{\min}$, equals $(f \cos \theta - \sin \theta)W$. To see how P^* varies with the angle α, we may conveniently divide P^* by $P^*_{\min}$:

$$\frac{P^*}{P^*_{\min}} = \frac{(f^2 \cos^2 \theta - \sin^2 \theta \sin^2 \alpha)^{1/2} - \sin \theta \cos \alpha}{f \cos \theta - \sin \theta}$$

$$= \frac{(f^2 \cot^2 \theta - \sin^2 \alpha)^{1/2} - \cos \alpha}{f \cot \theta - 1}.$$

Fig. 5.3-4 shows, schematically, the variation of P^* as a function of α in the range $0 \le \alpha \le \pi$.

Example 5.3-2

A uniform rod of weight W and length l lies on a rough table, the coefficient of limiting friction being f.   If the rod exerts a uniform pressure on the table, what force directed at right angles to the rod at one end will cause it to move ?

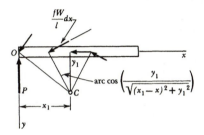

Fig. 5.3-5

Solution: Suppose that, initially, the rod breaks equilibrium by rotating about the point C whose coordinates are x_1, y_1, as shown in Fig. 5.3-5. Just before equilibrium is broken, there will be at each element, dx, of the bar a normal force on the bar $W\,dx/l$ exerted by the table and a frictional force $fW\,dx/l$.   These frictional forces will act at right angles to the radius

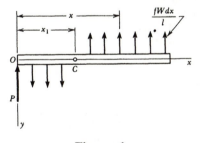

Fig. 5.3-6

vector from C, as shown.   Since they all have negative x-components, and since there is no external force in the x-direction to balance, it is obvious that y_1 must be zero and the point C must lie on the x-axis.   This also emerges from the first of the equations of equilibrium:

$$\sum F_x = -\frac{fW}{l}\,y_1 \int_0^l \frac{dx}{\sqrt{(x-x_1)^2 + y_1^2}} = 0.$$

We need not bother to work out the integral; it must be a positive number since the integrand is positive over the entire range of integration. The summation of forces in the x-direction can be zero only if $y_1 = 0$. An experienced hand would have started with this understanding. At any rate, the free-body diagram can now be redrawn, for the sake of simplicity, as shown in Fig. 5.3-6. The equation of equilibrium which asserts that the sum of the external forces in the x-direction must vanish is now identically satisfied. The other two equations of equilibrium determine x_1 and P when slipping impends.

$$\sum M_O = \frac{fW}{l} \int_0^{x_1} x \, dx - \frac{fW}{l} \int_{x_1}^{l} x \, dx = 0,$$

$$\sum F_y = \frac{fW}{l} \int_0^{x_1} dx - \frac{fW}{l} \int_{x_1}^{l} dx - P = 0.$$

The integrations are almost unnecessary, since the magnitude of the frictional force is uniform along the bar. From the moment equation

$$-x_1^2 + (l^2 - x_1^2) = 0, \qquad x_1 = \frac{l}{\sqrt{2}}.$$

From the force equation

$$P = \frac{fW}{l} [x_1 - (l - x_1)] = \frac{fW}{l} (2x_1 - l)$$

$$= fW(\sqrt{2} - 1).$$

This completes the analysis. It is interesting to note that the force required to start motion in this way is only about $0.41fW$—that is, about 41 per cent of what would be needed to make the bar move in simple translation. The point C about which the bar starts to rotate is about 71 per cent of the way along the bar from the end at which P is applied. In dynamics we treat the case that is, in a manner of speaking, the opposite of this one: the case in which P instead of being applied slowly is applied very rapidly, as a sharp blow or "impulse". It is found in the dynamical case that the bar begins to move about a point two-thirds of the way along the bar from the struck end. There is surprisingly little difference in the results for the two extreme cases, and in each the location of C is independent of the magnitude of the friction present.

Example 5.3-3

A uniform 100 lb beam 12 ft long rests on pads at each end. The horizontal surface on which it rests has a coefficient of friction of one-half. One end of the bar is restrained by a horizontal cable extending at right angles to the bar. Initially the cable is just taut. A force P is applied to the bar at a point 8 ft from the end at which the cable is attached, in a direction away from the cable at an angle of 45° with the bar, as shown in Fig. 5.3-7a. If P is gradually increased from zero, how large must it become before the bar moves?

Solution: The point C about which the bar will begin to move must be somewhere along the line of the cable (the y-axis in Fig. 5.3-7b). This follows from the fact that, since the cable is taut, equilibrium must be broken by motion of point A at right angles to the cable (i.e., along the x-axis). This means that the frictional force at A must, when motion impends, be in the negative x-direction; and, since this friction force must be at

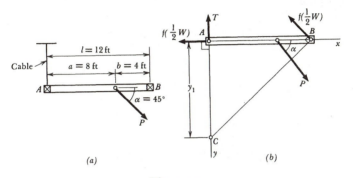

Fig. 5.3-7

right angles to the line *AC*, it follows that *C* must be on the *y*-axis. We denote the distance from the origin of coordinates at *A* to the point *C* by the symbol y_1, and the weight of the bar by the symbol *W*. Then the free-body diagram of the bar when motion impends is as shown in Fig. 5.3-7b. The equations of equilibrium are

$$\sum F_x = P \cos \alpha - \frac{1}{2} fW \frac{y_1}{\sqrt{y_1^2 + l^2}} - \frac{1}{2} fW = 0,$$

$$\sum F_y = -T - \frac{1}{2} fW \frac{l}{\sqrt{y_1^2 + l^2}} + P \sin \alpha = 0,$$

$$\sum M_A = -\frac{1}{2} fW \frac{l^2}{\sqrt{y_1^2 + l^2}} + Pa \sin \alpha = 0.$$

The second of these equations determines *T*, the tension in the cable—a quantity not of direct interest. We solve the first and third equations for the wanted quantities *P* and y_1:

$$y_1 = \frac{l^2 \cot^2 \alpha - a^2}{2a \cot \alpha}, \qquad P = \frac{1}{2} fW \frac{l^2}{a\sqrt{y_1^2 + l^2}} \operatorname{cosec} \alpha.$$

In the present case, $l = 12$ ft, $\cot \alpha = 1$, $a = 8$ ft, $f = \frac{1}{2}$, and $W = 100$ lb, so that

$$y_1 = \frac{144 - 64}{16} = 5 \text{ ft,}$$

$$P = \frac{(25)(12)^2}{(8)\sqrt{169}} \sqrt{2} = 49 \text{ lb.}$$

Finally, we must check the possibility that C coincides with one of the points of support, A. If this happens, the free-body diagram is as shown in Fig. 5.3-8.

$$\sum M_A = P'a \sin \alpha - \frac{1}{2}fWl = 0,$$

$$P' = \frac{1}{2}fW\frac{l}{a} \operatorname{cosec} \alpha.$$

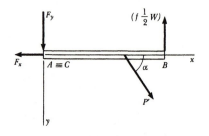

Fig. 5.3-8

We see from the general expression given above that this value of P can never exceed the one obtained previously. Also, if we solve for F_x we find

$$F_x = P' \sin \alpha = \frac{1}{2}fW\frac{l}{a},$$

which exceeds the largest permissible value of F_x, $\frac{1}{2}fW$.

5.4 Cable Friction

It is a common observation that by wrapping a rope around a post a man can hold in check a much larger force than he would ordinarily be able to exert. To see why this device is so effective, consider the case of a flexible cable carrying a force T_0 at one end and a force $T_1 > T_0$ at the other end, the cable being pressed for part of its length against a post. We suppose that T_1 is just large enough to produce impending motion. Then every point in contact with the post is in a state of impending motion. The free-body diagram of a typical element of the cable where it is in contact with the post is shown in Fig. 5.4-1. The post exerts on the cable a distributed normal pressure, p, and a distributed frictional tangential force, fp, units of force per unit of length. In general p must be regarded as a function of position along the cable. The equations of equilibrium

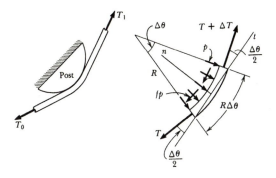

Fig. 5.4-1

for the typical element shown are found by summing forces in the normal and tangential directions. The normal equation of equilibrium is

$$\sum F_n = T \sin\left(\frac{\Delta\theta}{2}\right) + (T + \Delta T) \sin\left(\frac{\Delta\theta}{2}\right) - p(R\Delta\theta) = 0 \quad \textbf{5.4-1a}$$

or

$$2T \sin\left(\frac{\Delta\theta}{2}\right) + \Delta T \sin\left(\frac{\Delta\theta}{2}\right) - pR\,\Delta\theta = 0. \qquad \textbf{5.4-1b}$$

If the element is taken to be very small, then $\sin(\Delta\theta/2)$ is approximately equal to $\Delta\theta/2$, and $\Delta T \sin(\Delta\theta/2)$ will be negligible compared to the other terms. Therefore, if Eq. 5.4-1 is to hold for *any* $\Delta\theta$, however small (i.e., hold in the limit as $\Delta\theta\to0$), we must have

$$T = pR. \qquad \textbf{5.4-2}$$

Precisely the same conclusion can be reached from Eq. 4.2-12b if we recall that $d\theta/ds = 1/R$, the reciprocal of the radius of curvature, and that $q_n = -p$. Indeed, the situation pictured in Fig. 5.4-1 is only a special case of the flexible cable carrying a distributed load.

Equation 5.4-2, however, is an interesting result in its own right. It implies that for any given pull, T, the pressure, p, will be large if the radius of curvature, R, is small. No wonder that a cord tends to bite into a parcel where it goes round a corner! For a circular post R will be a constant, but there is nothing in the analysis that would prevent R from being a function of position along the curve. Turning now to the equation of tangential equilibrium, we have

$$\sum F_t = -T \cos \left(\frac{\Delta\theta}{2} \right) + (T + \Delta T) \cos \left(\frac{\Delta\theta}{2} \right) - fp(R\,\Delta\theta) = 0. \qquad \textbf{5.4-3}$$

In the limit, as $\Delta\theta$ is taken smaller and smaller, Eq. 5.4-3 implies that

$$dT = fpR\,d\theta \qquad \textbf{5.4-4}$$

or, in view of Eq. 5.4-2, that

$$\frac{dT}{T} = f\,d\theta. \qquad \textbf{5.4-5}$$

If we measure the angle θ from the point where the cable first makes contact with the post on the low-tension side, the tension at any other point is given by

$$\int_{T_0}^{T} \frac{dT}{T} = f \int_0^{\theta} d\theta, \qquad \log \left(\frac{T}{T_0} \right) = f\theta,$$

$$T = T_0 e^{f\theta}. \qquad \textbf{5.4-6}$$

Here θ is the angle through which the tangent to the post has turned in going from the element where the tension is T_0 to the element where the tension is T. This angle is measured in radians. In the case of a circular post, which is the one usually encountered in technology, θ is the same as the angle subtended at the center of the circle. Equation 5.4-4 is, of course, a special case of Eq. 4.2-12a with $ds = R\,d\theta$ and $q_t = -fp$. We see that T is a function of θ, growing exponentially with θ. It follows from Eq. 5.4-2 that if R is constant, p also increases exponentially with θ or with arclength, s. In the case of a smooth pulley, on the other hand, it was shown in Example 4.2-1 that $p =$ constant.

If the cable loses contact with the post at a point where $\theta = \theta_1$, we have

$$T_1 = T_0 e^{f\theta_1} \qquad \textbf{5.4-7}$$

or

$$T_1 - T_0 = T_0(e^{f\theta_1} - 1). \qquad \textbf{5.4-8}$$

Now we see why the load that can be held in this way is so large. The exponent $f\theta_1$ increases with the wrap angle θ_1 (once around the post, whatever its shape, makes $\theta_1 = 2\pi$) and, even if f is small, several complete circuits will make $e^{f\theta_1}$ a large number.

Example 5.4-1

A brake band consists of a wire band wound over half the circumference of a rotating drum and capable of being tightened by a force P, as shown in Fig. 5.4-2a. For the dimensions a, b, and R shown and with $f=0.12$, how

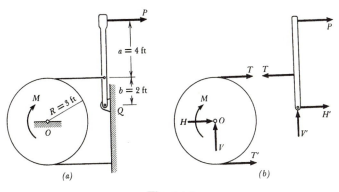

Fig. 5.4-2

large a force, P, is required to prevent the drum from rotating when a clockwise torque, M, of 100 lb-ft is applied to it?

Solution: The free-body diagrams of drum and brake lever are shown in Fig. 5.4-2b. Taking moments about the point, Q, at which the brake lever is hinged, we have

$$T = P\left(\frac{a+b}{b}\right).$$

Since T is on the "slack" or less taut side of the brake band and since rotation of the drum is impending, we have, from Eq. 5.4-7,

$$T' = Te^{0.12\pi} = P\left(\frac{a+b}{b}\right)e^{0.12\pi}.$$

The angle of wrap in this case is $\theta_1 = \pi$ radians. Now, on taking moments about the center, O, of the drum, we find

$$(T' - T)R - M = 0,$$

or

$$M = PR\left(\frac{a+b}{b}\right)(e^{0.12\pi} - 1)$$

$$= P(3)\left(\frac{6}{2}\right)(1.457 - 1) = 4.11P.$$

Therefore, with M given as 100 lb-ft, we find the necessary P to be

$$P = (0.243)(100) = 24.3 \quad \text{lb.}$$

It is interesting to note, before leaving the example, that this design is poorer in resisting counterclockwise torque than in resisting clockwise torque. If M is counterclockwise, T will be the tight and T' the slack side of the band. Then

$$T = T'e^{0.12\pi}, \qquad T' = P\left(\frac{a+b}{b}\right)e^{-0.12\pi},$$

and, on taking moments about the center of the drum, we obtain

$$-(T-T')R+M = 0, \qquad M = PR\left(\frac{a+b}{b}\right)(1-e^{-0.12\pi}) = 2.81P,$$

from which

$$P = 0.356M = (0.356)(100) = 35.6 \quad \text{lb.}$$

About 47 per cent more force is required to restrain counterclockwise motion than clockwise motion. The designer can make the brake equally effective in both directions by the design modification shown in one of the exercises.

Example 5.4-2

A light belt is used to transmit power from a 4-ft diameter driving pulley to a 2-ft diameter driven pulley, the angles of wrap being 210° and 150°, respectively, as shown in Fig. 5.4-3. If the coefficient of limiting friction is 0.25, and the ultimate tensile strength of the belt is 500 lb, how much torque can be transmitted to the smaller pulley before slip occurs?

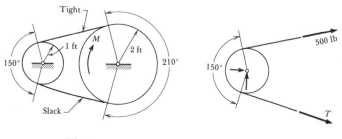

Fig. 5.4-3   **Fig. 5.4-4**

Solution: In this instance the upper side of the belt will be taut and the lower side will be under relatively less tension. To achieve the greatest possible torque, the tension in the taut side should be 500 lb. Slip will occur first on the smaller pulley since the angle of wrap is smaller there. The free-body diagram for the smaller pulley is shown in Fig. 5.4-4. Since slip impends, we have

$$500 = Te^{(0.25)(5\pi/6)} = Te^{0.654},$$

$$T = \frac{500}{1.92} = 260 \quad \text{lb},$$

and the torque transmitted to the pulley cannot exceed

$$(500)(1) - (260)(1) = 240 \quad \text{lb-ft}.$$

In order to transmit this torque, a moment of $(500-260)(2)=480$ lb-ft must be applied to the driven pulley. If more than this torque is applied the belt will slip. The designer who wishes to increase the amount of torque transmitted, without changing the radii of the pulleys (which controls the rotational speed of the smaller pulley), may introduce an idler pulley, as shown in Fig. 5.4-5, which serves to increase the angle of wrap.

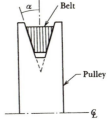

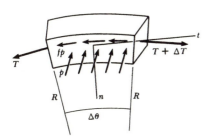

Fig. 5.4-5   Fig. 5.4-6

Example **5.4-3**

Explain the effectiveness of the commercial Vee-belt design used for power-transmission pulleys.

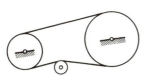

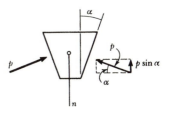

Fig. 5.4-7

Solution: For this pulley, the cross-section of the groove has the shape of a flat-bottomed *V*, as shown in Fig. 5.4-6. The belt need not have a trapezoidal cross-section—it could just as well be circular in cross-section—though the form shown is somewhat better adapted to reduce wear. We isolate an element of belt and draw its free-body diagram as shown in Fig. 5.4-7. The analysis now follows closely the lines of the text of Section 5.4, except that there are now two contact surfaces instead of one as in the case of the flat-faced pulley, and these faces are inclined to the vertical by an angle α. Resolving forces along the normal and tangential directions, we have

$$\sum F_n = T \sin\left(\frac{\Delta\theta}{2}\right) + (T+\Delta T)\sin\left(\frac{\Delta\theta}{2}\right) - (p\sin\alpha)(2R\,\Delta\theta) = 0,$$

$$\sum F_t = (T+\Delta T)\cos\left(\frac{\Delta\theta}{2}\right) - T\cos\left(\frac{\Delta\theta}{2}\right) - fp(2R\,\Delta\theta) = 0.$$

In the limit, as $\Delta\theta \to 0$, these become

$$T - 2pR\sin\alpha = 0$$

and

$$\frac{dT}{d\theta} = 2fpR = \frac{f}{\sin\alpha}T.$$

The last equation is the same as Eq. 5.4-5, with f replaced by $f/\sin\alpha$. We shall have

$$T = T_0 e^{(f\,\mathrm{cosec}\,\alpha)\theta}.$$

By making $\alpha = 30°$, for example, the coefficient of friction is effectively increased by a factor of 2. Of course the pressure p is also increased.

5.5 Friction in Machine Elements

For many of the fundamental machine elements the Amontons-Coulomb friction law provides a satisfactory estimate of the principal frictional effects encountered in engineering practice. We discuss here some of the commonest of these elements: the screw, the thrust bearing, and the simple journal bearing.

The load-carrying screw is essentially an inclined plane wound round a cylinder. The cross-section of the thread may take any one of a number of forms, depending on the function of the screw. Some of the commoner forms are pictured in Fig. 5.5-1; of these, the first three are most suitable for heavy loads such as are encountered in testing machines and in screw-jacks, while the V-shaped thread is suitable for a fastener. We shall initially consider the square thread, because its analysis is less complicated geometrically than the treatment of the others.

We suppose, then, that the screw is carrying an axial load W, which is balanced by a normal pressure p exerted by the nut on the underside of the threads.   The units of p are force per unit length, length being measured along the thread.   The screw is also subjected

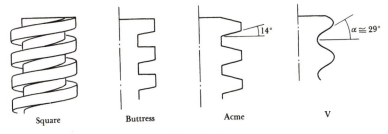

| Square | Buttress | Acme | V |

Fig. 5.5-1

to a torque, M, about its longitudinal axis, which is resisted partly by p and partly by a frictional shear stress exerted on the threads by the nut.   When M is just sufficient to produce motion of the screw

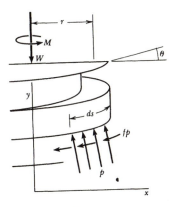

Fig. 5.5-2

in a direction opposite to that of the load W, this frictional effect is fp units of force per unit of thread length.   The free-body diagram of the screw is as shown schematically in Fig. 5.5-2.   Notice that in this figure only a small length of thread, ds, is shown loaded so as to avoid cluttering the figure unnecessarily.   The angle of the inclined plane, sometimes termed the *screw angle*, is denoted by the symbol θ. If we follow a thread round the circumference of the screw we find, after the tangent to the thread has turned through 360°, that we have moved a distance l along the axis from the point at which we started. Then $\theta = \arctan(l/2\pi r)$, where r is the radius of the screw. The distance l, which is the distance that the screw would advance relative to the nut in a single turn, is known as the *lead* of the screw. Returning now to the free-body diagram, the equations of equilibrium are

$$\sum F_y = \int_0^S (p \cos \theta - fp \sin \theta)\, ds - W = 0, \qquad \text{5·5-1}$$

$$\sum M_y = \int_0^S (fp \cos \theta + p \sin \theta)r\, ds - M = 0. \qquad \text{5·5-2}$$

Here p is a function of s, distance measured along the thread. The limits of integration are $s = 0$, where the thread first makes contact with the nut, and $s = S$, where the thread emerges from the nut. Since p is the only quantity in the integrand that depends on the variable of integration, s, these equations may be written

$$\left(\int_0^S p\, ds\right)(\cos \theta - f \sin \theta) = W, \qquad \text{5·5-3}$$

$$\left(\int_0^S p\, ds\right)(f \cos \theta + \sin \theta)r = M. \qquad \text{5·5-4}$$

The other equations of equilibrium are, of course, automatically satisfied on account of the axial symmetry of the screw. Taking the ratio of Eq. 5·5-4 to Eq. 5·5-3, we have

$$\frac{M}{W} = \frac{r(f \cos \theta + \sin \theta)}{\cos \theta - f \sin \theta} \qquad \text{5·5-5a}$$

or

$$M = rW \frac{f \cos \theta + \sin \theta}{\cos \theta - f \sin \theta}. \qquad \text{5·5-5b}$$

This can be put in a more easily remembered form if we introduce the angle of friction. Let $f = \tan \theta^*$. Then

$$M = rW \frac{\tan \theta^* \cos \theta + \sin \theta}{\cos \theta - \tan \theta^* \sin \theta}$$

$$= rW \frac{\sin \theta^* \cos \theta + \cos \theta^* \sin \theta}{\cos \theta^* \cos \theta - \sin \theta^* \sin \theta} = rW \frac{\sin(\theta + \theta^*)}{\cos(\theta + \theta^*)}$$

or

$$M = rW \tan(\theta + \theta^*). \qquad \text{5·5-6}$$

This is the torque required to raise the weight W. Of course, $\theta \le (\pi/2) - \theta^*$ for the torque to be positive.

The equations of equilibrium also fix the average normal pressure, p_{av}, exerted on the threads. This is rather important from a practical point of view because the threads will be stripped off if they are not strong enough to carry the load. Since

$$p_{av} = \frac{1}{S} \int_0^S p \, ds,$$

it follows at once from Eq. 5.5-4 that

$$p_{av} = \frac{M}{rS} (f \cos \theta + \sin \theta)^{-1} = \frac{M}{rS} \cos \theta^* \, \mathrm{cosec} \, (\theta + \theta^*). \quad \textbf{5.5-7}$$

Should the designer feel that this load per inch of thread is too great, he may increase the size of the thread so as to make it stronger, he may increase the length of the nut so as to increase S, or he may call for a reduced lead, also increasing S. Alternatively, a second spiral thread of the same lead may be cut on the cylindrical body of the screw with its threads half way between those of the first set. This doubles S while leaving the other quantities of the analysis unchanged. Such a screw is said to be *double-threaded*. When load-carrying capacity is essential, double- and even triple-threaded screws can be used. They are, of course, more expensive than the ordinary single-threaded screw. The *pitch* of a screw is the axial distance between successive threads; for a single-threaded screw, pitch and lead are identical; for a double-threaded screw the pitch is half the lead.*

What is known as the *efficiency* of a screw is the ratio of the force or torque which would be required in the complete absence of friction to the force or torque actually required to produce motion. In the case of a load-raising square screw it can be seen from Eq. 5.5-6 that this ratio is

$$\frac{\tan \theta}{\tan (\theta + \theta^*)}. \quad \textbf{5.5-8}$$

It follows that if the efficiency is to be a maximum we must choose θ so that

$$\frac{d}{d\theta} \left[\frac{\tan \theta}{\tan (\theta + \theta^*)} \right] = 0.$$

---

* This nomenclature is unfortunate. The thread of a screw is a helix, and the mathematical definition of the pitch of a helix is what is known as the lead of the screw.

To solve this equation, we perform the indicated differentiation of a quotient and then multiply both numerator and denominator of the resulting fraction by $2 \cos^2 \theta \cos^2 (\theta + \theta^*)$:

$$\frac{d}{d\theta}\left[\frac{\tan \theta}{\tan (\theta + \theta^*)}\right] = \frac{\tan (\theta + \theta^*) \sec^2 \theta - \tan \theta \sec^2 (\theta + \theta^*)}{\tan^2 (\theta + \theta^*)}$$

$$= \frac{2 \sin (\theta + \theta^*) \cos (\theta + \theta^*) - 2 \sin \theta \cos \theta}{2 \sin^2 (\theta + \theta^*) \cos^2 \theta}.$$

This last expression may be rewritten:

$$\frac{d}{d\theta}\left[\frac{\tan \theta}{\tan (\theta + \theta^*)}\right] = \frac{\sin 2(\theta + \theta^*) - \sin 2\theta}{2 \sin^2 (\theta + \theta^*) \cos^2 \theta}.$$

Setting this last expression for the derivative equal to zero, we find that the value of θ for maximum efficiency is determined by the trigonometric equation

$$\sin 2(\theta + \theta^*) - \sin 2\theta = 0$$

or [from $\sin A - \sin B = 2 \cos \frac{1}{2}(A + B) \sin \frac{1}{2}(A - B)$]

$$2 \cos (2\theta + \theta^*) \sin \theta^* = 0.$$

Since $\theta^* \neq 0$, the first root of this equation is given by $2\theta + \theta^* = \pi/2$, or

$$\theta = \frac{\pi}{4} - \frac{\theta^*}{2}. \qquad \textbf{5·5-9}$$

This value of θ is half of the permissible maximum value $(\pi/2) - \theta^*$ of θ found above. This value of θ substituted in 5.5-8 indicates that the maximum possible efficiency of a screw is

$$\frac{\tan \left(\dfrac{\pi}{4} - \dfrac{\theta^*}{2}\right)}{\tan \left(\dfrac{\pi}{4} + \dfrac{\theta^*}{2}\right)} = \frac{1 - \sin \theta^*}{1 + \sin \theta^*}. \qquad \textbf{5·5-10}$$

If $f = 0.20$, θ^* will be arctan $(0.2) = 11.3°$. It follows that the maximum efficiency of the screw will be 67.2 per cent. To attain this efficiency, θ should have the value $45° - 5.6° = 39.4°$. This means that for a one-inch radius screw the lead would have to be $2\pi r \tan \theta = 5.15$ inches. Most screw designs are not, in practice, intended to achieve maximum efficiency.

Rather more important than efficiency is the question of whether the release of the torque, M, which lifts the weight W, will result in the weight forcing the screw down again. To see whether or not this will happen we suppose that the moment, M', has the proper value to make downward motion impend. Then if the M' required turns out to be negative we can conclude that in the absence of any torque no motion will occur. The free-body diagram is the same as that shown in Fig. 5.5-2 except that the symbol M is replaced by the symbol M' and the directions of the frictional tractions are reversed. The equations of equilibrium are then the same as 5.5-1 and 5.5-2, except that M is replaced by M' and f by $-f$. From Eq. 5.5-5b we conclude that for downward motion

$$M' = rW \frac{-f \cos \theta + \sin \theta}{\cos \theta + f \sin \theta} = rW \tan (\theta - \theta^*). \qquad \textbf{5·5-11}$$

M' will be zero, or the screw will be on the verge of slipping back, when $f = \tan \theta$ or $\theta = \theta^*$. This result might have been anticipated from the fact that the screw is a form of inclined plane. It follows from Eq. 5.5-8 that if a load-bearing screw is to be non-reversible the efficiency cannot exceed $\tan \theta^* / \tan 2\theta^*$, a value never greater than 50 per cent.

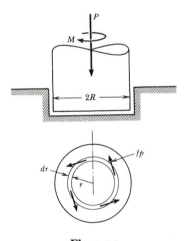

Fig. 5·5-3

Another common machine element in which friction plays an important role is the ordinary flat-collar thrust bearing or friction clutch. The simplest possible example of such a bearing is shown in Fig. 5.5-3. The shaft is supposed to be rotating at a constant rate so that an equilibrium treatment is appropriate.

Taking moments of the friction forces about the axis of the shaft, we have

$$M = 2\pi \int_0^R rfp(r\,dr).$$ 5.5-12

Here p is the intensity of normal pressure on the base of the shaft (note that its dimensions are force per unit area). Lacking further information, all we can really assert with regard to p is that it is radially symmetric. At any value of r it seems reasonable to say that p will have the same magnitude all round. Use has been made of this presumption in writing Eq. 5.5-12. It is painfully obvious, however, that progress in the analysis depends on a knowledge of the distribution of the pressure over the surface. This can really only be derived from deformable-body mechanics. Two assumptions are, however, common in machine design: (1) the normal pressure is uniform over the contact region, and (2) the wear is uniform over the contact region. On the first assumption p is a constant equal to the load carried by the shaft divided by the area of the bearing: $p = P/\pi R^2$. Then the required torque is given by the expression

$$M = 2\pi \frac{P}{\pi R^2} f \int_0^R r^2\,dr = \frac{2}{3} fPR.$$ 5.5-13

On the second assumption, if we say that wear is proportional to the pressure and to the speed, we conclude that the product of pressure and speed is constant. This implies that pr is constant, since the speed at any point on the bearing face is proportional to the radial distance, r. Call this constant C. Then $p = C/r$ and, from Eq. 5.5-12,

$$M = 2\pi Cf \int_0^R r\,dr = \pi CfR^2.$$ 5.5-14

To determine the constant C we write the equation of equilibrium obtained by summing forces in the axial direction.

$$-P + 2\pi \int_0^R p(r\,dr) = 0,$$

$$C \int_0^R dr = \frac{P}{2\pi}, \qquad C = \frac{P}{2\pi R}.$$ 5.5-15

It follows that

$$M = \frac{1}{2} fPR.$$ 5.5-16

Neither of these two common assumptions is exactly correct and the truth is probably somewhere between them. The torque M generally is less on the second assumption than on the first. The power required to rotate the shaft at a constant rate of ω radians per second is $M\omega$. This is also the rate at which heat is generated by friction at the bearing.

Another common machine element that can be analyzed with the aid of the friction law is the plain axle and bearing. When an axle turns in a very slightly larger cylindrical cavity it makes line contact with the hole along a generator of the cylinder. To fix ideas, let us compute the friction needed to restrain a body from

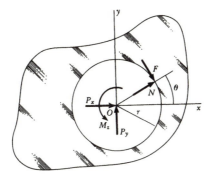

Fig. 5.5-4

rotating about a fixed axle. We suppose that the axle is in the z-direction and that the forces which act on the body, other than those exerted by the axle, can be reduced to an equipollent system consisting of a force $\mathbf{P}=P_x\mathbf{i}+P_y\mathbf{j}$ acting at O, the center of the axle, together with a couple, M_z, about the z-axis. The force exerted by the axle on the body will consist of a normal and a frictional reaction at the point, Q, where the axle presses on the circumference of the bearing. The free-body diagram is shown in Fig. 5.5-4. The point Q is located by the polar angle θ. The radius of the hole is denoted r; this is essentially the same as the radius of the axle. Now the equations of equilibrium are

$$P_x + F \sin \theta + N \cos \theta = 0,$$
$$P_y - F \cos \theta + N \sin \theta = 0, \qquad \textbf{5.5-17}$$
$$M_z - rF = 0.$$

These equations determine F, N, and θ, on the assumption that the friction is great enough to preserve equilibrium. To discover how much friction is actually needed to preserve equilibrium we suppose that counterclockwise rotation about the axle impends. Then $F = fN$ and is clockwise, as shown in Fig. 5.5-4. The first two of Eqs. 5.5-17 become

$$N(\cos \theta + f \sin \theta) = -P_x,$$
$$N(\sin \theta - f \cos \theta) = -P_y. \qquad \textbf{5.5-18}$$

Taking the ratio of these two expressions, we have

$$\frac{\sin \theta - f \cos \theta}{\cos \theta + f \sin \theta} = \frac{P_y}{P_x} \qquad \textbf{5.5-19}$$

or, what is equivalent,

$$\tan (\theta + \theta^*) = \frac{P_y}{P_x}. \qquad \textbf{5.5-20}$$

On squaring each of Eqs. 5.5-18 and adding, we have

$$(1 + f^2)N^2 = P_x^2 + P_y^2$$

or

$$N = \sqrt{P_x^2 + P_y^2} \cos \theta^*. \qquad \textbf{5.5-21}$$

Equations 5.5-20 and 5.5-21 serve to determine N and θ for this limiting case. But the last of Eqs. 5.5-17 must also be satisfied. This means that f must be such as to satisfy the equation

$$M_z - fr\sqrt{P_x^2 + P_y^2} \cos \theta^* = 0$$

or

$$\sin \theta^* = \frac{M_z}{r\sqrt{P_x^2 + P_y^2}}. \qquad \textbf{5.5-22}$$

The value of θ^* given by the last expression, then, is the smallest value of friction that will prevent rotation about the axle with the given set of forces, $\mathbf{P}$ and M_z.

As is so often the case, it is instructive to look at the situation from another point of view. The force components F and N are equivalent to a single force of magnitude $(P_x^2 + P_y^2)^{1/2}$ applied at a point on the rim of the bearing. The moment of this force about the center of the axle, when motion impends, is fNr, which, substituting for N from Eq. 5.5-21, is equal to $fr(P_x^2 + P_y^2)^{1/2} \cos \theta^*$. This means that the resultant of F and N has a moment arm about the center of the axle of magnitude $fr \cos \theta^*$, or, since $f = \tan \theta^*$, a moment arm $r \sin \theta^*$. To summarize: if we draw a circle with center at the center of the axle and radius $r \sin \theta^*$, the reaction at the rim of the axle must be tangent to this circle when motion impends. The circle of radius $r \sin \theta^*$ is known as the friction circle. Its use is usually the most expedient way to cope with axle friction analysis.

Many other instances of machine-element combinations could be adduced for which the effect of friction on behavior can be anticipated with the aid of the Amontons-Coulomb friction law. Some of these are best understood by means of particular examples, such as the ones which follow. The reader should be alert to recognize that in many cases an inherently statically indeterminate situation is made amenable to analysis either by the fact that motion impends or by that species of reasonable approximation to the existing pressure distribution which is so common in engineering.

Example 5.5-1

A screw jack is to be used to raise a load of 1500 lb. The user of the jack is to exert a force of 20 lb at the end of a 1 ft handle. If a single-threaded square screw of 1 in. radius is used, what should its pitch be? What will be the efficiency of the screw? Will the weight slip down when the force is released from the handle? Take $f = 0.10$.

Solution: For raising the load we have (Eq. 5.5-6)

$$M = rW \tan(\theta + \theta^*).$$

In the present case $M = 240$ lb-in., $r = 1$ in., and $W = 1500$ lb, so that

$$\tan(\theta + \theta^*) = \frac{240}{(1)(1500)} = 0.16 \quad \text{or} \quad \theta + \theta^* = 9.1°.$$

Since $\theta^* = \arctan(0.10) = 5.7°$ we must have a screw angle $\theta = 3.4°$. This implies that the pitch (or lead) of the screw will be given by

$$\tan \theta = \frac{l}{2\pi r}, \qquad l = 2\pi r \tan \theta,$$

$$l = (2\pi)(1)(0.059) = 0.37 \text{ in.}$$

We will therefore have about 3 threads per inch along the body of the screw. Before leaving this part of the question we note that if f were 0.16 or more, no pitch, however small, would serve to make this jack function. A higher torque would be needed. We also note that for design purposes $l = 0.37$ in. is a maximum value. We could make l smaller and the screw jack would still function properly; in fact it would require even less torque. It would, however, be slower in raising the weight.

The efficiency of the screw is given by Eq. 5.5-8.

$$\text{Efficiency} = \frac{\tan \theta}{\tan(\theta + \theta^*)} = \frac{0.059}{0.16} = 0.37.$$

Finally, we note that the weight W will not slip down when the force is released from the handle, because $f > \tan \theta$.

Although it is not a part of the question asked, a designer would have to examine the load carried by the thread so as to be certain that it was not excessive. We see from Eq. 5.3-7 that

$$p_{av} = \frac{M}{rS} \frac{\cos \theta^*}{\sin (\theta + \theta^*)} = \frac{240}{(1)(S)} \frac{0.9953}{0.158} = \frac{1580}{S} \quad \text{lb/in.}$$

Since there are about three threads per inch on screw and nut, each thread will have a thickness of no more than 0.16 in. at the root. The average shear stress at the root of the thread will be about $1580/(0.16S) = 9880/S$ pounds per square inch when S is measured in inches. The allowable shear stress, if steel is used, is about 10^4 lb/in.^{2}, so that only a short engagement with the nut, say $S = 1$ in., is required for a safe design from this point of view.

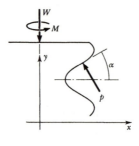

Fig. 5.5-5

Example 5.5-2

Find the torque required to exert a force W when a V-shaped screw thread is used.

Solution: We take the cross-section to be defined by the angle α, the semi-vertical angle of the V, as shown in Fig. 5.5-5. Now the normal pressure p per unit length of thread has a y-component $p \cos \alpha \cos \theta$ and a z-component $p \cos \alpha \sin \theta$. The frictional shear (directed out of the figure, toward the reader) has a y-component $-fp \sin \theta$ and a z-component $fp \cos \theta$. The equations of equilibrium analogous to Eqs. 5.5-1,2 are

$$\sum F_y = \int_0^S (p \cos \theta \cos \alpha - fp \sin \theta) \, ds - W = 0,$$

$$\sum M_y = \int_0^S r(p \sin \theta \cos \alpha + fp \cos \theta) \, ds - M = 0.$$

Just as in the case of Eqs. 5.5-3,4 we separate out $\int_0^S p \, ds$ and take the ratio of these expressions to find

$$\frac{M}{rW} = \frac{f \cos \theta + \cos \alpha \sin \theta}{\cos \theta \cos \alpha - f \sin \theta},$$

$$M = rW \frac{(f \sec \alpha) \cos \theta + \sin \theta}{\cos \theta - (f \sec \alpha) \sin \theta}.$$

This is the same as Eq. 5.5-5b, with f replaced by $f \sec \alpha$. It follows that all the conclusions previously reached for the case of a square thread apply to a V-shaped thread if we replace f by $f \sec \alpha$. In other words, the influence of the V-shape is to make friction more effective. The moment required to exert a force W is

$$M = rW \frac{\tan \theta + \sec \alpha \tan \theta^*}{1 - \sec \alpha \tan \theta \tan \theta^*}.$$

We see why, apart from cost, V-shaped threads are preferred for fasteners while square threads are preferred for power screws. In most commercial V-screws $\alpha \cong 29°$ so that friction is effectively increased by a factor 1.14.

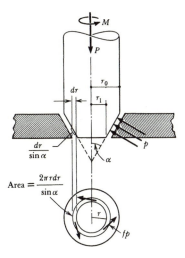

Fig. 5.5-6

Example 5.5-3

Find the torque required to rotate a conical pivot bearing of semi-vertical angle α, outer radius r_0, inner radius r_1, carrying a load P if the coefficient of *friction is f. Do this on the assumption* (1) *that the normal pressure is uniform;* (2) *that the wear is uniformly distributed.*

Solution: (1) The surface area of the bearing is

$$\frac{\pi(r_0^2 - r_1^2)}{\sin \alpha}.$$

Summing forces vertically, we have

$$p \sin \alpha \, \frac{\pi(r_0^2 - r_1^2)}{\sin \alpha} = P,$$

$$p = \frac{P}{\pi(r_0^2 - r_1^2)}.$$

The moment of the frictional forces carried on any ring element of the bearing extending from r to $r + dr$ is

$$\frac{rfp(2\pi r \, dr)}{\sin \alpha},$$

so, taking moments about the vertical axis, we have

$$\left(\frac{2\pi f}{\sin \alpha}\right) \frac{P}{\pi(r_0^2 - r_1^2)} \int_{r_1}^{r_0} r^2 \, dr - M = 0,$$

$$M = \frac{2fP}{3 \sin \alpha} \frac{r_0^3 - r_1^3}{r_0^2 - r_1^2} = \frac{2fP}{3 \sin \alpha} \frac{r_0^2 + r_0 r_1 + r_1^2}{r_0 + r_1}.$$

(2) If wear is uniform, $p = C/r$. We determine C by summing forces vertically:

$$P = \int_{r_1}^{r_0} p \sin \alpha \, \frac{2\pi r \, dr}{\sin \alpha} = 2\pi C(r_0 - r_1)$$

or

$$C = \frac{P}{2\pi(r_0 - r_1)}.$$

The moment of the friction forces about the axis of the bearing is

$$M' = \int_{r_1}^{r_0} rfp \, \frac{2\pi r \, dr}{\sin \alpha} = \frac{2\pi fC}{\sin \alpha} \int_{r_1}^{r_0} r \, dr = \frac{\pi fC}{\sin \alpha} (r_0^2 - r_1^2).$$

Substituting for C, we obtain

$$M' = \frac{fP}{2 \sin \alpha} (r_0 + r_1).$$

It is interesting to note that if we take $\alpha = \pi/2$ and $r_1 = 0$ these results for M and M' reduce to the previous case of a flat-ended bearing.

Now let us compare the torques for uniform pressure and uniform wear:

$$\frac{M}{M'} = \frac{4}{3} \frac{r_0^2 + r_0 r_1 + r_1^2}{(r_0 + r_1)^2} = \frac{4}{3} \left[1 - \frac{r_0 r_1}{(r_0 + r_1)^2} \right].$$

Since the geometric mean of two different positive numbers is always less than their arithmetic mean, it follows that $(r_0 r_1)^{1/2} < \frac{1}{2}(r_0 + r_1)$ and hence that $(r_0 r_1)/(r_0 + r_1)^2 < \frac{1}{4}$. We conclude that M is always greater than M', i.e., the torque requirement is reduced when the bearing is "run in," or worn.

Example 5.5-4

A friction clutch (Fig. 5.5-7) connecting two coaxial shafts consists of two circular plates, which, when pressed together, can exert a force of up to 400 lb on one another. The plates have an annulus of brake lining of inside

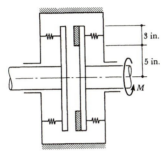

3 in.

5 in.

M

Fig. 5.5-7

radius $r_1 = 5$ in., outside radius $r_0 = 8$ in., and $f = 0.25$. For this material, how much torque can be transmitted before the clutch slips?

Solution: Assuming that p is uniform over the friction material, we have $p = 400/\pi (64 - 25) = 3.27$ lb/in^2.
Therefore,

$$M = 2\pi \int_{r_1}^{r_0} fpr^2 \, dr = \frac{2\pi f p}{3} (r_0^3 - r_1^3)$$
$$= \left(\frac{1}{3}\right)(6.28)(0.25)(3.27)(512 - 125) = 660 \text{ lb-in.}$$

On the other hand, when the clutch is worn we have $p = C/r$ so that

$$P = 2\pi \int_{r_1}^{r_0} pr \, dr = 2\pi C \int_{r_1}^{r_0} dr = 2\pi C(r_0 - r_1),$$
$$C = \frac{P}{2\pi(r_0 - r_1)};$$
$$M' = 2\pi \int_{r_1}^{r_0} fpr^2 \, dr = 2\pi f \frac{P}{2\pi(r_0 - r_1)} \int_{r_1}^{r_0} r \, dr = \frac{1}{2} fP(r_0 + r_1),$$
$$M' = \left(\frac{1}{2}\right)\left(\frac{1}{4}\right)(400)(8 + 5) = 650 \text{ lb-in.}$$

and there is a slight drop in the torque which can be transmitted.

Example 5.5-5

A 200 lb weight hangs from a cable wound on a 10 in. radius drum. The drum can rotate on a horizontal axle of 3 in. radius, the coefficient of friction at the bearing being 0.10. On the same axle is a wheel of 20 in. radius around which another cable is fastened. Find the minimum force P needed to raise the weight in each of the two cases shown in Fig. 5.5-8a and b.

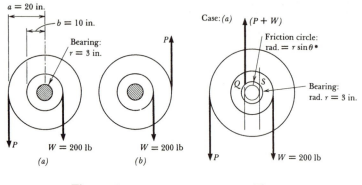

Fig. 5.5-8   **Fig. 5.5-9**

Solution: For case (a), we draw the free-body diagram of the drum and wheel as shown in Fig. 5.5-9. Notice the friction circle (which is drawn to an enlarged scale). The force exerted by the axle on the drum must be vertical and directed upward so as to balance P and W. In fact we can see that its magnitude must be $P+W$. It can have no horizontal component since there is no other external force to balance such a component. This force exerted by the axle on the drum must act at a point on the rim of the bearing and must be tangent to the friction circle. It must therefore act either at point Q or at point S. To see which of these points gives us the correct tangent we note that the drum is on the point of rotating counterclockwise. The friction force acting on the drum must therefore produce a clockwise moment about the geometrical center. This means that the force exerted by the axle on the wheel must act at Q and not at S. Now it is a simple matter to take moments about Q:

$$P(a - r \sin \theta^*) - W(b + r \sin \theta^*) = 0,$$

$$P = W \frac{b + r \sin \theta^*}{a - r \sin \theta^*} = 200 \frac{10 + (3)(0.1)}{20 - (3)(0.1)} = (200)(0.523) = 105 \text{ lb.}$$

If the weight were being lowered, the axle would bear on the drum at S.

For case (b), the free-body diagram is as shown in Fig. 5.5-10. Since $W > P$, it follows that the bearing reaction on the drum is upward and of magnitude $W - P$. It must again act at point Q in order to be tangent

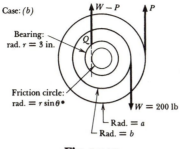

Fig. 5.5-10

to the friction circle, bear against the drum, and have a frictional component which opposes upward motion of W. Now taking moments about Q, we obtain

$$P(a+r \sin \theta^*) - W(b+r \sin \theta^*) = 0,$$

$$P = W \frac{b+r \sin \theta^*}{a+r \sin \theta^*} = 200 \frac{10+0.3}{20+0.3} = (200)(0.507) = 101 \text{ lb.}$$

Of course if friction were neglected we should have $P=100$ lb. The arrangement (b) requires less force to raise the weight than does (a) because the reaction at the bearing is smaller ($W-P$ rather than $W+P$) and this implies a smaller frictional component.

Example 5.5-6

A railway freight car has journal bearings of radius 2.5 in. and coefficient of friction 0.10. If the wheels themselves have a radius of 22 in., how much pull is needed to move a 20,000 lb load?

Solution: The freight car is shown schematically in Fig. 5.5-11 and a free-body diagram of one of the wheels in Fig. 5.5-12. The friction circle is shown enlarged for clarity. Two forces act on the wheel, one at S where the wheel is in contact with the rail and one at a point on the bearing

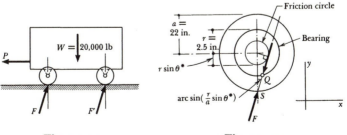

Fig. 5.5-11   **Fig. 5.5-12**

where the axle touches it. This latter force is tangent to the friction circle when the traction P is just great enough to make the wheel roll to the left. But the two forces at wheel and bearing must be "equal and opposite"; therefore the force at S must be tangent to the friction circle also, as shown. Of the two possible tangents the one that intersects the bearing circle at Q is the proper one because the corresponding force exerted by the axle produces a clockwise rotation about the center of the wheel as it should, since the wheel is about to rotate counterclockwise. The horizontal component of the reaction, F, at point S is

$$F\frac{r}{a}\sin\theta*$$

and the vertical component is

$$F\left(1-\frac{r^2}{a^2}\sin^2\theta*\right)^{\frac{1}{2}}.$$

The situation at the other wheel is the same, except that the reaction is not necessarily the same. Call it F'. We have

$$(F+F')\left(1-\frac{r^2}{a^2}\sin^2\theta*\right)^{\frac{1}{2}}=W$$

and

$$(F+F')\frac{r}{a}\sin\theta*=P,$$

so that

$$P=W\frac{\dfrac{r}{a}\sin\theta*}{\left(1-\dfrac{r^2}{a^2}\sin^2\theta*\right)^{\frac{1}{2}}}.$$

It appears from this result that it is desirable to have large wheels and small axles. With the numerical values of this example, $\theta*=5.7°$ and

$$P=20{,}000\frac{\left(\dfrac{2.5}{22}\right)(0.1)}{\left[1-\left(\dfrac{2.5}{22}\right)^2(0.1)^2\right]^{\frac{1}{2}}}=(20{,}000)(0.011)=220\text{ lb.}$$

This is equivalent to an over-all coefficient of friction of 0.011. We see why it is much easier to pull a load in a wagon than to slide it along the ground.

Exercises

5.2-1: (a) A 30 lb block is placed on a plane inclined at an angle of 30° to the horizontal. The coefficient of limiting static friction is 0.30 and of kinetic friction is 0.25. Will the block be in equilibrium? What is the magnitude of the friction force acting on the block?

Ans.: 6.49 lb.

(b) If the slope angle is halved, to 15°, what are the answers to the questions of part (a)?

Ans.: 7.76 lb.

5.2-2: A force, P, making an angle, ϕ, with the inclined plane is applied to the 30 lb block of the previous exercise. For each of the two slope angles, (i) what is the minimum value of P and (ii) what is the corresponding ϕ to initiate motion of the block up the plane?

Ans.: For both slope angles, $\phi = 16.7°$; for case (a), $P_{min} = 21.8$ lb; for case (b), $P_{min} = 15.8$ lb.

5.2-3: The 30 lb uniform block on the 15° slope of Exercise 5.2-1b has dimensions 2.5 in. by 5 in. by 10 in. It is placed on the slope in each of the three positions shown. Will the block be in equilibrium for all three positions? Compute the distance from the lower corner of the block to the resultant normal reaction if the block is in equilibrium.

Ans.: (a) 4.66 in.

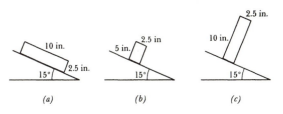

(a)   (b)   (c)

Exer. 5.2-3

5.2-4: The three boxes of Example 5.2-2 are shown here again, with an additional vertical load $Q = kW$ applied to the top box. Show that this load may be made so large that slipping can occur only at the floor, and

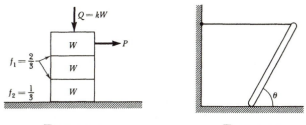

Exer. 5.2-4   **Exer. 5.2-5**

find the critical value of k. If $k = 3$, what force P is needed to slip the stack of boxes?

Ans.: $P = 2W$ when $k = 3$.

5.2-5: The uniform rod is supported by the horizontal cable and the rough floor (coefficient of limiting friction 0.5). What is the smallest angle θ at which the rod can be in equilibrium?

Ans.: 45°.

5.2-6: A light wedge is used to raise a 2500 lb weight, as shown. The coefficient of friction at all contact surfaces is 0.5. What is the minimum force, P, needed to raise the weight?

Ans.: 6230 lb.

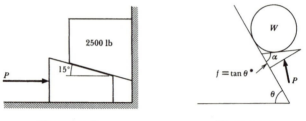

Exer. 5.2-6   Exer. 5.2-7

5.2-7: The cylinder shown is held on the inclined plane by a wedge and a force, P. The cylinder has a weight W but the weight of the wedge may be ignored. The coefficient of friction between wedge and plane is $f = \tan \theta^*$ but all other contacts may be regarded as smooth. Supposing that the angle of inclination of the plane, θ, is less than θ^* and that θ^* is less than the wedge angle, α, find the smallest value of P which will prevent the cylinder and wedge from sliding down the plane. (Note that P is not necessarily parallel to the inclined plane.)

Ans.: $P = W \sin \theta \sin (\alpha - \theta^*)/\sin \alpha$.

5.2-8: A horizontal rod of 1 in. radius and weighing 50 lb rests on a V-shaped support, the coefficient of friction at the points of contact being 0.1. The apical angle of the V is 90° and the entire arrangement is symmetrical about a vertical line through the center of the cylinder and the apex of the V-support. A light arm 2 in. long extends horizontally from the cylinder and at right angles to it; the arm and the V-support do not interfere with one another. What downward force must be applied at the end of this arm in order to make the cylinder rotate?

Ans.: 2.45 lb.

5.2-9: An automobile is parked on an incline. The vertical through the mass center meets the ground at a point, P. Supposing that either the front wheels or the rear wheels (but not both sets) can be braked, show that it is best to brake the set closest to the point P.

5.2-10: The pin connections between the 13 in. light connecting rod and both the 5 in. radius flywheel and the small 3 lb piston are smooth, as is the shaft bearing at the center of the flywheel. The coefficient of limiting

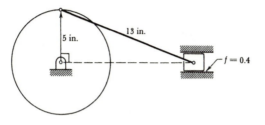

Exer. 5.2-10

static friction between the piston and the rough wall of the horizontal cylinder is $f = 0.4$.

(a) What counterclockwise torque M_1 supplied to the flywheel is needed to cause motion to impend?

Ans.: $M_1 = 5.14$ lb-in.

(b) What clockwise torque M_2 is needed for impending motion?

Ans.: $M_2 = 7.2$ lb-in.

5.2-11: A load W is to be raised by two light wedges, each of angle α. Which of the pictured arrangements, (a) or (b), is to be preferred? Consider both the load, P, required on each wedge for raising W and the self-locking characteristics of each mechanism. What would happen in case (b) if there were no vertical guide for W?

Ans.: System (a) is to be preferred for lifting the weight.

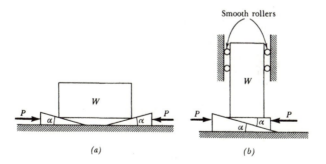

Exer. 5.2-11

5.2-12: The lifting device shown is to be designed so that the load will not slip, however heavy it may be. Show that this design criterion will be satisfied provided the dimensions are such that the angle α is less than the friction angle θ^*. Ignore friction at the pivot joints.

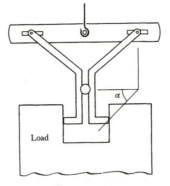

Exer. 5.2-12

5.2-13: What load, P, on the handle of the light brake lever is required to prevent rotation of the flywheel under the applied counterclockwise torque, M? The coefficient of friction is f. Can friction locking occur for proper choices of the system parameters? What happens if the direction of the torque is reversed?

Ans.: $P = \dfrac{M}{r}\dfrac{(a-fd)}{cf}$; locking occurs when $a=fd$; for clockwise M,

$P = \dfrac{M}{r}\dfrac{(a+fd)}{cf}$.

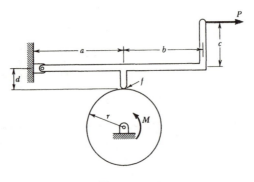

Exer. 5.2-13

5.2-14: A light collar 4 in. long slides on a vertical 3 in. diameter rod, the coefficient of friction at all points of contact being 0.2. Show that if the dimension c is greater than 8.5 in. the collar will not slip down, no matter how heavy the weight W.

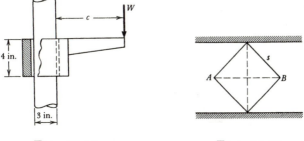

Exer. 5.2-14   **Exer. 5.2-15**

5.2-15: A light square block of side s is clamped tightly at two corners in the jaws of a vise, as shown. The coefficient of friction is f at each contact. If the clamping force is N, what force, P, along the diagonal AB is necessary to cause the block to slip? What torque, M, about an axis perpendicular to the plane is needed to cause slipping?

5.2-16: Body A weighs 200 lb, body B weighs 300 lb, and the weight of the brake may be ignored. The coefficients of friction are 0.30 between brake and A; 0.20 between A and B; and 0.10 between B and the plane. Find the minimum force, P, that will cause B to have impending motion to the right.

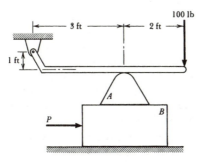

Exer. 5.2-16

Ans: 111 lb.

5.2-17: The force P applied to the bottom block is increased until motion of that block impends.

(a) What is the relationship between the coefficients of friction, the dimensions a and b, and the weights of the blocks that determines whether motion impends at contacts 1 and 3 or contacts 2 and 3 ?

Ans.: Motion impends at 1 and 3 if $bf_2 > af_1$.

(b) If $W = 2500$ lb, $w = 1500$ lb, $f_1 = 2f_2 = 3f_3 = 0.6$, and $b/a = 3/2$, what is P for slip impending ?

Ans.: 1050 lb.

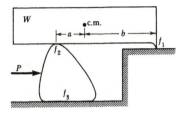

Exer. 5.2-17

5.2-18: A tapered cotter pin holds two rods which carry a tension, T. If the friction angle is θ^*, what is the largest taper angle, α, permissible ?

Ans.: $\alpha = 2\theta^*$.

Exer. 5.2-18

5.2-19: A man of weight W pushes on the smooth side of a crate of weight W'. If the coefficient of friction on the floor for both man and crate is f, and if the mass-center of the man is at his mid-height, show that when $W > W'$ he can incline his body to the vertical by any angle less than $\arctan (2fW'/W)$ and that when $W' > W$ he can incline his body to the vertical by any angle less than $\arctan 2f$.

5.2-20: A window is counterbalanced by sash weights. Show that it can be raised by a single vertical force if and only if that force is applied at a distance from the centerline of the window less than the height of the window divided by twice the coefficient of friction between window and moulding.

5.2-21: A uniform rod of length $2a$ and weight W rests on a horizontal cylinder of radius r, the axis of the rod being at right angles to the axis of the cylinder. If the angle of friction is θ^* and if the rod is initially horizontal, what is the largest weight that can be placed on the end of the rod without causing it to slip?

Ans.: $Wr\theta*/(a-r\theta*)$.

5.3-1: A beam rests on supports at its midpoint and at two equidistant points, their distances from the center support being chosen so that each of the three points of support carries a load of 100 lb (the total weight of the beam being, of course, 300 lb). The three points of support are at the same horizontal level. The coefficient of friction at each point of support is $\frac{1}{4}$. If a horizontal force, gradually increasing from zero, is applied to the beam at right angles to its length at one of the outer points of support, how large can this force grow before equilibrium becomes impossible?

Ans.: 37.5 lb.

5.3-2: A heavy circular disk is supported on three bosses, A, B, and C, which form an equilateral triangle as shown. The weight of the disk is W and the coefficient of friction at each boss is f. A gradually increasing horizontal force is applied to one of the bosses in a direction parallel to the

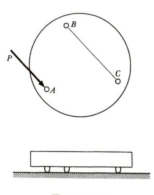

Exer. 5.3-2

line joining the other two. (a) How large can P become before equilibrium is broken? (b) Locate the axis about which the disk begins to rotate.

Ans.: (a) $P = \frac{2}{3}fW$; (b) initial rotation is about axis through the angle bisectors.

5.3-3: A horizontal cylinder of weight W rests in a V-shaped groove whose sides are inclined at 45° to the vertical. The coefficient of friction between groove and cylinder is 0.6. What is the minimum force, P, that will disturb equilibrium, when applied to the cylinder?

Ans.: $0.35W$.

5.3-4: Two identical uniform rods, *AB* and *BC*, of weight *W*, are smoothly pinned at *B* and are laid in a straight line on a rough horizontal table. The length of each rod is *l* and the coefficient of friction is *f*. A gradually increasing force, *P*, is applied to point *A* at right angles to the bars. Show that equilibrium will be broken when $P = fW[\sqrt{4\sqrt{2}-2} - \sqrt{2}]$.

5.3-5: A particle is held on a rough inclined plane by a smooth light rod which is smoothly pivoted at a point of the plane. The acute angle, α, between this rod and the line of steepest slope on the plane is so large that, in the absence of any further constraint, the particle would not be in

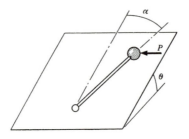

Exer. 5.3-5

equilibrium. Let *W* be the weight of the particle, *f* be the coefficient of friction between it and the plane, and θ be the angle of inclination of the plane to the horizontal. Show that the minimum additional horizontal force, *P*, needed to preserve equilibrium is given by the expression $W(\sin \theta \sin \alpha - f \cos \theta)/\cos \alpha$.

5.3-6: A uniform rod of length $3l$ and weight *W* rests on a rough horizontal table, being supported at one end, *A*, and at a point distant $2l$ from *A* by small projections. At the end of the rod remote from *A* it is subjected to a gradually increasing force, *P*. When equilibrium is broken the rod begins to rotate about a point distant $2l$ from the points of support. Find (a) the magnitude of *P* at this instant and (b) the angle it makes with the bar.

Ans.: (a) $P = 0.9fW$; (b) $16°$.

5.3-7: The rod of the previous problem is subjected to a couple about a vertical axis through the end of the rod remote from *A*. How large can this couple become before equilibrium is broken? How does the rod begin to move?

Ans.: $M = \frac{1}{2}fWl$.

5.3-8: A square table of weight *W* with a leg at each corner stands on a horizontal floor whose coefficient of friction is *f*. Find the minimum horizontal force which, applied at a corner, will make the table slip.

Ans.: $\frac{1}{4}(1 + \sqrt{2})fW$.

5.4-1: A pulley of 9 in. diameter drives a second pulley of 25 in. diameter in the same plane, their centers being 4 ft apart, by means of a flat belt. (a) If the minimum tension is 100 lb, what is the maximum tension when the belt is on the point of slipping? (b) How much torque can be transmitted in this way? Take $f = 0.3$ and assume that the belt is arranged as in the figure.

Ans.: (a) 232 lb; (b) 595 lb-in.

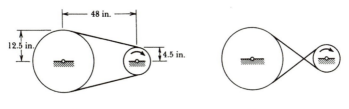

Exer. 5.4-1	**Exer. 5.4-2**

5.4-2: If, in Exercise 5.4-1, the belt is crossed, as in the revised figure, but all other conditions remain the same, answer the same questions.

Ans.: (a) 313 lb; (b) 960 lb-in.

5.4-3: The band brake shown is a modification of one described in the text. It is designed to be equally effective in resisting counterclockwise and clockwise torque applied to the drum. Find the maximum torque that can be restrained by this arrangement and verify that this value is in fact independent of the direction of rotation which impends.

Ans.: $M = \dfrac{Pr(a+b)}{b} \left(\dfrac{1 - e^{-\frac{3}{2}\pi f}}{1 + e^{-\frac{3}{2}\pi f}} \right).$

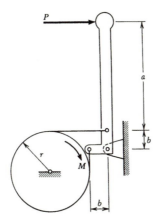

Exer. 5.4-3

5.4-4: How large a load can be restrained by a man exerting a pull of 100 lb on a rope wrapped for one and one-half turns around a post if $f = 0.4$?

Ans.: 4320 lb.

5.4-5: A circular disk of weight W is pressed against a perfectly rough wall by a cord attached to the wall and carrying a weight P. The cord makes an angle α with the wall. The coefficient of friction between cord and disk is f. Show that the minimum value of P which will maintain equilibrium is $W/[(1 + \cos \alpha)e^{\alpha f} - 2]$.

5.4-6: A rough peg has a horizontal axis and a cross-section in the form of an ellipse, the major axis, a, being horizontal and the minor axis, b, being vertical. Two weights, W_1 and W_2, hang from the ends of a string that passes over the peg. Initially W_1 is about to slip downwards. How much weight can be added to W_2 before it begins to descend?

Ans.: $(W_1^2 - W_2^2)/W_2$.

5.4-7: A flexible steel wire is threaded around three posts of radius $r = 1.5$ in., spaced as shown. (a) What tension, T, is needed to overcome a resistance of 100 lb at the end of the wire if $f = 0.32$? (b) What is the maximum pressure exerted by the wire on the post?

Exer. 5.4-5

Ans.: (a) $T = 272$ lb; (b) $p_{max} = 181$ lb/in.

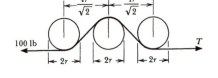

Exer. 5.4-7

5.4-8: A belt wraps half way around a fixed pulley having a $45°$ total apical angle V-groove in which the coefficient of friction is 0.3. One end of the belt carries a 1000 lb weight while the other is pulled with a force, P. (a) What is the minimum P to raise the weight? (b) At what value of P will the weight descend slowly?

Ans.: (a) 11,740 lb; (b) 85 lb.

5.4-9: A cord hangs in a vertical plane over a fixed horizontal peg and carries weights W, W' at its ends. If the cord is on the point of slipping and $W' > W$, show that the coefficient of friction is given by the expression

$$f = \frac{1}{\pi} \log (W'/W).$$

5.4-10: A device for applying torque to a shaft of radius r consists of a bar, AC, to which is attached a strap, BDA. The bar AC is normal to the surface of the cylinder, on which it bears at A. Show that no matter how large is the force P applied at C at right angles to the bar, no slipping will occur, provided f exceeds the value determined by the equation

$$4e^{-(5\pi f)/3} - 2\sqrt{3}f = 1.$$

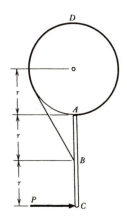

5.5-1: A screw jack has a square-threaded screw of mean diameter 2 in. and pitch 0.5 in. If the coefficient of friction between screw and nut is 0.15, what force must be exerted at the end of a lever 25 in. long in order to raise a 4000 lb load? **Ans.:** 38 lb.

5.5-2: What force is necessary to lower the load of the previous problem, and what is the efficiency of the screw jack? **Ans.:** 11 lb in direction opposite to that used in raising load.

Exer. 5.4-10

5.5-3: A square-threaded screw is tightened by a moment of 1500 in-lb to a tension, T. The screw has a diameter of 1.5 in. and a pitch of 0.3 in. The base of the nut has an outside diameter of 3.2 in. and an inside diameter of 1.6 in. Find the pull, T, (a) on the assumption that all friction is negligible, (b) on the assumption that $f = 0.2$ for the contact of screw in nut,

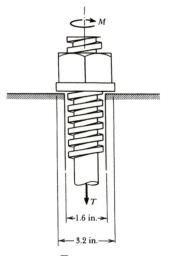

Exer. 5.5-3

but can be neglected elsewhere, and (c) on the assumption that $f=0.2$ at *all* surfaces where motion impends.

Ans.: (a) 31,400 lb; (b) 7470 lb; (c) 3400 lb.

5.5-4: If the nut in the previous problem has a depth of 1 in., what will be the mean load per in. on the threads when they are tightened to an axial load $T=2000$ lb?

Ans.: 130 lb/in.

5.5-5: Estimate the tension in a quarter-inch screw having 32 V-shaped threads per in. when a torque of 50 lb-in. is applied. Take $f=0.15$ and the semi-apical angle as $29°$.

Ans.: 1840 lb.

5.5-6: Criticize the logical structure of the following line of reasoning: A machine is used to raise a weight, W, through a height of 1 ft by means of an applied force. Suppose the work put in is cW. Then the forward efficiency is $1/c$. The energy lost in friction is $cW-W=(c-1)W$. When the machine runs freely back under gravity to its original position, the work done by the weight is W and the work lost in friction is again $(c-1)W$. The surplus energy is $W-(c-1)W=(2-c)W$. If $c=2$ there is no surplus energy and so the machine will not automatically run back when the applied force is removed. Therefore if the forward efficiency is less than 50 per cent a machine will not automatically run back when the applied load is removed. Why is the argument only approximately true?

5.5-7: The tapered shank of a drill has a total apical angle of $12°$ and a mean diameter of $\frac{3}{4}$ in. It is entered lightly in the chuck and the drill is then operated with an axial force of 50 lb. If the coefficient of friction is 0.17, how much torque can the drill exert on the workpiece before it begins to slip in the chuck? Assume that the bearing of the drill in the chuck is "worn."

Ans.: 30.5 lb-in.

5.5-8: When a ball bearing is pressed into a steel plate by a force, P, it deforms and makes contact with the plate over a circle of radius a. Within this circle the normal pressure is governed by the law $p=(3P/2\pi a^3)\sqrt{a^2-r^2}$, where r denotes distance from the center of the contact circle. Using this expression for p, estimate the torque required to spin such a ball bearing about a vertical axis.

Ans.: $3\pi fPa/16$.

5.5-9: When a hard steel, flat-ended circular shaft is pressed axially into a somewhat softer elastic material, the pressure exerted on the base of the cylinder is given by the expression $p=(P/2\pi a)(a^2-r^2)^{-\frac{1}{2}}$. Assuming this distribution of pressure and a coefficient of friction f, find the torque required to rotate the shaft of a plain-thrust bearing. Compare your results with the text expressions for p constant and for p proportional to r^{-1}.

Ans.: $(\pi/4)fPa$.

5.5-10: A marine propulsion shaft carries a thrust of 6700 lb. This load is held by collars on which the pressure may be considered uniform. The collars are 8 in. *OD* and 5 in. *ID*. For $f=0.08$ compute the torque and horsepower lost in friction when the shaft rotates at 120 rpm.

Ans.: 1740 lb-in; 3.3 hp.

5.5-11: A plain thrust bearing is recessed as shown. Find expressions for the torque required to produce uniform rotation if the friction coefficient is f. Assume (a) uniform pressure and (b) uniform wear.

Ans.: (a) $M = \dfrac{2fP}{3}\left(\dfrac{r_0^3-r_1^3}{r_0^2-r_1^2}\right)$; (b) $M' = \dfrac{1}{2}fP(r_0+r_1)$.

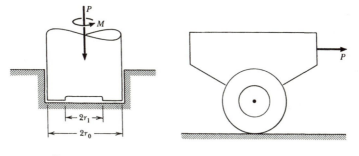

Exer. 5.5-11   **Exer. 5.5-12**

5.5-12: A two-wheeled cart carries a load, W, centered over the axle. (a) If the wheels are of 2.5 ft radius, if the axle is of 2 in. radius, and if the coefficient of friction at the journal bearing is 0.05, how heavy a load can be pulled by a horizontal force $P=100$ lb? (b) How large must the coefficient of friction be so that the wheel will not slip where it is in contact with the road?

Ans.: (a) 30,000 lb; (b) 0.0033.

5.5-13: A light cord passes over a wheel mounted on a bearing carried on a horizontal axle. The cord carries weights W and W', which are so related that the wheel is on the point of turning around the axle. Assuming the cord does not slip, show that the friction angle of the bearing is given by the expression $\theta^* = \arcsin\left(\dfrac{W-W'}{W+W'}\dfrac{R}{r}\right)$, where R and r are the radii of the wheel and axle, respectively. Compare this result with that for Exercise 5.4-9.

5.5-14: A shaft 6 in. in diameter carries a vertical load of 10,000 lb and a horizontal load of 15,000 lb. For $f=0.04$ find the horsepower lost in friction at the journals when the shaft is driven at 120 rpm by a torque.

Ans.: 4.31 hp.

5.5-15: The bell-crank shown is carried in a fixed axle at O, the radius of the axle being 2 in. and the coefficient of friction being 0.3.   Arm OA is 10 in. long and makes an angle of 45° with the horizontal.   Arm OB is 2 ft long and makes an angle of 30° with the horizontal.   A 100 lb downward

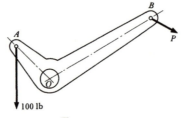

Exer. 5.5-15

force acts at A.   (a) What force, P, acting at B in a direction at an angle of 60° with the vertical, is needed to make the bell-crank rotate in a clockwise direction ? (b) Compare this with the force that would be required in the absence of friction.

Ans.: (a) 37 lb; (b) 34 lb.

APPENDIX

Centers of Mass

A.1 **Definition of the Mass Center of a Body**

In this appendix we summarize the results of Section 2.7 on mass centers, discuss methods of locating them, and provide a table of positions of the mass centers of some uniform bodies.

(a) *System of Discrete Particles.* Suppose, for each of a system of N particles, we know the mass m_i, $i = 1, 2, 3, \ldots N$, and the position vector $\mathbf{r}_i$ with respect to a fixed point O. The mass center of the system is the point in space with position vector $\mathbf{r}^*$ defined by

$$\mathbf{r}^* = \frac{\displaystyle\sum_{i=1}^{N} m_i \mathbf{r}_i}{\displaystyle\sum_{i=1}^{N} m_i} = \frac{\displaystyle\sum_{i=1}^{N} m_i \mathbf{r}_i}{m}, \qquad \textbf{A.1-1}$$

where

$$m = \sum_{i=1}^{N} m_i \qquad \textbf{A.1-2}$$

is the total mass of the system. If a particular cartesian coordinate system (x, y, z) with origin at O is introduced, so that the i-th particle has position coordinates (x_i, y_i, z_i), i.e., $\mathbf{r}_i = x_i \mathbf{i} + y_i \mathbf{j} + z_i \mathbf{k}$, then the mass center has position coordinates (x^*, y^*, z^*)—or $\mathbf{r}^* = x^*\mathbf{i} + y^*\mathbf{j} + z^*\mathbf{k}$—given by the scalar equivalents to Eq. A.1-1:

$$x^* = \frac{\sum\limits_{i=1}^{N} m_i x_i}{\sum\limits_{i=1}^{N} m_i}, \qquad y^* = \frac{\sum\limits_{i=1}^{N} m_i y_i}{\sum\limits_{i=1}^{N} m_i}, \qquad z^* = \frac{\sum\limits_{i=1}^{N} m_i z_i}{\sum\limits_{i=1}^{N} m_i}. \qquad \textbf{A.1-3}$$

   (b) *Continuous Systems.* Suppose a body with distributed mass is given. The mass center position $\mathbf{r}^*$ is then defined by an integral equivalent to Eq. A.1-1. Formally, we divide the body into N parts, denoting the mass of the i-th part by Δm_i, $i = 1, 2, \ldots N$, and the position vector to an arbitrary point in the i-th part by $\mathbf{r}_i$, $i = 1, 2, \ldots N$. We then form the Riemann sums

$$\sum_{i=1}^{N} \Delta m_i, \qquad \sum_{i=1}^{N} \mathbf{r}_i \, \Delta m_i,$$

and consider the limit processes

$$\lim_{\substack{N \to \infty \\ \Delta m_i \to 0}} \left(\sum_{i=1}^{N} \Delta m_i \right), \qquad \lim_{\substack{N \to \infty \\ \Delta m_i \to 0}} \left(\sum_{i=1}^{N} \mathbf{r}_i \, \Delta m_i \right).$$

If these limits exist whatever the mode of subdivision may be and whatever the choice of the points $\mathbf{r}_i$ may be, then the Riemann integrals

$$m = \int dm \quad \text{and} \quad \int \mathbf{r} \, dm$$

exist, with m being the total mass of the body again. The mass center has position vector

$$\mathbf{r}^* = \frac{1}{m} \int \mathbf{r} \, dm \qquad\qquad \textbf{A.1-4}$$

with respect to the origin O.

   If a cartesian coordinate system (x, y, z) is introduced, with $\mathbf{r} = x\mathbf{i} + y\mathbf{j} + z\mathbf{k}$, then the scalar equivalents to A.1-4 give the position coordinates (x^*, y^*, z^*) of the mass center:

$$x^* = \frac{1}{m} \int x \, dm, \qquad y^* = \frac{1}{m} \int y \, dm, \qquad z^* = \frac{1}{m} \int z \, dm. \qquad \textbf{A.1-5}$$

   The integrals of Eqs. A.1-4 and A.1-5 are evaluated usually as line, surface, or volume integrals over the spatial region occupied by the body. A mass density function is introduced in order to transform the integrals into the desired form. If the body is curvilinear in form (thin rods or beams, for example), a line density

$\lambda(s)$ with dimensions mass per unit length may be defined: $\lambda = \lim_{\Delta s \to 0} (\Delta m/\Delta s)$, where Δm is the mass of a length Δs of the body. Similarly, surface and volume densities $\sigma = \lim_{\Delta A \to 0} (\Delta m/\Delta A)$ and $\rho = \lim_{\Delta V \to 0} (\Delta m/\Delta V)$ may be defined for bodies having the appropriate shapes: thin plates and shells are examples of the former, solid cylinders and thick slabs are examples of the latter. The integrals A.1-5 then become of the form, say for the case of a volume density,

$$x^* = \frac{1}{m} \int \int \int_V x\rho \, dV, \qquad y^* = \frac{1}{m} \int \int \int_V y\rho \, dV,$$

$$z^* = \frac{1}{m} \int \int \int_V z\rho \, dV, \qquad\qquad \textbf{A.1-6}$$

where the total mass m is given by

$$m = \int \int \int_V \rho \, dV. \qquad\qquad \textbf{A.1-7}$$

The mass center is of importance in statics primarily because of its usual identification with the center of gravity and hence with a point on the line of action of the resultant weight force on a body. In dynamics, it appears as the favored point in many ways: as the point which has known acceleration when the external forces are known (principle of motion of the mass center), as an appropriate base point for the principle of moment of momentum, and as the fundamental base point for moment of inertia computations.

The form of the computation for the position of the mass center is one of common occurrence in mathematics and physics. The integrals $\int x \, dm$, etc., are examples of the concept of the *first moment* of a scalar quantity, in this case mass. In general, $\int x \, df$ is the first moment of f with respect to the yz-plane. If f is, say, volume, then the computation leads to the coordinates (x^*, y^*, z^*) of the *centroid* of the volume. If $df = p \, dA$, where p is the pressure on the plane area dA, then we obtain first moment integrals of the density form—with the pressure p replacing the surface mass density σ—and the resulting point coordinates are those of the *center of pressure* (Section 4.4). In particular, if the mass density of a body is constant, then it disappears from the computations of Eqs. A.1-6, since it is a common factor in both the numerator and the denominator (m) of each equation. In this circumstance, the mass center and the geometrical centroid of the body are the same point. Such a body of constant density is said to be *uniform*.

A.2 Translation and Rotation of Axes

In this section we shall compute the position of the mass center with respect to another coordinate system, given its position with respect to one coordinate system. This computation is straightforward, involving no more than vectorial addition and the scalar product for finding components of vectors.

Suppose we are given one set of cartesian axes with origin O: $Oxyz$, and a parallel set with origin P: $Pxyz$ (Fig. A.2-1). The latter

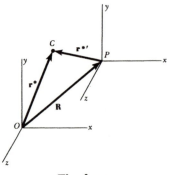

Fig. A.2-1

set is said to be translated with respect to the first (and conversely); that is, there has been a simple shift of origin from O to P, with no change in the coordinate directions. Let $\mathbf{R} = X\mathbf{i} + Y\mathbf{j} + Z\mathbf{k}$ denote the position of P relative to O, $\mathbf{r}^* = x^*\mathbf{i} + y^*\mathbf{j} + z^*\mathbf{k}$ denote the position of the mass center C of a system relative to O, and $\mathbf{r}^{*\prime} = x^{*\prime}\mathbf{i} + y^{*\prime}\mathbf{j} + z^{*\prime}\mathbf{k}$ denote the position of the mass center C relative to P. The triangle law of addition results in

$$\mathbf{r}^* = \mathbf{r}^{*\prime} + \mathbf{R} \qquad\qquad \textbf{A.2-1}$$

or

$$x^* = x^{*\prime} + X, \qquad y^* = y^{*\prime} + Y, \qquad z^* = z^{*\prime} + Z. \qquad \textbf{A.2-2}$$

Given $\mathbf{R}$ and either $\mathbf{r}^*$ or $\mathbf{r}^{*\prime}$, these serve to compute the third vector.

Suppose we are given the two sets of cartesian coordinates $Oxyz$ and $Ox'y'z'$ with the same origin, and the position vectors $\mathbf{r}^* = x^*\mathbf{i} + y^*\mathbf{j} + z^*\mathbf{k}$ and $\mathbf{r}^{*\prime} = x^{*\prime}\mathbf{i}' + y^{*\prime}\mathbf{j}' + z^{*\prime}\mathbf{k}'$ of the mass center C of the system in the two sets (Fig. A.2-2). In fact, $\mathbf{r}^* \equiv \mathbf{r}^{*\prime}$, since the vector from O to the mass center C is a given directed line segment; only its *representation* in scalar components differs. Either set of axes is said to be rotated from the other, or to be obtainable by a rotation (see *Dynamics*, Chapter VI).   Since $x^{*\prime}$ is the x'-component of $\mathbf{r}^*$, we must have $x^{*\prime}=\mathbf{r}^*\cdot\mathbf{i}'$, where $\mathbf{i}'$ is the unit vector in the x'-direction.   The (x, y, z) components of $\mathbf{i}'$ are the direction cosines of the x'-direction with respect to the (x, y, z) axes; therefore,

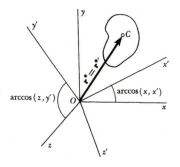

Fig. A.2–2

$$x^{*\prime} = x^* \cos (x, x')+y^* \cos (y, x')+z^* \cos (z, x'),$$

and, similarly,

$$y^{*\prime} = x^* \cos (x, y')+y^* \cos (y, y')+z^* \cos (z, y'), \qquad \textbf{A.2-3}$$

$$z^{*\prime} = x^* \cos (x, z')+y^* \cos (y, z')+z^* \cos (z, z').$$

Indeed, these typify the transformation of the components (A_x, A_y, A_z) into $(A_{x'}, A_{y'}, A_{z'})$ of any vector $\mathbf{A}$ under such a rotational transformation of coordinates (see Section 4.5).

A.3 **Further Methods of Locating the Mass Center**

(a) *Bodies with a Plane of Mass Symmetry.*   If a body has a plane of mass symmetry, then the mass center must lie in that plane.   A plane of mass symmetry is one such that the mass density is the same at points which are mirror images of one another in the plane.   A plane of geometrical symmetry is not necessarily a plane of mass symmetry; it will be also a plane of mass symmetry for uniform bodies, however.

(b) *Bodies with Multiple Symmetries.*   If a body has two planes of mass symmetry, then the mass center must lie on their line of intersection; if a body has three or more planes of mass symmetry, then their common point of intersection is the mass center.   The case when three or more planes of symmetry intersect along a common line, of course, leads only to the conclusion that the mass center lies on that line. If every plane through such a line is a plane of mass symmetry, then the axis is an axis of symmetry and the body is axially symmetric —a uniform circular cylinder, for example. A uniform sphere is an example of a radially symmetric body, with every plane through a single point (which must be the mass center) being a plane of symmetry. Symmetry arguments are often useful in locating the mass center (and centroid) of geometrically regular uniform bodies.

(c) *Method of Decomposition.* The mass center of complex bodies may be located easily if the body can be decomposed into parts, for each of which the mass center is known. The composite body has its mass center at the position given by treating the body as a system of discrete particles, each particle having mass equal to the mass of one of the subdivisions and position the same as the position of the mass center of the subdivision.

For example, the uniform isosceles trapezoid of base angle 60°, base lengths l and L, and total mass m shown in Fig. A.3-1a, has an x-coordinate of mass center given by $x^* = L/2$, by symmetry. The

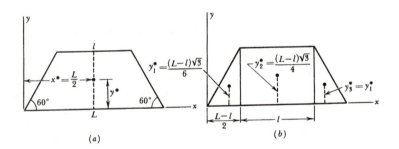

Fig. A.3-1

y-coordinate may be found by decomposing the trapezoid into two triangles and a rectangle, as in Fig. A.3-1b. The altitude of the trapezoid is $h = (L-l)\tan 60°/2$; the y-coordinates of the mass centers of the parts are, therefore, as given in the figure. Since the trapezoid is uniform, the mass of each part is equal to the total mass, m, multiplied by the ratio of the area of the part to the total area:

$$m_1 = m_3 = \left(\frac{L-l}{L+l}\right)\left(\frac{m}{2}\right), \qquad m_2 = \left(\frac{2l}{L+l}\right)m.$$

Therefore,

$$my^* = m_1 y_1^* + m_2 y_2^* + m_3 y_3^* = 2m_1 y_1^* + m_2 y_2^*$$

$$= \left(\frac{L-l}{L+l}\, m\right)\left(\frac{L-l}{6}\,\sqrt{3}\right) + \left(\frac{2l}{L+l}\, m\right)\left(\frac{L-l}{4}\,\sqrt{3}\right)$$

$$= \frac{m(L-l)\sqrt{3}}{12(L+l)}\,[2L-2l+6l],$$

or

$$y^* = \frac{(L-l)(L+2l)\sqrt{3}}{6(L+l)}.$$

The same conclusion can be reached in another way, by using a "negative" mass and removing part of a larger body to obtain the one we wish to consider. The trapezoid considered before is part of a large equilateral triangle of side L, obtained by cutting off a triangle of side l (Fig. A.3-2).

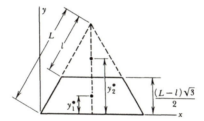

Fig. A.3-2

If the large triangle is body 1 and the small triangle is body 2, we have:

$$m_1 = \frac{L^2}{L^2-l^2}\, m, \qquad m_2 = \frac{l^2}{L^2-l^2}\, m, \qquad m = m_1 + (-m_2);$$

$$y_1^* = \frac{L\sqrt{3}}{6}, \qquad y_2^* = \frac{(L-l)\sqrt{3}}{2} + \frac{l\sqrt{3}}{6};$$

$$my^* = m_1 y_1^* + (-m_2)y_2^* = \frac{mL^3\sqrt{3}}{6(L^2-l^2)} - \frac{ml^2(3L-2l)\sqrt{3}}{6(L^2-l^2)}$$

$$= \frac{m(L^3-3Ll^2+2l^3)\sqrt{3}}{6(L^2-l^2)} = \frac{m(L-l)^2(L+2l)\sqrt{3}}{6(L^2-l^2)}.$$

Therefore,

$$y^* = \frac{(L-l)(L+2l)\sqrt{3}}{6(L+l)},$$

the same, of course, as the previous result.

A.4 Center of Mass of Uniform Bodies

The table contains the following information for uniform mass distributions:

(1) Name of body, picture relative to a coordinate system $Oxyz$.

(2) The length (L), area (A), and volume (V) of one-, two-, and three-dimensional mass distributions, respectively; the total mass is the appropriate constant mass density multiplied by L, A, or V.

(3) The position $\mathbf{r}^* = x^*\mathbf{i} + y^*\mathbf{j} + z^*\mathbf{k}$ of the mass center relative to the given origin O and the given axis system.

Table of Centers of Mass for Uniform Bodies

Body	Axis System	$L-A-V$ $\mathbf{r}^*$
Thin Rod, Straight		$L = l$ $x^* = \dfrac{l}{2}$ $y^* = z^* = 0$
Thin Rod, Circular Arc		$L = 2r\theta$ $x^* = \dfrac{r\sin\theta}{\theta}$ $y^* = z^* = 0$

Body	Axis System	$L - A - V$ r^*
Thin Rod, Circular Hoop	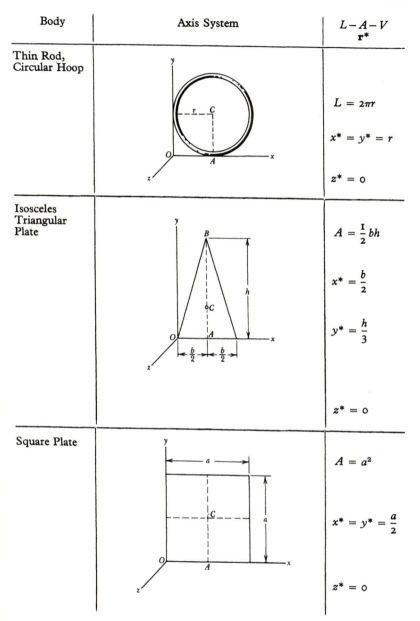	$L = 2\pi r$ $x^* = y^* = r$ $z^* = 0$
Isosceles Triangular Plate		$A = \frac{1}{2} bh$ $x^* = \frac{b}{2}$ $y^* = \frac{h}{3}$ $z^* = 0$
Square Plate		$A = a^2$ $x^* = y^* = \frac{a}{2}$ $z^* = 0$

Body	Axis System	$L-A-V$ $\mathbf{r}^*$
Rectangular Plate	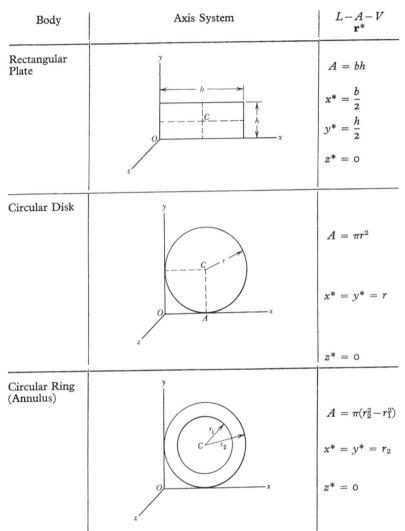	$A = bh$ $x^* = \dfrac{b}{2}$ $y^* = \dfrac{h}{2}$ $z^* = 0$
Circular Disk		$A = \pi r^2$ $x^* = y^* = r$ $z^* = 0$
Circular Ring (Annulus)		$A = \pi(r_2^2 - r_1^2)$ $x^* = y^* = r_2$ $z^* = 0$

Body	Axis System	$L-A-V$ $\mathbf{r}^*$
Circular Sector	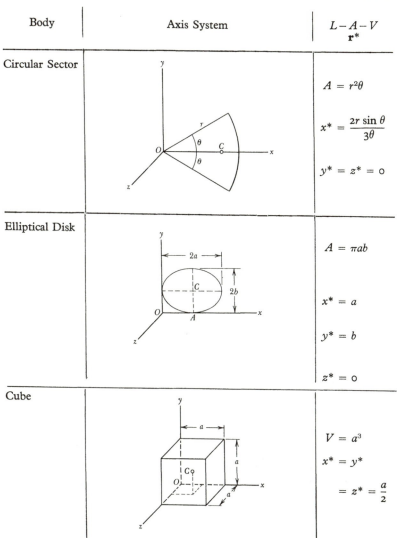	$A = r^2\theta$ $$x^* = \frac{2r \sin \theta}{3\theta}$$ $y^* = z^* = 0$
Elliptical Disk		$A = \pi ab$ $x^* = a$ $y^* = b$ $z^* = 0$
Cube		$V = a^3$ $x^* = y^*$ $= z^* = \dfrac{a}{2}$

356

Body	Axis System	$L - A - V$ $\mathbf{r}^*$
Rectangular Parallelepiped		$V = abc$ $x^* = \dfrac{a}{2}$ $y^* = \dfrac{b}{2}$ $z^* = \dfrac{c}{2}$
Right Circular Cone		$V = \dfrac{1}{3}\pi r^2 h$ $x^* = z^* = 0$ $y^* = \dfrac{h}{4}$
Right Circular Cylinder, Solid		$V = \pi r^2 h$ $x^* = z^* = 0$ $y^* = \dfrac{h}{2}$

Body	Axis System	$L-A-V$ $\mathbf{r}^*$
Right Circular Cylinder, Hollow	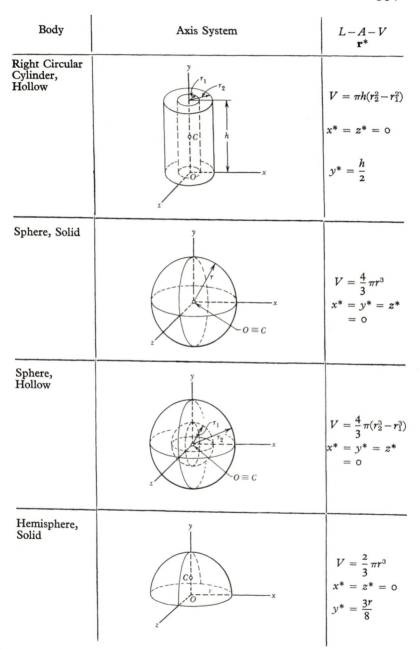	$V = \pi h(r_2^2 - r_1^2)$ $x^* = z^* = 0$ $y^* = \dfrac{h}{2}$
Sphere, Solid		$V = \dfrac{4}{3}\pi r^3$ $x^* = y^* = z^*$ $= 0$
Sphere, Hollow		$V = \dfrac{4}{3}\pi(r_2^3 - r_1^3)$ $x^* = y^* = z^*$ $= 0$
Hemisphere, Solid		$V = \dfrac{2}{3}\pi r^3$ $x^* = z^* = 0$ $y^* = \dfrac{3r}{8}$

Body	Axis System	$L - A - V$ $\mathbf{r}^*$
Ellipsoid	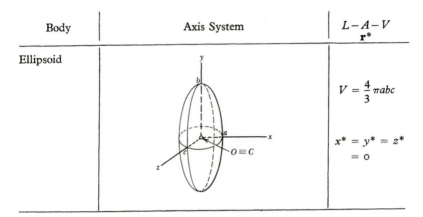	$V = \dfrac{4}{3}\pi abc$ $x^* = y^* = z^*$ $= 0$

INDEX

INDEX

Efficiency (of screw), 317
Encastré joint, 87
Engineering unit of mass, 64
Equality of vectors, 3
Equation:
  of a line, 9
  of a plane, 27
  of state for fluid, 259
    for perfect gas, 261
Equations, vector, 48
Equilibrating vector, 52
Equilibrium, 61–62
  convective, 279
  equations of, 92–95
    for beam, 236
    for cable, 222–223
    for continuum, 270
    for fluid, 256
  moment, 95
  of particle, 61–62, 64, 92
  of rigid body, 98–101, 146
  of systems, 92–95
Equipollent force sets, 102, 104
Equivalent force sets, 102
External force, 73

F

First moment (of scalar), 347
Fixed joint, 87
Flexibility of cable, 78, 270
Flexure of beam, 123, 234
Fluid, 251
  equilibrium equation of, 256
Force, 60–61 (see also Couple;
    Distributed forces; Force set)
  axial, 234
  at beam support, 238
  buoyant, 262
  center of, 116
  central, 116
  contact, 76, 77, 87
  distributed, 114, 120, 215
  external, 73
  frictional (see Friction)
  gravitational, 74
  internal, 73
  inverse-square law, 75
  moment of, 64

Force (*continued*)
  normal, 77
  restoring, 81
  resultant, 62, 93, 102
  shear, 234
  transmissibility of, 101
  units of, 63–64
  vector nature of, 61
  weight, 75, 118
Force density, 114
Force set, 102 ff.
  coplanar, 106
  equipollent, 102, 104
  equivalent, 102
  parallel, 106
    center of, 107
  reduction of, 104
  resultant, 104
Force system (see Force set)
Four-bar linkage, 186
Frame, 148, 184
Frame of reference, inertial, 61
Free-body diagram, 73
Free end, 238
Free vector, 264
Friction, 77, 280
  angle of, 287
  bearing, 319
  cable, 308
  circle of, 323
  coefficient of kinetic, 286
  coefficient of (limiting), 282, 286
    table, 284
  cone of, 288
  effect of deformability on, 300
  internal, 285
  laws of, 286
  limiting, 282
  in lubricated bearings, 281, 286
  in machine elements, 314 ff.
  rolling, 281, 285
  screw, 314
Friction clutch, 319
Friction locking, 289, 319

G

Gas, 251
  perfect, 260

A CATALOG OF SELECTED
DOVER BOOKS
IN SCIENCE AND MATHEMATICS

A CATALOG OF SELECTED

DOVER BOOKS

IN SCIENCE AND MATHEMATICS

Astronomy

BURNHAM'S CELESTIAL HANDBOOK, Robert Burnham, Jr. Thorough guide to the stars beyond our solar system. Exhaustive treatment. Alphabetical by constellation: Andromeda to Cetus in Vol. 1; Chamaeleon to Orion in Vol. 2; and Pavo to Vulpecula in Vol. 3. Hundreds of illustrations. Index in Vol. 3. 2,000pp. 6⅛ x 9¼.
23567-X, 23568-8, 23673-0 Pa., Three-vol. set $46.85

THE EXTRATERRESTRIAL LIFE DEBATE, 1750–1900, Michael J. Crowe. First detailed, scholarly study in English of the many ideas that developed between 1750 and 1900 regarding the existence of intelligent extraterrestrial life. Examines ideas of Kant, Herschel, Voltaire, Percival Lowell, many other scientists and thinkers. 16 illustrations. 704pp. 5⅜ x 8½. 40675-X Pa. $19.95

A HISTORY OF ASTRONOMY, A. Pannekoek. Well-balanced, carefully reasoned study covers such topics as Ptolemaic theory, work of Copernicus, Kepler, Newton, Eddington's work on stars, much more. Illustrated. References. 521pp. 5⅜ x 8½.
65994-1 Pa. $15.95

AMATEUR ASTRONOMER'S HANDBOOK, J. B. Sidgwick. Timeless, comprehensive coverage of telescopes, mirrors, lenses, mountings, telescope drives, micrometers, spectroscopes, more. 189 illustrations. 576pp. 5⅝ x 8¼. (Available in U.S. only.)
24034-7 Pa. $13.95

STARS AND RELATIVITY, Ya. B. Zel'dovich and I. D. Novikov. Vol. 1 of *Relativistic Astrophysics* by famed Russian scientists. General relativity, properties of matter under astrophysical conditions, stars, and stellar systems. Deep physical insights, clear presentation. 1971 edition. References. 544pp. 5⅝ x 8¼.
69424-0 Pa. $14.95

Chemistry

CHEMICAL MAGIC, Leonard A. Ford. Second Edition, Revised by E. Winston Grundmeier. Over 100 unusual stunts demonstrating cold fire, dust explosions, much more. Text explains scientific principles and stresses safety precautions. 128pp. 5⅜ x 8½. 67628-5 Pa. $5.95

THE DEVELOPMENT OF MODERN CHEMISTRY, Aaron J. Ihde. Authoritative history of chemistry from ancient Greek theory to 20th-century innovation. Covers major chemists and their discoveries. 209 illustrations. 14 tables. Bibliographies. Indices. Appendices. 851pp. 5⅜ x 8½. 64235-6 Pa. $24.95

CATALYSIS IN CHEMISTRY AND ENZYMOLOGY, William P. Jencks. Exceptionally clear coverage of mechanisms for catalysis, forces in aqueous solution, carbonyl- and acyl-group reactions, practical kinetics, more. 864pp. 5⅜ x 8½.
65460-5 Pa. $19.95

A HISTORY OF MECHANICS, René Dugas. Monumental study of mechanical principles from antiquity to quantum mechanics. Contributions of ancient Greeks, Galileo, Leonardo, Kepler, Lagrange, many others. 671pp. 5⅜ x 8½.
65632-2 Pa. $18.95

STATISTICAL MECHANICS: Principles and Applications, Terrell L. Hill. Standard text covers fundamentals of statistical mechanics, applications to fluctuation theory, imperfect gases, distribution functions, more. 448pp. 5⅜ x 8½.
65390-0 Pa. $14.95

THE VARIATIONAL PRINCIPLES OF MECHANICS, Cornelius Lanczos. Graduate level coverage of calculus of variations, equations of motion, relativistic mechanics, more. First inexpensive paperbound edition of classic treatise. Index. Bibliography. 418pp. 5⅜ x 8½.
65067-7 Pa. $14.95

THE VARIOUS AND INGENIOUS MACHINES OF AGOSTINO RAMELLI: A Classic Sixteenth-Century Illustrated Treatise on Technology, Agostino Ramelli. One of the most widely known and copied works on machinery in the 16th century. 194 detailed plates of water pumps, grain mills, cranes, more. 608pp. 9 x 12.
28180-9 Pa. $24.95

ORDINARY DIFFERENTIAL EQUATIONS AND STABILITY THEORY: An Introduction, David A. Sánchez. Brief, modern treatment. Linear equation, stability theory for autonomous and nonautonomous systems, etc. 164pp. 5⅜ x 8¼.
63828-6 Pa. $6.95

ROTARY WING AERODYNAMICS, W. Z. Stepniewski. Clear, concise text covers aerodynamic phenomena of the rotor and offers guidelines for helicopter performance evaluation. Orignially prepared for NASA. 537 figures. 640pp. 6⅛ x 9¼.
64647-5 Pa. $16.95

INTRODUCTION TO SPACE DYNAMICS, William Tyrrell Thomson. Comprehensive, classic introduction to space-flight engineering for advanced undergraduate and graduate students. Includes vector algebra, kinematics, transformation of coordinates. Bibliography. Index. 352pp. 5⅜ x 8½.
65113-4 Pa. $10.95

HISTORY OF STRENGTH OF MATERIALS, Stephen P. Timoshenko. Excellent historical survey of the strength of materials with many references to the theories of elasticity and structure. 245 figures. 452pp. 5⅜ x 8½.
61187-6 Pa. $14.95

CONSTRUCTIONS AND COMBINATORIAL PROBLEMS IN DESIGN OF EXPERIMENTS, Damaraju Raghavarao. In-depth reference work examines orthogonal Latin squares, incomplete block designs, tactical configuration, partial geometry, much more. Abundant explanations, examples. 416pp. 5⅜ x 8¼.
65685-3 Pa. $10.95

ESSAYS ON THE THEORY OF NUMBERS, Richard Dedekind. Two classic essays by great German mathematician: on the theory of irrational numbers; and on transfinite numbers and properties of natural numbers. 115pp. 5⅜ x 8½.
21010-3 Pa. $7.95

APPLIED COMPLEX VARIABLES, John W. Dettman. Step-by-step coverage of fundamentals of analytic function theory–plus lucid exposition of five important applications: Potential Theory; Ordinary Differential Equations; Fourier Transforms; Laplace Transforms; Asymptotic Expansions. 66 figures. Exercises at chapter ends. 512pp. 5⅜ x 8½. 64670-X Pa. $14.95

INTRODUCTION TO LINEAR ALGEBRA AND DIFFERENTIAL EQUATIONS, John W. Dettman. Excellent text covers complex numbers, determinants, orthonormal bases, Laplace transforms, much more. Exercises with solutions. Undergraduate level. 416pp. 5⅜ x 8½. 65191-6 Pa. $12.95

MATHEMATICAL METHODS IN PHYSICS AND ENGINEERING, John W. Dettman. Algebraically based approach to vectors, mapping, diffraction, other topics in applied math. Also generalized functions, analytic function theory, more. Exercises. 448pp. 5⅜ x 8¼. 65649-7 Pa. $12.95

CALCULUS OF VARIATIONS WITH APPLICATIONS, George M. Ewing. Applications-oriented introduction to variational theory develops insight and promotes understanding of specialized books, research papers. Suitable for advanced undergraduate/graduate students as primary, supplementary text. 352pp. 5⅜ x 8½.
64856-7 Pa. $9.95

COMPLEX VARIABLES, Francis J. Flanigan. Unusual approach, delaying complex algebra till harmonic functions have been analyzed from real variable viewpoint. Includes problems with answers. 364pp. 5⅜ x 8½. 61388-7 Pa. $10.95

AN INTRODUCTION TO THE CALCULUS OF VARIATIONS, Charles Fox. Graduate-level text covers variations of an integral, isoperimetrical problems, least action, special relativity, approximations, more. References. 279pp. 5⅜ x 8½.
65499-0 Pa. $10.95

CATASTROPHE THEORY FOR SCIENTISTS AND ENGINEERS, Robert Gilmore. Advanced-level treatment describes mathematics of theory grounded in the work of Poincaré, R. Thom, other mathematicians. Also important applications to problems in mathematics, physics, chemistry and engineering. 1981 edition. References. 28 tables. 397 black-and-white illustrations. xvii + 666pp. 6⅛ x 9¼.
67539-4 Pa. $17.95

INTRODUCTION TO DIFFERENCE EQUATIONS, Samuel Goldberg. Exceptionally clear exposition of important discipline with applications to sociology, psychology, economics. Many illustrative examples; over 250 problems. 260pp. 5⅜ x 8½.
65084-7 Pa. $10.95

NUMERICAL METHODS FOR SCIENTISTS AND ENGINEERS, Richard Hamming. Classic text stresses frequency approach in coverage of algorithms, polynomial approximation, Fourier approximation, exponential approximation, other topics. Revised and enlarged 2nd edition. 721pp. 5⅜ x 8½. 65241-6 Pa. $17.95

THEORY OF MATRICES, Sam Perlis. Outstanding text covering rank, nonsingularity and inverses in connection with the development of canonical matrices under the relation of equivalence, and without the intervention of determinants. Includes exercises. 237pp. 5⅜ x 8½.   66810-X Pa. $8.95

INTRODUCTION TO ANALYSIS, Maxwell Rosenlicht. Unusually clear, accessible coverage of set theory, real number system, metric spaces, continuous functions, Riemann integration, multiple integrals, more. Wide range of problems. Undergraduate level. Bibliography. 254pp. 5⅜ x 8½.   65038-3 Pa. $11.95

MODERN NONLINEAR EQUATIONS, Thomas L. Saaty. Emphasizes practical solution of problems; covers seven types of equations. ". . . a welcome contribution to the existing literature...."–*Math Reviews*. 490pp. 5⅜ x 8½.   64232-1 Pa. $13.95

MATRICES AND LINEAR ALGEBRA, Hans Schneider and George Phillip Barker. Basic textbook covers theory of matrices and its applications to systems of linear equations and related topics such as determinants, eigenvalues and differential equations. Numerous exercises. 432pp. 5⅜ x 8½.   66014-1 Pa. $12.95

MATHEMATICS APPLIED TO CONTINUUM MECHANICS, Lee A. Segel. Analyzes models of fluid flow and solid deformation. For upper-level math, science and engineering students. 608pp. 5⅜ x 8½.   65369-2 Pa. $18.95

ELEMENTS OF REAL ANALYSIS, David A. Sprecher. Classic text covers fundamental concepts, real number system, point sets, functions of a real variable, Fourier series, much more. Over 500 exercises. 352pp. 5⅜ x 8½.   65385-4 Pa. $11.95

AN INTRODUCTION TO MATRICES, SETS AND GROUPS FOR SCIENCE STUDENTS, G. Stephenson. Concise, readable text introduces sets, groups, and most importantly, matrices to undergraduate students of physics, chemistry, and engineering. Problems. 164pp. 5⅜ x 8½.   65077-4 Pa. $7.95

SET THEORY AND LOGIC, Robert R. Stoll. Lucid introduction to unified theory of mathematical concepts. Set theory and logic seen as tools for conceptual understanding of real number system. 496pp. 5⅜ x 8¼.   63829-4 Pa. $14.95

TENSOR CALCULUS, J.L. Synge and A. Schild. Widely used introductory text covers spaces and tensors, basic operations in Riemannian space, non-Riemannian spaces, etc. 324pp. 5⅜ x 8¼.   63612-7 Pa. $13.95

ORDINARY DIFFERENTIAL EQUATIONS, Morris Tenenbaum and Harry Pollard. Exhaustive survey of ordinary differential equations for undergraduates in mathematics, engineering, science. Thorough analysis of theorems. Diagrams. Bibliography. Index. 818pp. 5⅜ x 8½.   64940-7 Pa. $19.95

INTEGRAL EQUATIONS, F. G. Tricomi. Authoritative, well-written treatment of extremely useful mathematical tool with wide applications. Volterra Equations, Fredholm Equations, much more. Advanced undergraduate to graduate level. Exercises. Bibliography. 238pp. 5⅜ x 8½.   64828-1 Pa. $8.95

Math–Decision Theory, Statistics, Probability

ELEMENTARY DECISION THEORY, Herman Chernoff and Lincoln E. Moses. Clear introduction to statistics and statistical theory covers data processing, probability and random variables, testing hypotheses, much more. Exercises. 364pp. 5⅜ x 8½. 65218-1 Pa. $12.95

STATISTICS MANUAL, Edwin L. Crow et al. Comprehensive, practical collection of classical and modern methods prepared by U.S. Naval Ordnance Test Station. Stress on use. Basics of statistics assumed. 288pp. 5⅜ x 8½. 60599-X Pa. $8.95

SOME THEORY OF SAMPLING, William Edwards Deming. Analysis of the problems, theory and design of sampling techniques for social scientists, industrial managers and others who find statistics important at work. 61 tables. 90 figures. xvii +602pp. 5⅜ x 8½. 64684-X Pa. $16.95

STATISTICAL ADJUSTMENT OF DATA, W. Edwards Deming. Introduction to basic concepts of statistics, curve fitting, least squares solution, conditions without parameter, conditions containing parameters. 26 exercises worked out. 271pp. 5⅜ x 8½. 64685-8 Pa. $9.95

LINEAR PROGRAMMING AND ECONOMIC ANALYSIS, Robert Dorfman, Paul A. Samuelson and Robert M. Solow. First comprehensive treatment of linear programming in standard economic analysis. Game theory, modern welfare economics, Leontief input-output, more. 525pp. 5⅜ x 8½. 65491-5 Pa. $17.95

DICTIONARY/OUTLINE OF BASIC STATISTICS, John E. Freund and Frank J. Williams. A clear concise dictionary of over 1,000 statistical terms and an outline of statistical formulas covering probability, nonparametric tests, much more. 208pp. 5⅜ x 8½. 66796-0 Pa. $8.95

PROBABILITY: An Introduction, Samuel Goldberg. Excellent basic text covers set theory, probability theory for finite sample spaces, binomial theorem, much more. 360 problems. Bibliographies. 322pp. 5⅜ x 8½. 65252-1 Pa. $11.95

GAMES AND DECISIONS: Introduction and Critical Survey, R. Duncan Luce and Howard Raiffa. Superb nontechnical introduction to game theory, primarily applied to social sciences. Utility theory, zero-sum games, n-person games, decision-making, much more. Bibliography. 509pp. 5⅜ x 8½. 65943-7 Pa. $14.95

FIFTY CHALLENGING PROBLEMS IN PROBABILITY WITH SOLUTIONS, Frederick Mosteller. Remarkable puzzlers, graded in difficulty, illustrate elementary and advanced aspects of probability. Detailed solutions. 88pp. 5⅜ x 8½. 65355-2 Pa. $5.95

PROBABILITY THEORY: A Concise Course, Y. A. Rozanov. Highly readable, self-contained introduction covers combination of events, dependent events, Bernoulli trials, etc. 148pp. 5⅜ x 8¼. 63544-9 Pa. $8.95

CURVATURE AND HOMOLOGY: Enlarged Edition, Samuel I. Goldberg. Revised edition examines topology of differentiable manifolds; curvature, homology of Riemannian manifolds; compact Lie groups; complex manifolds; curvature, homology of Kaehler manifolds. New Preface. Four new appendixes. 416pp. 5⅜ x 8½.
40207-X Pa. $14.95

TOPOLOGY, John G. Hocking and Gail S. Young. Superb one-year course in classical topology. Topological spaces and functions, point-set topology, much more. Examples and problems. Bibliography. Index. 384pp. 5⅝ x 8¼. 65676-4 Pa. $13.95

LECTURES ON CLASSICAL DIFFERENTIAL GEOMETRY, Second Edition, Dirk J. Struik. Excellent brief introduction covers curves, theory of surfaces, fundamental equations, geometry on a surface, conformal mapping, other topics. Problems. 240pp. 5⅜ x 8½. 65609-8 Pa. $9.95

Math–History of

A SHORT ACCOUNT OF THE HISTORY OF MATHEMATICS, W. W. Rouse Ball. One of clearest, most authoritative surveys from the Egyptians and Phoenicians through 19th-century figures such as Grassman, Galois, Riemann. Fourth edition. 522pp. 5⅜ x 8½. 20630-0 Pa. $13.95

THE HISTORICAL ROOTS OF ELEMENTARY MATHEMATICS, Lucas N. H. Bunt, Phillip S. Jones, and Jack D. Bedient. Fundamental underpinnings of modern arithmetic, algebra, geometry and number systems derived from ancient civilizations. 320pp. 5⅜ x 8½. 25563-8 Pa. $9.95

GAMES, GODS & GAMBLING: A History of Probability and Statistical Ideas, F. N. David. Episodes from the lives of Galileo, Fermat, Pascal, and others illustrate this fascinating account of the roots of mathematics. Features thought-provoking references to classics, archaeology, biography, poetry. 1962 edition. 304pp. 5⅜ x 8½. (Available in U.S. only.) 40023-9 Pa. $9.95

HISTORY OF MATHEMATICS, David E. Smith. Nontechnical survey from ancient Greece and Orient to late 19th century; evolution of arithmetic, geometry, trigonometry, calculating devices, algebra, the calculus. 362 illustrations. 1,355pp. 5⅜ x 8½. Two-vol. set.
Vol. I: 20429-4 Pa. $13.95
Vol. II: 20430-8 Pa. $14.95

A CONCISE HISTORY OF MATHEMATICS, Dirk J. Struik. The best brief history of mathematics. Stresses origins and covers every major figure from ancient Near East to 19th century. 41 illustrations. 195pp. 5⅜ x 8½. 60255-9 Pa. $8.95

THE HISTORY OF THE CALCULUS AND ITS CONCEPTUAL DEVELOPMENT, Carl B. Boyer. Origins in antiquity, medieval contributions, work of Newton, Leibniz, rigorous formulation. Treatment is verbal. 346pp. 5⅜ x 8½.
60509-4 Pa. $9.95

THE PHYSICS OF WAVES, William C. Elmore and Mark A. Heald. Unique overview of classical wave theory. Acoustics, optics, electromagnetic radiation, more. Ideal as classroom text or for self-study. Problems. 477pp. 5⅜ x 8½.
64926-1 Pa. $14.95

PHYSICAL PRINCIPLES OF THE QUANTUM THEORY, Werner Heisenberg. Nobel Laureate discusses quantum theory, uncertainty, wave mechanics, work of Dirac, Schroedinger, Compton, Wilson, Einstein, etc. 184pp. 5⅜ x 8½.
60113-7 Pa. $8.95

ATOMIC SPECTRA AND ATOMIC STRUCTURE, Gerhard Herzberg. One of best introductions; especially for specialist in other fields. Treatment is physical rather than mathematical. 80 illustrations. 257pp. 5⅜ x 8½.   60115-3 Pa. $11.95

AN INTRODUCTION TO STATISTICAL THERMODYNAMICS, Terrell L. Hill. Excellent basic text offers wide-ranging coverage of quantum statistical mechanics, systems of interacting molecules, quantum statistics, more. 523pp. 5⅜ x 8½.
65242-4 Pa. $14.95

THEORETICAL PHYSICS, Georg Joos, with Ira M. Freeman. Classic overview covers essential math, mechanics, electromagnetic theory, thermodynamics, quantum mechanics, nuclear physics, other topics. First paperback edition. xxiii + 885pp. 5⅜ x 8½.   65227-0 Pa. $24.95

PROBLEMS AND SOLUTIONS IN QUANTUM CHEMISTRY AND PHYSICS, Charles S. Johnson, Jr. and Lee G. Pedersen. Unusually varied problems, detailed solutions in coverage of quantum mechanics, wave mechanics, angular momentum, molecular spectroscopy, more. 280 problems plus 139 supplementary exercises. 430pp. 6½ x 9¼.   65236-X Pa. $14.95

THEORETICAL SOLID STATE PHYSICS, Vol. 1: Perfect Lattices in Equilibrium; Vol. II: Non-Equilibrium and Disorder, William Jones and Norman H. March. Monumental reference work covers fundamental theory of equilibrium properties of perfect crystalline solids, non-equilibrium properties, defects and disordered systems. Appendices. Problems. Preface. Diagrams. Index. Bibliography. Total of 1,301pp. 5⅜ x 8½. Two volumes.   Vol. I: 65015-4 Pa. $16.95
Vol. II: 65016-2 Pa. $16.95

A TREATISE ON ELECTRICITY AND MAGNETISM, James Clerk Maxwell. Important foundation work of modern physics. Brings to final form Maxwell's theory of electromagnetism and rigorously derives his general equations of field theory. 1,084pp. 5⅜ x 8½. Two-vol. set.   Vol. I: 60636-8 Pa. $14.95
Vol. II: 60637-6 Pa. $14.95

OPTICKS, Sir Isaac Newton. Newton's own experiments with spectroscopy, colors, lenses, reflection, refraction, etc., in language the layman can follow. Foreword by Albert Einstein. 532pp. 5⅜ x 8½.   60205-2 Pa. $13.95

THEORY OF ELECTROMAGNETIC WAVE PROPAGATION, Charles Herach Papas. Graduate-level study discusses the Maxwell field equations, radiation from wire antennas, the Doppler effect and more. xiii + 244pp. 5⅜ x 8½.
65678-0 Pa. $9.95

INTRODUCTION TO QUANTUM MECHANICS With Applications to Chemistry, Linus Pauling & E. Bright Wilson, Jr. Classic undergraduate text by Nobel Prize winner applies quantum mechanics to chemical and physical problems. Numerous tables and figures enhance the text. Chapter bibliographies. Appendices. Index. 468pp. 5⅜ x 8½. 64871-0 Pa. $13.95

METHODS OF THERMODYNAMICS, Howard Reiss. Outstanding text focuses on physical technique of thermodynamics, typical problem areas of understanding, and significance and use of thermodynamic potential. 1965 edition. 238pp. 5⅜ x 8½. 69445-3 Pa. $8.95

TENSOR ANALYSIS FOR PHYSICISTS, J. A. Schouten. Concise exposition of the mathematical basis of tensor analysis, integrated with well-chosen physical examples of the theory. Exercises. Index. Bibliography. 289pp. 5⅜ x 8½. 65582-2 Pa. $13.95

RELATIVITY IN ILLUSTRATIONS, Jacob T. Schwartz. Clear nontechnical treatment makes relativity more accessible than ever before. Over 60 drawings illustrate concepts more clearly than text alone. Only high school geometry needed. Bibliography. 128pp. 6⅛ x 9¼. 25965-X Pa. $7.95

THE ELECTROMAGNETIC FIELD, Albert Shadowitz. Comprehensive undergraduate text covers basics of electric and magnetic fields, builds up to electromagnetic theory. Also related topics, including relativity. Over 900 problems. 768pp. 5⅝ x 8¼. 65660-8 Pa. $19.95

GREAT EXPERIMENTS IN PHYSICS: Firsthand Accounts from Galileo to Einstein, edited by Morris H. Shamos. 25 crucial discoveries: Newton's laws of motion, Chadwick's study of the neutron, Hertz on electromagnetic waves, more. Original accounts clearly annotated. 370pp. 5⅜ x 8½. 25346-5 Pa. $12.95

RELATIVITY, THERMODYNAMICS AND COSMOLOGY, Richard C. Tolman. Landmark study extends thermodynamics to special, general relativity; also applications of relativistic mechanics, thermodynamics to cosmological models. 501pp. 5⅜ x 8½. 65383-8 Pa. $15.95

LIGHT SCATTERING BY SMALL PARTICLES, H. C. van de Hulst. Comprehensive treatment including full range of useful approximation methods for researchers in chemistry, meteorology and astronomy. 44 illustrations. 470pp. 5⅜ x 8½. 64228-3 Pa. $14.95

STATISTICAL PHYSICS, Gregory H. Wannier. Classic text combines thermodynamics, statistical mechanics and kinetic theory in one unified presentation of thermal physics. Problems with solutions. Bibliography. 532pp. 5⅜ x 8½. 65401-X Pa. $14.95

Prices subject to change without notice.

Available at your book dealer or online at **www.doverpublications.com**. Write for free Dover Mathematics and Science Catalog (59065-8) to Dept. GI, Dover Publications, Inc., 31 East 2nd St., Mineola, NY 11501. Dover publishes more than 400 books each year on science, elementary and advanced mathematics, biology, music, art, literature, history, social sciences, and other subjects.